Microbial Engineering: Concepts and Applications

Microbial Engineering: Concepts and Applications

Edited by **Lucy Phillip**

SYRAWOOD
PUBLISHING HOUSE

New York

Published by Syrawood Publishing House,
750 Third Avenue, 9th Floor,
New York, NY 10017, USA
www.syrawoodpublishinghouse.com

Microbial Engineering: Concepts and Applications
Edited by Lucy Phillip

International Standard Book Number: 978-1-68286-134-9 (Hardback)

Printed in the United States of America.

Contents

VI   Contents

Preface

The main aim of this book is to educate learners and enhance their research focus by presenting diverse topics covering this vast field. This is an advanced book which compiles significant studies by distinguished experts. This book addresses successive solutions to the challenges arising in the area of application, along with it; the book provides scope for future developments.

Microbial engineering is an important branch of applied biological sciences and finds applications across a wide range of disciplines, such as agriculture, pharmaceutics, etc. This book contains some path-breaking studies in the field of microbial engineering. Researches revolving around biocatalysis, biotransformation, biosynthesis, etc. have been presented in this book. This book focuses on the applications of microbial engineering for industrial purposes. Students, researchers, experts and all associated with microbial engineering and allied fields will benefit alike from it.

It was a great honour to edit this book, though there were challenges, as it involved a lot of communication and networking between me and the editorial team. However, the end result was this all-inclusive book covering diverse themes in the field.

Finally, it is important to acknowledge the efforts of the contributors for their excellent chapters, through which a wide variety of issues have been addressed. I would also like to thank my colleagues for their valuable feedback during the making of this book.

Editor

Influences of 3-hydroxypropionaldehyde and lactate on the production of 1,3-propanediol by *Klebsiella pneumoniae*

Zhihui Zhong, Longfei Liu, Jiajia Zhou, Lirong Gao, Jiajie Xu, Shuilin Fu and Heng Gong[*]

Abstract

Background: 1,3-Propanediol is the starting point of a new-generation polymer with superior properties which can be used in many industrial fields. 3-Hydroxypropionaldehyde and lactate have been identified as two important metabolites in the biological route of 1,3-propanediol bioconversion from glycerol. Here, influence of lactate on the inhibition caused by 3-hydroxypropionaldehyde of 1,3-propanediol fermentation by *Klebsiella pneumoniae* is reported.

Methods: The influences of 3-hydroxypropionaldehyde and lactate on 1,3-propanediol production were investigated in normal and lactate pathway deficient strains with different fermentation conditions.

Results: By using the strains KG1 and L-type lactate dehydrogenase-deficient mutant (KG1Δldh), the results indicated that an early accumulation of 3-hydroxypropionaldehyde directly inhibited the 1,3-propanediol production rather than through lactate accumulation during the late stage of fermentation. Then, the influence of extra addition of lactate on the late stage of fermentation was investigated, and the inhibitory effect of lactate did not appear. At last, it was found that by reducing 3-hydroxypropionaldehyde accumulation in the early stage of fermentation, the concentration and yield of 1,3-propanediol increased by 18% and 16%, respectively, over the initial experimental levels.

Conclusions: An early accumulation of 3-hydroxypropionaldehyde directly decreased the final 1,3-propanediol concentration rather than through lactate accumulation during the late stage of fermentation.

Keywords: Fermentation; Glycerol; 3-Hydroxypropionaldehyde; Lactate; 1,3-Propanediol

Background

1,3-Propanediol (1,3-PD) can be formulated into numerous industrial products such as composites, adhesives, and laminates. With the increasing emergence of novel applications of 1,3-PD, the demand for its biological production has been on a rise [1]. The conventional chemical conversion of acrolein (also known as propenal) into 1,3-PD has several drawbacks [2]. Nevertheless, the bioconversion of 1,3-PD from glycerol by *Klebsiella pneumoniae* has attracted much attention worldwide, and numerous efforts have been made to enhance the 1,3-PD fermentation, such as construction of recombinant *K. pneumoniae* strains [1-8]. Due to the tremendous growth of the biofuel industry, a glycerol surplus has

been created, resulting in a decrease in glycerol prices. Compared with the recombinant *E. coli* producing 1,3-PD from glucose described in a patent by Nakamura and Whited [7], the *K. pneumoniae* strain producing 1,3-PD from glycerol as substrate has lots of advantages and is being refocused [8].

The production of 1,3-PD by *K. pneumoniae* is generally performed under anaerobic conditions with glycerol as the sole carbon source. In the metabolic reactions, glycerol is dissimilated through coupled oxidative and reductive pathways [1,2,4-6,9]. The reductive branch consists of two steps: glycerol is first dehydrated to 3-hydroxypropionaldehyde (3-HPA) and then 3-HPA is reduced to 1,3-PD under the consumption of reducing power (NADH). Reducing power and various byproducts were produced in the oxidative branch.

Among the intermediates and by-products, 3-HPA has been identified as an important intermediate during the

* Correspondence: gongheng@ecust.edu.cn
State Key Laboratory of Bioreactor Engineering, East China University of Science and Technology, 130 Meilong Road, Shanghai 200237, People's Republic of China

fermentation course and lactate is the main byproduct in the final fermentation result, accumulation of either can decrease the 1,3-PD concentration [9,10]. If lactate accumulates to high levels, it impedes cellular activity [10] and decreases the final 1,3-PD concentration by consuming NADH [6]. Hao et al. [9] reported that a very high concentration of 3-HPA could inhibit the growth of, and lead to the death of, cells. When the concentration of 3-HPA increases to a higher level in the medium (ac 10 mM), the activity of the 1,3-propanediol oxidoreductase in the cells decreases correspondingly, which leads to a decrease of the 3-HPA conversion rate. Then, the accumulation of 3-HPA is accelerated furthermore. 3-HPA accumulation in culture medium is triggered by this positive feedback mechanism.

However, the mechanism of inhibition caused by moderate concentrations of 3-HPA has not been studied systematically. As an intermediate product, 3-HPA accumulates in the early stage of fermentation. The accumulation of 3-HPA in the culture medium can significantly decrease most of the enzymatic activities in *K. pneumoniae* strains with the exception of lactate dehydrogenase (the key enzyme in the interconversion of lactate to pyruvate) [11]. If the concentration of 3-HPA in the early stage of fermentation is high, then lactate will accumulate to reach a high concentration during the late stage of fermentation. Hence, it is not clear whether the decrease of final fermentation concentration of 1,3-PD is caused by 3-HPA accumulation in the early stage of fermentation or due to lactate accumulation in the late stage of fermentation. Clarifying the mechanism of inhibition caused by 3-HPA concentrations could help researchers to 'fine-tune' the entire bioprocess.

In the present study, influences of 3-HPA and lactate on 1,3-PD fermentation was investigated using two strains (KG1 and KG1Δldh (lactate dehydrogenase-deficient strain engineered from KG1)), and then 1,3-PD fermentation process was optimized by strategic glycerol feeding to keep the 3-HPA below the inhibitory level [9].

Methods
Construction of mutants using Red recombination
The KG1Δldh strain was constructed by homologous recombination. Gene (*ldhA*) was selected for knockout by Red recombination [12]. The disruption cassette was amplified by two-step PCR from the chromosomal DNA of *K. pneumoniae* and pKD4, in which the upstream and downstream homologous extensions of target gene were amplified with two pair of primers (5′-ATGAAAATCGCGG TTTAT-3′, 5′-AAGCAGCTCCAGCCTACACATGGGCC TTCAGCTCTTCC-3′; 5′-AGGAGGATATTCATATGGA CGCGATCTGTTCTTTGAAG-3′, 5′-TTAGACGATGGC GTTCGG-3′) and kanamycin resistance gene was amplified with one pair of primers (5′-TGTGTAGGCTG-GAGCTGCTT-3′, 5′-GTCCATATGAATATCCTCCT-3′).

For Red facilitated homologous recombination in *K. pneumoniae*, the pKD46-Tc was first transformed into the parent cell. Next, the disruption cassette of the target gene was further transformed into the parent cell with pKD46-Tc. The recombined strain was screened on LB plates containing kanamycin and grown at 37°C. To delete the kanamycin resistance gene in the recombined strain, pCP20-Tc was transformed into the recombined strain. Cells were then cultured overnight at 42°C, diluted and plated onto solid LB to form single colonies, which were further screened for loss of kanamycin and tetracycline hydrochloride resistance.

Microorganisms and culture conditions
KG1 was obtained by the screening of wild-type *K. pneumoniae*, and the KG1Δldh strain was constructed by homologous recombination [12]. These strains have been grown on a pre-culture medium containing (per liter) 3.4 g $K_2HPO_4 \cdot 3H_2O$, 2 g $(NH_4)_2SO_4$, 1.3 g KH_2PO_4, 0.2 g $MgSO_4$, 1 g yeast extract, and 0.3 mL trace element solution composed of (per liter) 70 g $ZnCl_2$, 3.2 g $NiCl_2 \cdot 6H_2O$, 10 g $MnCl_2 \cdot 4H_2O$, 20 g $CoCl_2 \cdot 2H_2O$, 0.6 g H_3BO_3, and 0.45 g $Na_2MoO_4 \cdot 2H_2O$. Each liter of fermentation medium comprised 1 g KCl, 2.15 g $KH_2PO_4 \cdot 2H_2O$, 4 g $(NH_4)_2SO_4$, 0.4 g $MgSO_4 \cdot 7H_2O$, 2 g yeast extract, and 0.4 mL trace element solution.

The seed cells were incubated in 500-mL flasks containing 100 mL seed medium, at 37°C, 200 rpm for 10 h. Fed-batch fermentation in a 5-l stirred reactor (NC Bio, Shanghai, China) was carried out under anaerobic condition with a 5% (*v/v*) inoculation at 140 rpm at 37°C for 30 h. The initial glycerol concentration was 40 g/L and was maintained around 20 g/L (±5 g/L) by feeding glycerol during the entire fermentation process. The pH value was maintained 6.8 by the automatic addition of 50% (*w/w*) KOH.

Analytical methods
The cell optical density was related to the dry cell weight (DCW) with an experimentally determined calibration curve. The optical density (OD) of the cultures was monitored at 650 nm after appropriate dilution. The broth components were measured using a HPLC system equipped with a 2487 Dual-Wavelength Absorbance Detector (Waters, Milford, MA, USA) and a Platisil ODS column (5 μm, 250 M × 4.6 M; Dikma Technologies, Beijing, China). The column temperature was 65°C, and the mobile phase was 5 mM H_2SO_4 at 0.8 mL/min. The 3-HPA concentration was measured using the tryptophan colorimetric method [13].

Results and discussion
Accumulation of 3-HPA and lactate at different stages of fermentation
The cell growth, glycerol consumption, and metabolites production during the fed-batch fermentation using

KG1 with an initial glycerol concentration of 40 g/L were illustrated in Figure 1. In terms of 3-HPA and lactate formation, fed-batch fermentation can be divided into two stages: stage I (0 to 12.5 h) was the early stage of fermentation, in which cells were growing and the biomass and 3-HPA concentration reached a peak. Stage II (12.5 to 30 h) was the late stage of fermentation, in which the concentrations of lactate and 1,3-PD increased gradually. In stage II, glycerol was fed and maintained around 20 g/L (217 mM) till the end of the fermentation. Due to dilution of the medium, a slight decrease in biomass was observed.

Effect of the accumulation of 3-HPA on lactate formation

Figure 2A,B illustrated the relationship between 3-HPA accumulation in stage I with lactate formation in stage II. Despite the initial concentration of glycerol, no conditions were changed during the entire fed-batch fermentation by KG1. With an initial glycerol concentration of 40 g/L, a peak concentration (14.6 mM) of 3-HPA at 12 h in stage I was observed (Figure 2A). The peak concentration of 3-HPA decreased by 25% to 10.9 mM (Figure 2A) with the initial concentration of glycerol lowered to 30 g/L. Compared with initial glycerol concentration of 40 g/L, the final concentration of lactate decreased by 39% to 31.4 mM (Figure 2B), and the cell growth and 1,3-PD production at the end of fermentation increased

by 10% and 9%, respectively, at the condition of initial glycerol concentration at 30 g/L (Table 1, Figure 2C).

These results suggested that 3-HPA concentration vary with different initial concentrations of glycerol: the higher the initial concentration of glycerol, the higher the 3-HPA concentration in the early stage of fermentation. And there was a close relationship between 3-HPA accumulation in the early stage of fermentation and lactate accumulation in the late stage of fermentation: a higher concentration of 3-HPA resulted in a higher concentration of lactate. A higher concentration of 3-HPA during the early stage of fermentation significantly affected the final concentrations of the biomass and 1,3-PD, as did the lactate concentration in the late stage of fermentation.

Effect of lactate on the inhibition caused by 3-HPA in 1,3-PD fermentation

To investigate whether 3-HPA accumulation during the early stage affected cell growth and 1,3-PD production indirectly through lactate accumulation during the late stage of fermentation, a fed-batch fermentation with initial glycerol concentration of 40 g/L by using a lactate dehydrogenase-deficient strain (KG1Δldh) was examined. Although the concentration of lactate was very much lower due to the deficiency of lactate dehydrogenase [5] (Figure 2B), the variations of 3-HPA was almost the same

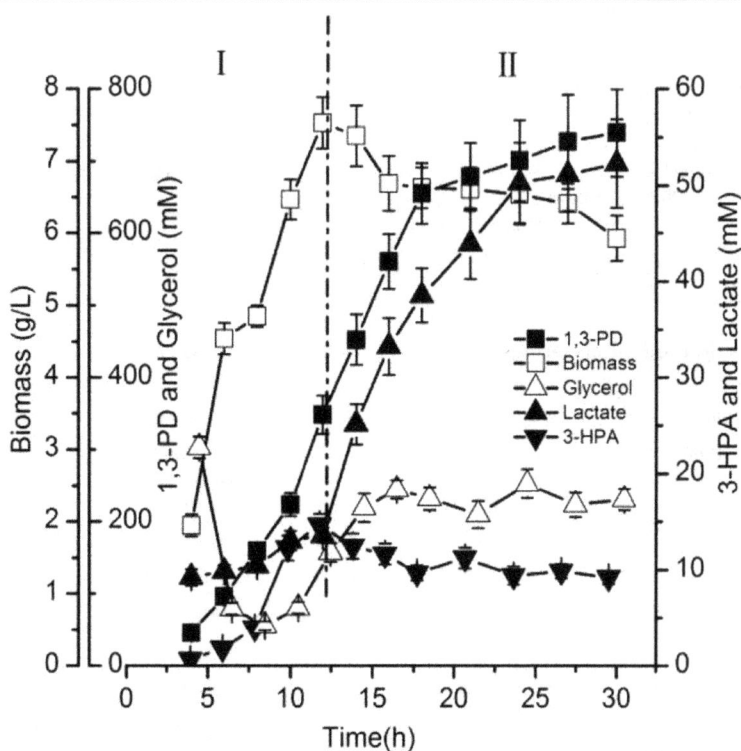

Figure 1 Time course of 1,3-propanediol fed-batch fermentation by KG1. With an initial concentration of glycerol of 40 g/L. Open squares, biomass; closed squares, 1,3-PD; open triangles, glycerol; closed triangles, lactate; closed inverted triangles, 3-hydroxypropionaldehyde (3-HPA).

as that of KG1 (Figure 2A). The final concentrations of biomass and 1,3-PD by KG1Δldh were almost identical to that by KG1 (Table 1, Figure 2C).

The above results suggested that lactate accumulation during the late stage of fermentation do not have an obvious influence on 1,3-PD formation in cells. So, 3-HPA accumulation during the early stage of fermentation decreased the 1,3-PD concentration directly rather than through lactate accumulation during the late stage of fermentation. Maybe this was because of the 3-HPA accumulation during the early stage of fermentation inhibited the central carbon pathway at the post translation level of the cell growth, which could result in irreversible damage to cells in the fermentation broth and would cause an irreversible cessation of the fermentation process until the end of fermentation [11,14].

Lactate synthesized by the oxidative pathway in *K. pneumoniae*, which competed NADH with 1,3-PD formation [15]. It was reported that NADH derived from blocking lactic acid synthesis pathway was used to produce 2,3-BD instead of 1,3-PD [16]. Our experiments data also showed that, compared with KG1, the final concentrations of 2,3-BD increased by 17% while the final concentration and yield of 1,3-PD were almost unchanged by KG1Δldh.

Effect of addition of lactate on the late stage of fermentation

When KG1Δldh was cultured at an initial glycerol concentration of 30 g/L, lactate was added in stage II to maintain a concentration around 50 mM. Compared with the culture of KG1 at an initial glycerol concentration of 40 g/L, the final cell density and 1,3-PD concentration were increased with a decreased 3-HPA in stage I, even though the lactate concentration was similar (Table 1).

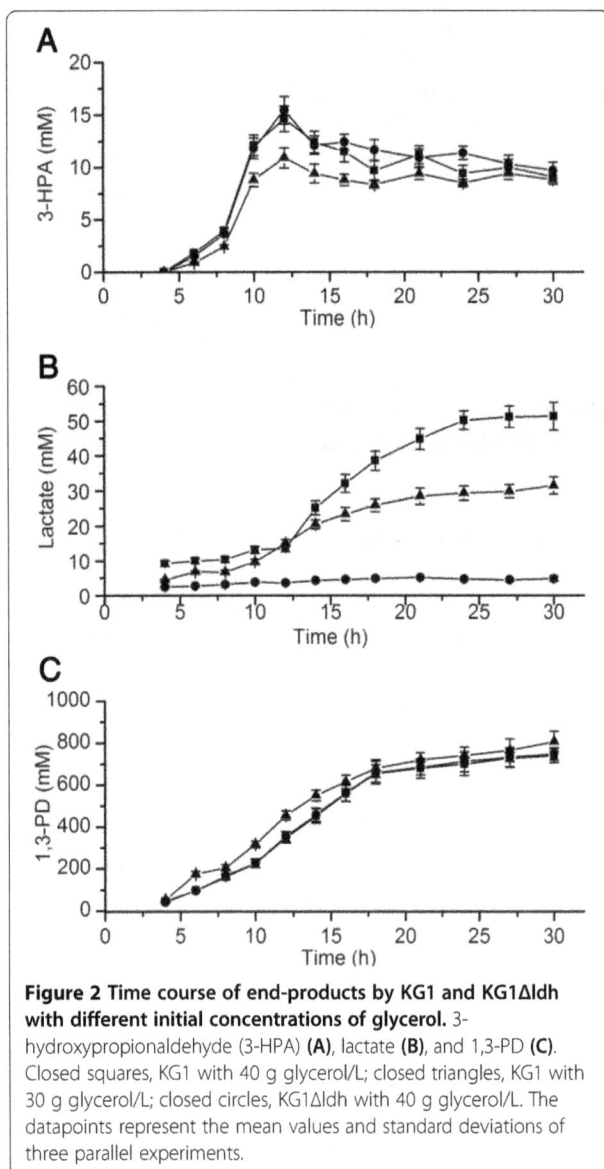

Figure 2 Time course of end-products by KG1 and KG1Δldh with different initial concentrations of glycerol. 3-hydroxypropionaldehyde (3-HPA) **(A)**, lactate **(B)**, and 1,3-PD **(C)**. Closed squares, KG1 with 40 g glycerol/L; closed triangles, KG1 with 30 g glycerol/L; closed circles, KG1Δldh with 40 g glycerol/L. The datapoints represent the mean values and standard deviations of three parallel experiments.

Table 1 Results of 1,3-propanediol fed-batch fermentation by KG1 and KG1Δldh with different initial concentrations of glycerol

Strain and initial glycerol concentration (g/L)		Lactate (mM)	3-HPA (12 h) (mM)	Biomass (g/L)	2,3-BD (mM)	1,3-PD (mM)	1,3-PD yield (%)
KG1	20	18.9 ± 1.1	<1.5	6.9 ± 0.6	242 ± 13	872 ± 43	64 ± 2
KG1Δldh	20	4.9 ± 0.5	<2.0	7.1 ± 0.7	261 ± 9	884 ± 38	65 ± 2
KG1	30	31.4 ± 2.5	10.9 ± 1.0	6.5 ± 0.3	213 ± 17	808 ± 48	59 ± 2
KG1Δldh	30	55.3 ± 5.6 (+50)	11.9 ± 1.3	6.6 ± 0.3	240 ± 20	804 ± 48	59 ± 2
KG1	40	51.3 ± 4.0	14.6 ± 1.2	5.9 ± 0.4	190 ± 12	740 ± 32	55 ± 3
KG1	40	101.3 ± 9.0 (+50)	14.5 ± 1.2	5.3 ± 0.4	195 ± 11	742 ± 45	52 ± 3
KG1Δldh	40	4.8 ± 0.7	15.4 ± 1.3	5.7 ± 0.3	222 ± 19	748 ± 30	56 ± 2
KG1Δldh	40	54.8 ± 4.2 (+50)	13.5 ± 1.0	5.9 ± 0.4	225 ± 15	747 ± 54	55 ± 3

(+50) = 50 mM lactate adding into the fermentation media in the late stage of fermentation. Yield of 1,3-propanediol is on basis of mass. The date are final values except 3-HPA; the data of 3-HPA are peak values in 12 h of fermentation. Values are the mean values and standard deviations of three parallel experiments.

To further investigate the effect of lactate in the late stage on 1,3-PD production, 50 mM of lactate was added to the late stage of the 1,3-PD fed-batch fermentation using KG1 with an initial glycerol concentration of 40 g/L. The result was shown in Figure 3A,B. With the addition of 50 mM lactate in stage II, the growth of cells was slightly inhibited, the final biomass concentration was 5.3 g/L and was 10% lower than that without lactate addition. 1,3-PD concentration was 742 mM, which was nearly unchanged with that without adding 50 mM lactate (Table 1). In 1,3-PD production by *K. pneumoniae*,

lactate is main byproducts and can even reach 450 mM in the broth [5,6]. In our experiments, within the concentration of 100 mM, lactate has no inhibitory effect on 1,3-PD production, indicating that 3-HPA, rather than lactate, inhibits 1,3-PD production.

Effect of a lower concentration of 3-HPA on 1,3-PD fermentation by *K. pneumoniae*

Based on the results shown above, further experiments were executed using an initial concentration of glycerol of 20 g/L. The 3-HPA concentration curve using KG1ΔΙdh was almost the same as that of KG1 (Figure 4A) and they were the lowest among all of experiments due to the lowest initial concentration of glycerol. The final lactate concentration by KG1 was much lower than that with 40 g/L glycerol, whereas, by KG1ΔΙdh, the concentration was below the detection limit (Figure 4B). The concentrations of biomass and 1,3-PD were virtually identical (Table 1). The concentration and yield of 1,3-PD increased by 18% and 16%, respectively, compared to the original experimental method with KG1.

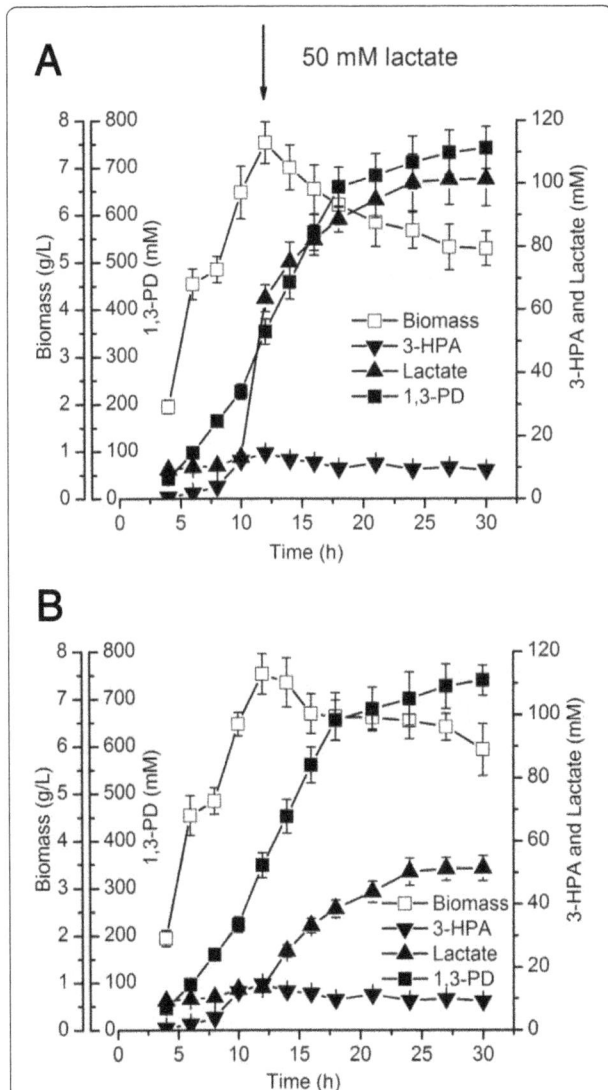

Figure 3 The effect of extra addition of lactate on the late stage of fermentation. 1,3-propanediol fed-batch fermentation by KG1 with an initial concentration of glycerol of 40 g/L, in (A), 50 mM lactate were added into the fermentation media in the late stage of fermentation. And in (B), there is no addition of lactate. Open squares, biomass; closed triangles, lactate; closed inverted triangles, 3-hydroxypropionaldehyde (3-HPA). The datapoints represent the mean values and standard deviations of three parallel experiments.

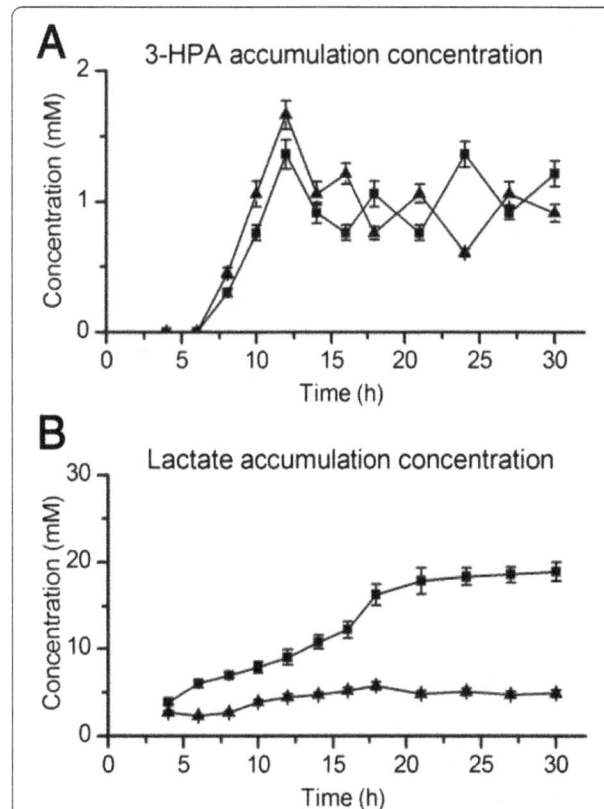

Figure 4 Time course of end-products by KG1 and KG1ΔΙdh with lower initial concentration of glycerol. 3-hydroxypropionaldehyde (3-HPA) (A) and lactate (B). Closed squares, KG1 with 20 g glycerol/L; closed triangles, KG1ΔΙdh with 20 g glycerol/L. The datapoints represent the mean values and standard deviations of three parallel experiments.

Conclusions

In this study, several 1,3-PD fed-batch fermentations had been carried out by using the strains KG1 and KG1Δldh to clarify the effects of inhibition caused by moderate 3-HPA concentrations during fermentation. The results indicated that 3-HPA accumulation during the early stage of fermentation was the main inhibitor in the fermentation and the inhibitory effect was unaffected by lactate accumulation during the late stage of fermentation. The influences of extra lactate being added into the fermentation media in the late stage of fermentation on the cell growth and 1,3-PD production was also investigated, and a concentration of 100 mM lactate was observed to have no inhibitory effect on 1,3-PD production.

Based on the above results, a new fed-batch fermentation process using an initial glycerol concentration of 20 g/L by KG1 was developed, the concentration and yield of 1,3-PD reached 872 mM and 64%, respectively, which were increased by 18% and 16%, respectively, over the initial experimental levels. The information gained in this study could be helpful to enhance 1,3-PD production by finely controlling the entire bioprocess.

Abbreviations
3-HPA: 3-hydroxypropionaldehyde; 2,3-BD: 2,3-butanediol; HPLC: high-performance liquid chromatography; *K. pneumoniae*: *Klebsiella pneumoniae*; NADH: nicotinamide adenine dinucleotide reduced form; 1,3-PD: 1,3-propanediol.

Competing interests
The authors declare that they have no competing interests.

Authors' contributions
ZZ designed and drafted the manuscript. JZ carried out the molecular genetic studies, participated in the sequence alignment, and drafted the manuscript. LG participated in the sequence alignment. LL and SF participated in the design of the study and performed the statistical analysis. JX conceived of the study, participated in its design and coordination and helped to draft the manuscript. HG guided our design, draft and revise of the study during the whole processing. All authors read and approved the final manuscript.

References
1. Kaur G, Srivastava AK, Chand S (2012) Advances in biotechnological production of 1,3-propanediol. Biochem Eng J 64:106–118, doi:10.1016/j.bej.2012.03.002.
2. Maervoet VET, Mey MD, Beauprez J, Maeseneire SD, Soetaert WK (2011) Enhancing the microbial conversion of glycerol to 1,3-propanediol using metabolic engineering. Org Process Res Dev 15:189–202, doi:10.1021/op1001929.
3. Zhu JG, Li S, Ji XJ, Huang H, Hu N (2009) Enhanced 1,3-propanediol production in recombinant *Klebsiella pneumoniae* carrying the gene yqhD encoding 1,3-propanediol oxidoreductase isoenzyme. World J Microbiol Biotechnol 25:1217–1223, doi:10.1007/s11274-009-0005-7.
4. Biebl H, Menzel K, Zeng AP, Deckwer WD (1999) Microbial production of 1,3-propanediol. Appl Microbiol Biotechnol 52:289–297, doi:10.1007/s002530051523.
5. Xu YZ, Guo NN, Zheng ZM, Ou XJ, Liu HJ, Liu DH (2009) Metabolism in 1,3-propanediol fed-batch fermentation by a D-lactate deficient mutant of *Klebsiella pneumoniae*. Biotechnol Bioeng 104:965–972, doi:10.1002/bit.22455.
6. Yang G, Tian JS, Li JL (2007) Fermentation of 1,3-propanediol by a lactate deficient mutant of *Klebsiella oxytoca* under microaerobic conditions. Appl Microbiol Biotechnol 73:1017–1024, doi:10.1007/s00253-006-0563-7.
7. Nakamura CE, Whited GM (2003) Metabolic engineering for the microbial production of 1,3-propanediol. Curr Opin Biotechnol 14:454–459, doi:10.1016/j.copbio.2003.08.005.
8. Seo JW, Seo MY, Oh BR, Heo SY, Baek JO, Rairakhwada D, Luo LH, Hong WK, Kim CH (2010) Identification and utilization of a 1,3-propanediol oxidoreductase isoenzyme for production of 1,3-propanediol from glycerol in *Klebsiella pneumoniae*. Appl Microbiol Biotechnol 85:659–666, doi:10.1007/s00253-009-2123-4.
9. Hao J, Lin RH, Zheng ZM, Sun Y, Liu DH (2008) 3-Hydroxypropionaldehyde guided glycerol feeding strategy in aerobic 1,3-propanediol production by *Klebsiella pneumoniae*. J Ind Microbiol Biotechnol 35:1615–1624, doi:10.1007/s10295-008-0405-y.
10. Cheng KK, Liu HJ, Liu DH (2005) Multiple growth inhibition of *Klebsiella pneumoniae* in 1,3-propanediol fermentation. Biotechnol Lett 27:19–22, doi:10.1007/s10529-004-6308-8.
11. Zheng ZM, Wang TP, Xu YZ, Dong CQ, Liu DH (2011) Inhibitory mechanism of 3-hydroxypropionaldehyde accumulation in 1,3-propanediol synthesis with *Klebsiella pneumoniae*. Afr J Biotechnol 10:6794–6798, doi:10.5897/AJB11.825.
12. Yamamoto S, Izumiya H, Morita M, Arakawa E, Watanabe H (2009) Application of (lambda) Red recombination system to Vibrio cholerae genetics: simple methods for inactivation and modification of chromosomal genes. Gene 438:57–64, doi:10.1016/j.gene.2009.02.015.
13. Cirde SJ, Stone L, Boruff CS (1945) Acrolein determination by means of tryptophane. Ind Eng Chem Anal Ed 17:259–262, doi:10.1021/i560140a021.
14. Hao J, Wang W, Tian JS, Li JL, Liu DH (2008) Decrease of 3-hydroxypropionaldehyde accumulation in 1,3-propanediol production by over-expressing dhaT gene in *Klebsiella pneumonia* TUAC01. J Ind Microbiol Biotechnol 35:735–741, doi:10.1007/s10295-008-0340-y.
15. Kaur G, Srivastava AK, Chand S (2012) Advances in biotechnological production of 1,3-propanediol. Biochem Eng J 64:106–118, doi:10.1016/j.bej.2012.03.002.
16. Celińska E (2010) Debottlenecking the 1,3-propanediol pathway by metabolic engineering. Biotechnol Adv 28:519–530, doi:10.1016/j.biotechadv.2010.03.003.

2

Microbial transformation of quinic acid to shikimic acid by *Bacillus megaterium*

Saptarshi Ghosh, Harish Pawar, Omkar Pai and Uttam Chand Banerjee[*]

Abstract

Background: Biotransformation of quinic acid to shikimic acid was attempted using whole cells of Bacillus megaterium as a biocatalyst.

Results: Physico-chemical parameters such as temperature (37°C), pH (7.0), agitation (200 rpm), substrate (5 mM) and cell mass concentrations (200 kg/m 3) and reaction time (3 h) were found optimum to enhance the bioconversion. Maximum conversion (89%) of quinic acid to shikimic acid was achieved using the above optimized parameters. Shikimic acid was extracted from the reaction mixture by a pH-dependent method and maximum recovery (76%) was obtained with petroleum ether.

Conclusions: Biotransformation of quinic acid to shikimic acid seems to be a better alternative over its fermentative production.

Keywords: Shikimic acid; Quinic acid; Biotransformation

Background

Shikimic acid, a key intermediate of the aromatic amino acid synthesis pathway [1] has immense pharmaceutical importance. This pathway is present in animals, plants and even in microorganisms as the common route for the synthesis of aromatic compounds [2–4]. Shikimic acid, a highly functionalized six-membered carbocyclic ring with three asymmetric centres, is a potent chiral building block for the synthesis of several biologically important compounds [5]. Three aromatic amino acids, more than ten different antibiotics and many herbicides and pesticides have been synthesized from shikimic acid. The molecule has come under limelight in the recent years as it is an important precursor for the synthesis of Oseltamivir (Tamiflu), the only drug against avian flu caused by H5N1 virus [6,7]. Synthesis of Tamiflu is solely dependent on the supply of shikimic acid. It is estimated that nearly two-thirds of the requirement of shikimic acid is still being sourced from plants, with the remaining one-third only obtained from genetically engineered *Escherichia coli* [8,9]. Several genetically modified strains have been used for over production of shikimic acid by fermentation,

based on alteration in the central carbon metabolism [10–15]. Still, there is a need to develop cost-effective and better alternative method for shikimic acid production. Microbial biotransformation has been used for the production of shikimic acid from quinic acid in recent few years as a potent alternative of the available methods [16,17]. The production of shikimic acid from quinic acid (1.4 mM) using *Gluconobacter oxydans* as biocatalyst has been reported [18,19] with both whole cell catalyst and immobilized cell system (57% to 77% bioconversion). In the biotransformation, quinate dehydrogenase (QDH), a classical membrane-bound quinoprotein containing pyrroloquinoline quinine (PQQ) as the coenzyme functions as the primary enzyme in quinate oxidation converting quinic acid to 3-dehydroquinate (3-DHQ) (Figure 1) [9,18]. This 3-DHQ further gets converted to 3-dehydroshikimate (3-DHS) by 3-dehydroquinate dehydratase (DQD). Finally, 3-DHS gets converted to shikimic acid by the action of nicotinamide adenine dinucleotide phosphate (NADP)-dependent shikimate dehydrogenase (SKDH). Entry into the shikimate pathway from quinic acid seems to be advantageous over the classical pathway starting from glucose as there is less number of steps involved. From quinic acid, it is converted to shikimate by three enzymatic steps using microbial biotransformation [19]. Bioconversion of dehydroshikimate (0.2 mM) to

* Correspondence: ucbanerjee@niper.ac.in
Department of Pharmaceutical Technology (Biotechnology), National Institute of Pharmaceutical Education and Research, Sector- 67, S.A.S. Nagar, Punjab 160062, India

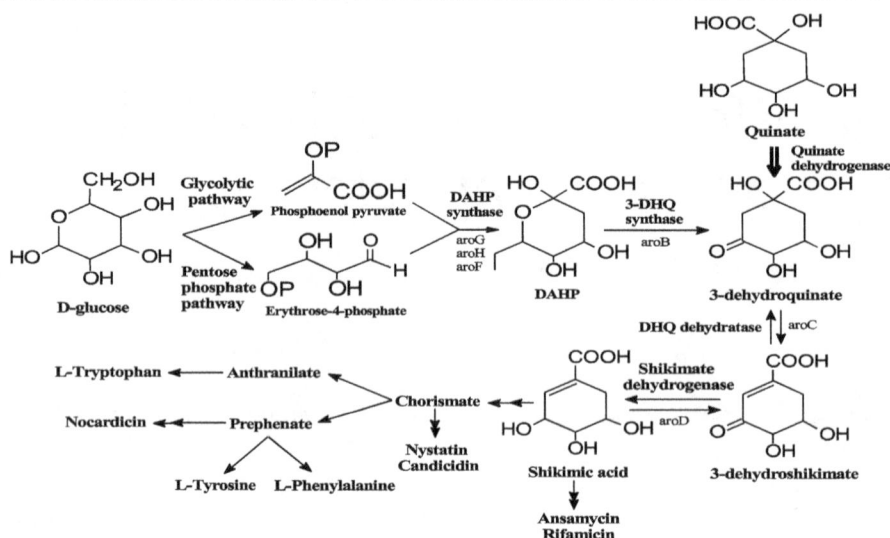

Figure 1 Metabolic map of shikimate pathway. Entrance to the shikimate pathway from quinate is indicated by a fat arrow [9,18].

shikimate using purified shikimate dehydrogenase with 100% yield was also reported in literature [20]. In this study, whole cells of *Bacillus megaterium* MTCC 428 were used as biocatalyst for the transformation of quinic acid to shikimic acid. Various physico-chemical parameters of the biochemical reaction were optimized to enhance the yield of shikimic acid.

Material and methods

Materials
Quinic acid, shikimic acid and dehydroshikimic acid were obtained from Sigma Aldrich chemical company (Milawukee, WI, USA). Growth media components were purchased from Hi-Media Inc. (Mumbai, India). All other chemicals were of analytical grade.

Microorganisms and culture conditions
Bacillus megaterium MTCC 428 was obtained from Microbial Type Culture Collection, Institute of Microbial Technology, Chandigarh, India. The organism was cultured at 37°C in modified nutrient broth (NB) medium (pH 7.0) for 24 h. The media components were as follows (kg/m^3): peptone 10, beef extract 1.5, yeast extract 1.5, and glucose 10.

Biocatalyst preparation
The strain was revived from glycerol stock and grown in a 20 mL medium and transferred (5%, v/v) to 100 mL production medium at 37°C (200 rpm) in an incubator shaker (Kuhner shaker, Germany). After 24 h, cells were harvested by centrifugation at 10,000 × g (Sigma 6 K15, GmbH, Germany). Cells were thoroughly washed with 100 mM phosphate buffer (pH 7.0), and resuspended in

the same buffer with a cell concentration of 100 kg/m^3 and used for the biotransformation reaction.

Optimization of biotransformation parameters
Various physico-chemical parameters for the biotransformation of quinic acid to shikimic acid were optimized in 5 mL reaction mixture by varying one parameter at a time. Temperature was varied between 25°C to 45°C and pH values were adjusted between pH 4.0 to 9.0. An increased substrate concentration (2 to 20 mM) in the reaction mixture was used. The cell mass concentration was varied between 100 to 400 kg/m^3. To optimize the cofactor recycling, different glucose concentrations (0.05 to 0.4 M) were used in the reaction mixture. The effect of reaction time on the yield of shikimic acid was observed by carrying out the reaction up to 6 h.

Analytical methods
High-performance liquid chromatography
An analytical HPLC method was developed using high-performance liquid chromatography system (Shimadzu 10 AD VP, Kyoto, Japan). Shikimic acid was analyzed on an Alltech OA-2000 organic acid column (100 × 6.5 mm, 6.5 μm) (Grace Davison Discovery Science, Deerfield, IL, USA) using 2.5 mM H$_2$SO$_4$ as mobile phase at a flow rate of 0.3 mL/min and detected at 215 nm by UV detector. A standard curve of shikimic acid was made to quantify the biotransformation.

LC-MS analysis
Reaction mixture was analyzed by LC-MS (Model: LTQ-XL, Thermo Scientific, USA) to further confirm the product formation. Alltech OA-2000 organic acid column (100 × 6.5 mm, 6.5 μm) (Grace Davison Discovery

Science, Deerfield, IL, USA) was used with 5 mM formic acid as mobile phase at a flow rate of 0.5 mL/min and detected by UV detector at 215 nm.

MALDI-TOF analysis

Samples were analyzed in MALDI-TOF/TOF mass spectrometer (Bruker Ultraflex-TOF/TOF, Madison, WI, USA) for detecting the molecular mass of the product. After the reaction, sample was extracted by petroleum ether with pH adjustment and dried under vacuum. The dried sample was solubilized in methanol and used for the analysis.

NMR analysis

After the completion of the reaction, the mixture was extracted with petroleum ether by pH adjustment and dried under vacuum. The dried sample was solubilized in DMSO and ^{1}H NMR and ^{13}C NMR spectra were obtained with Bruker AVANCE 400 MHz (^{1}H 400 and ^{13}C 100 MHz). Chemical shifts were expressed in δ units relative to the tetramethylsilane (TMS) signal as an internal reference in DMSO.

Extraction of shikimic acid from reaction mixture

Being a highly polar compound, it is very hard to isolate shikimic acid from the reaction mixture by solvent extraction. Therefore, pH-dependent extraction method was selected where, at a particular pH, the compound gets deionized and it comes into the organic non-polar phase during solvent extraction. Before the extraction, the pH of the solution was adjusted to 4.48 and extracted with different solvents such as ethyl acetate, dichloromethane, n-butanol, petroleum ether, etc.

Results and discussion

Biotransformation of quinic acid to shikimic acid using whole cells of B. megaterium

Quinic acid is converted to shikimic acid by two enzymatic systems in Gluconobacter oxydans, as reported by Adachi et al. [19,20]. Along with the classical shikimate pathway, starting from glucose, quinic acid can also be a potential entry point with the formation of 3-dehydroquinate by the enzyme quinate dehydrogenase. The reaction was carried out with quinic acid as the substrate and whole cells of B. megaterium as the catalyst. The reaction conditions were as follows: quinic acid, 2.5 mM; cell mass concentration, 100 kg/m^3; temperature, 30°C; mixing rate 200 rpm. Under the un-optimized conditions 42% conversion of quinic acid to shikimic acid was obtained (Additional file 1: Figure 1-4 of Supporting Information).

Optimization parameters

Various physico-chemical parameters for the biotransformation of quinic acid to shikimic acid using whole cells of B. megaterium were optimized.

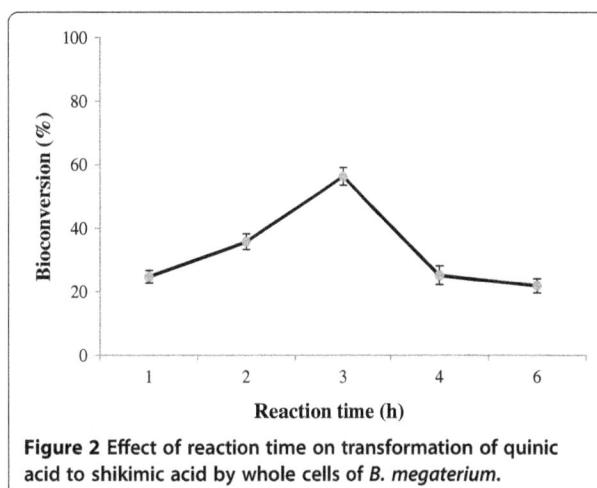

Figure 2 Effect of reaction time on transformation of quinic acid to shikimic acid by whole cells of B. megaterium.

Effect of reaction time

To determine the effect of reaction time, the reaction mixture (5 mL) containing whole cells of B. megaterium (200 kg/m^3), quinic acid (5 mM), and glucose (5%) was incubated at 30°C (200 rpm). Samples were taken at 1 h interval upto 6 h. It is evident from Figure 2 that maximum conversion took place in 3 h while it decreased with the incubation time (Additional file 1: Figure 5 of Supporting Information). It might be due to the oxidation of shikimic acid to dehydroshikimate on long incubation. Hence, the subsequent experiments were carried out for 3 h only which is better than the previous reports of having 1.5 h as reaction time with 1.4 mM substrate [19] and 5 h as reaction time with 0.2 mM substrate [20].

Effect of reaction temperature

To investigate the effect of reaction temperature, the reaction mixtures as mentioned above were incubated at various temperatures (20°C to 45°C). It is evident from Figure 3 that percentage conversion increased with the increase of incubation temperature and maximum

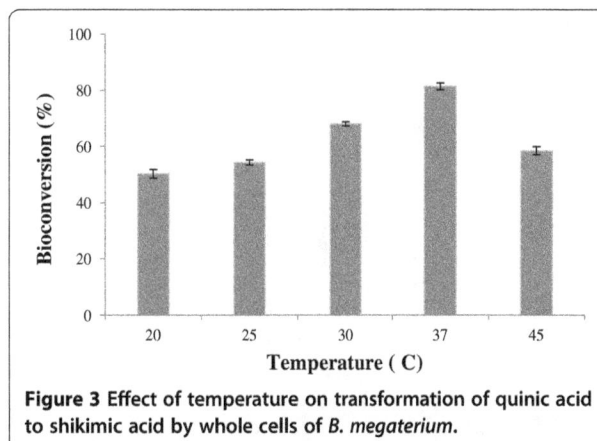

Figure 3 Effect of temperature on transformation of quinic acid to shikimic acid by whole cells of B. megaterium.

conversion was obtained at 37°C (Additional file 1: Figure 6 of Supporting Information). At higher temperature (45°C), the percentage conversion decreased. It might be due to the deactivation of the enzyme at higher temperature or oxidation of shikimic acid to dehydroshikimate by the reverse reaction. Biotransformation of quinic acid to shikimic acid by whole cells and pure enzyme was also reported in literature and in both the cases the reactions were carried out at 25°C [19,20].

Effect of reaction pH

To determine the effect of reaction pH, the reaction mixtures as mentioned above were incubated at 37°C having different initial pHs. The initial pHs of the reaction mixtures were adjusted between 4.0 to 9.0 using various buffers. It is evident from Figure 4 that percentage conversion increased with the increase in pH and maximum conversion was obtained at pH 7.0 which was similar to the previous reports [19,20] (Additional file 1: Figure 7 of Supporting Information). It was seen that at higher pH (7.5 to 9.0) there was a decrease in percentage conversion. It might be due to the oxidation of shikimic acid to dehydroshikimate by reverse reaction.

Effect of substrate concentration

To determine the effect of substrate concentration, reactions were carried out with the increasing concentrations of quinic acid (2 to 20 mM). It is evident from Figure 5 that maximum conversion took place with 5 mM quinic acid and it decreased with the increasing substrate concentration (Additional file 1: Figure 8 of Supporting Information). The low conversion of quinic acid to shikimic acid at higher substrate concentration might be due to the substrate inhibition. In our case, biotransformation with the whole cells of B. megaterium, a maximum of 80% conversion with 5 mM quinic acid was obtained while 55% to 77% conversion with the whole cells of Gluconobacter oxydans with 1.4 mM substrate was reported by Adachi et al.

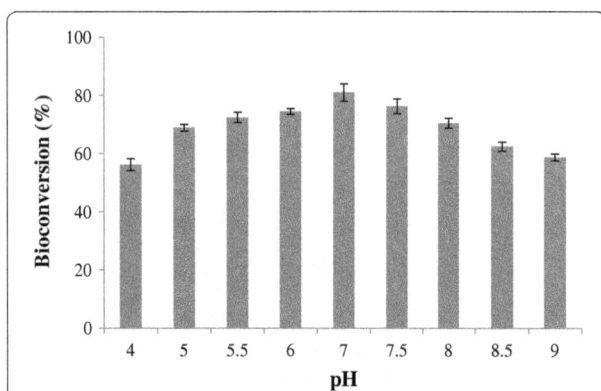

Figure 5 Effect of substrate concentration on transformation of quinic acid to shikimic acid by whole cells of B. megaterium.

[19]. At higher substrate concentration (above 5 mM), quinate dehydrogenase, the foremost enzyme in the biotransformation pathway might be inhibited.

Effect of cell mass concentration

To determine the effect of cell mass (biocatalyst) concentration, reactions were carried out with increasing cell mass concentrations (100 to 400 kg/m^3). It is seen from Figure 6 that maximum conversion was achieved with a cell mass concentration of 200 kg/m^3 and decreased thereafter (Additional file 1: Figure 9 of Supporting Information). Higher viscosity of the reaction mixture at higher cell mass concentration might be responsible for the lower bioconversion. Here, 84% bioconversion was achieved with 200 kg/m^3 cell mass at 3 h, while previous report mentioned a conversion of 57% to 77% at 20 h with 25 kg/m^3 cell mass [19] (Additional file 1: Figure 11 of Supporting Information).

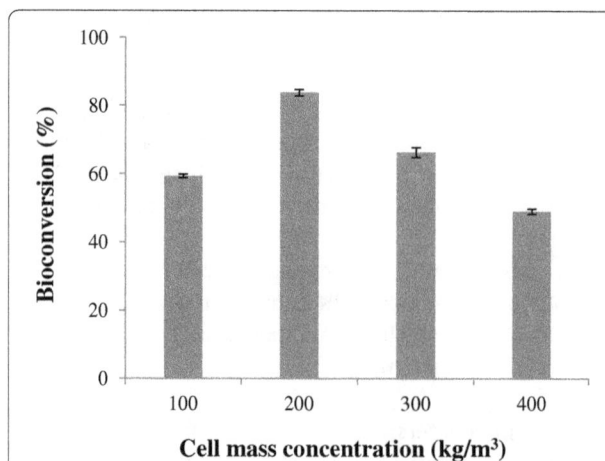

Figure 4 Effect of pH on transformation of quinic acid to shikimic acid by whole cells of B. megaterium.

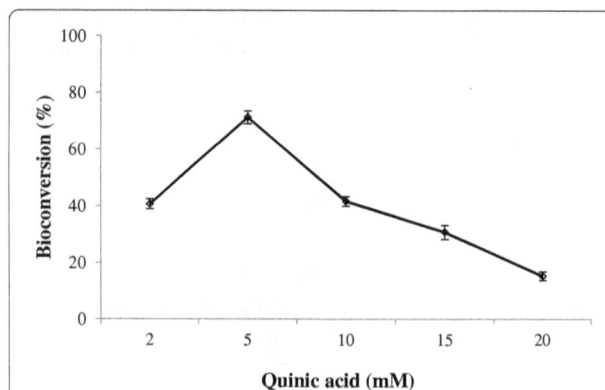

Figure 6 Effect of cell mass concentration on transformation of quinic acid to shikimic acid by whole cells of B. megaterium.

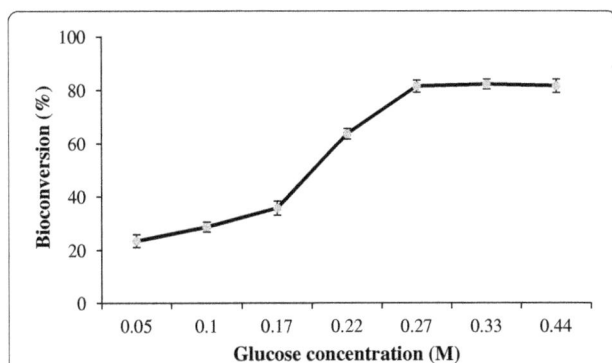

Figure 7 Effect of glucose concentration on transformation of quinic acid to shikimic acid by whole cells of *B. megaterium*.

Effect of glucose concentration

Cofactor recycling plays a key role in the NADPH-dependent enzyme biocatalysis; therefore, it is necessary to optimize the reaction condition for proper cofactor recycling [21,22]. To study the effect of cofactor generation, different concentrations of glucose (0.05 to 0.4 M) were used in the reaction mixture. It is evident from Figure 7 that percentage conversion increased with increase in glucose concentration and maximum conversion (81%) was achieved with 0.3 M glucose (Additional file 1: Figure 10 of Supporting Information). The conversion remained more or less same with the further increase of glucose concentration in the reaction mixture. Under the optimized conditions (37°C, 200 rpm, pH 7.0) 89% conversion was achieved at 3 h (Figure 8) while 57% to 77% conversion was reported using whole cells of *Gluconobacter oxydans* by Adachi et al. [19] (Additional file 1: Figure 11 of Supporting Information).

Optimization of downstream processing of shikimic acid from reaction mixture

The pH-dependent method was selected for the extraction of shikimic acid from the reaction mixture. After workup, the pH of the reaction mixture was adjusted to

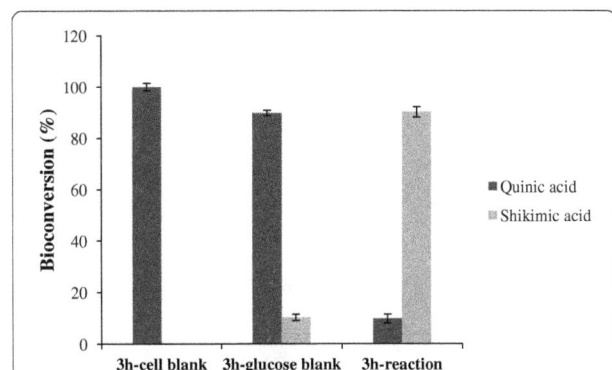

Figure 8 Reaction under optimized condition.

Table 1 Extraction of shikimic acid from reaction mixture using different solvents

Number	Solvent	Amount recovered (mM)	Recovery (%)
1	Ethyl acetate	1.15	25.55
2	Dichloromethane	0.1	2.2
3	*n*-Butanol	3.2	71.11
4	Petroleum ether	3.42	76

its pKa 4.48. Extraction was carried out using different solvents like ethyl acetate, *n*-butanol, petroleum ether and dichloromethane. Post extraction, the samples were concentrated on rotavapor and analyzed by HPLC. Maximum recovery (76%) of shikimic acid was obtained with petroleum ether (Table 1). The sample was then analyzed by matrix-assisted laser desorption/ionisation (MALDI) and nuclear magnetic resonance (NMR).

Identification of shikimic acid

After the completion of the reaction, product was extracted by pH-dependent method. The product was characterized as shikimic acid by ^{1}H and ^{13}C NMR and MALDI. Analytical data were as follows: ^{1}H NMR (400 MHz, MeOD): δ 2.18 to 2.22 (days, 1H), 2.69 to 2.73 (days, 1H), 3.69 to 3.70 (days, 1H), 4.01 (s, 1H), 4.38 (s, 1H), 6.81 (s, 1H); ^{13}C NMR (100 MHz, MeOD): δ 31.69, 67.35, 68.41, 72.78, 130.82, 138.81, 170.20; MALDI-TOF-TOF: *m/z* 197.32, 213.32 (Additional file 1: Figure 12-16 of Supporting Information).

Conclusion

A potential biotransformation process is reported for the production of shikimic acid using whole cells of *B. megaterium* as biocatalyst. Various reaction parameters (reaction time, temperature, pH, substrate, biocatalyst concentration, etc.) were optimized. Maximum conversion (89%) was achieved under the optimized condition. Biotransformation of quinic acid to shikimic acid using purified enzymes are also reported; however, it has many limitations such as the use of co-factors (NADH/NAD$^+$) in the reaction mixture and the quick denaturation of the enzyme during the course of reaction. Although a major advantage of mass transfer limitations are omitted in a cell-free system, due to inherent problems of enzyme denaturation, which makes this method difficult. The main limitation of transformation of quinic acid to shikimic acid is the use of higher substrate concentration. The enzyme titre may be increased by the well-developed techniques of recombinant DNA technology, directed evolution, etc. Shikimic acid production from quinate seems to be a better alternative over its fermentative production.

Additional file

Additional file 1: Supporting information.

Abbreviations

DAHP: 3-deoxy-arabino-heptulosonate-7-phosphate; DHQ: 3-dehydroquinate; DHS: 3-dehydroshikimate; DQD: 3-dehydroquinate dehydratase; GDH: glucose dehydrogenase; HPLC: high-performance liquid chromatography; MALDI: matrix assisted laser desorption/ionisation; MTCC: microbial type culture collection; NADP: nicotinamide adenine dinucleotide phosphate; NADPH: reduced form of nicotinamide adenine dinucleotide phosphate; NMR: nuclear magnetic resonance; QDH: quinate dehydrogenase; rpm: revolutions per minute; SKDH: Shikimate dehydrogenase.

Competing interests

The authors declare that they have no competing interests.

Authors' contributions

SG carried out experiments and data analysis during the study and drafted the manuscript. HP participated in the design of the study and performed experiments. OP conceived of the study and participated in designing experiments. UCB has been involved in drafting the manuscript and revising it critically for important intellectual content and also given final approval of the version to be published. All authors read and approved the final manuscript.

Acknowledgement

One of the authors SG acknowledges the financial support by Indian Council of Medical Research, India.

References

1. Herrmann KM, Weaver LM (1999) The shikimate pathway. Annu Rev Plant Physiol Plant Mol Biol 50:473–503
2. Ganem B (1978) Shikimate-derived metabolites: from glucose to aromatics – recent developments in natural-products of shikimic acid pathway. Tetrahedron 34:3353–3383
3. Pittard AJ (1996) Biosynthesis of aromatic amino acids. In: Neidhardt FC, Curtiss R III, Ingraham JL, Lin ECC, Low KB, Magasanic B, Reznikoff WS, Riley M, Schaechter M, Umbarger HE (eds) *Escherichia coli* and *Salmonella*. Cellular and Molecular Biology. American Society of Microbiology, Washington, DC, pp 458–484
4. Herrmann KM (1995) The shikimate pathway: early steps in the biosynthesis of aromatic compounds. Plant Cell 7:907–919
5. Krämer M, Bongaerts J, Bovenberg R, Kremer S, Müller U, Orf S, Wubbolts M, Raeven L (2003) Metabolic engineering for microbial production of shikimic acid. Metab Eng 5:277–283
6. Farina V, Brown JD (2006) Tamiflu: the supply problem. Angew Chem Int Ed 45:7330–7334
7. Raghavendra TR, Vaidyanathan P, Swathi HK, Ramesha BT, Ravikanth G, Ganeshaiah KN, Srikrishna A, Shaanker RU (2009) Prospecting for alternate sources of shikimic acid, a precursor of Tamiflu, a bird-flu drug. Curr Sci 96(6):771–772
8. Payne R, Edmonds M (2005) Isolation of shikimic acid from star anise seeds. J Chem Educ 82:599–600
9. Ghosh S, Chisti Y, Banerjee UC (2012) Production of shikimic acid. Biotechnol Adv 30:1425–1431
10. Knop DR, Draths KM, Chandran SS, Barker JL, Frost JW (2001) Hydroaromatic equilibrium during biosynthesis of shikimic acid. J Am Chem Soc 123:10173–10182
11. Chandran SS, Yi J, Draths KM, Von Daeniken R, Weber W, Frost JW (2003) Phosphoenolpyruvate availability and the biosynthesis of shikimic acid. Biotechnol Prog 19:808–814
12. Johansson L, Lindskog A, Silfversparre G, Cimander C, Nielsen KF, Lidén G (2005) Shikimic acid production by a modified strain of *E. coli* (W3110.shik1) under phosphate-limited and carbon-limited conditions. Biotechnol Bioeng 92:541–552
13. Ahn JO, Lee HW, Saha R, Park MS, Jung JK, Lee DY (2008) Exploring the effects of carbon sources on the metabolic capacity for shikimic acid production in *Escherichia coli* using in silico metabolic predictions. J Microbiol Biotechnol 18:1773–1784
14. Escalante A, Calderón R, Valdivia A, de Anda R, Hernández G, Ramírez OT, Gosset G, Bolívar F (2010) Metabolic engineering for the production of shikimic acid in an evolved *Escherichia coli* strain lacking the phosphoenolpyruvate: carbohydrate phosphotransferase system. Microb Cell Fact 9:21
15. Cui YY, Ling C, Zhang YY, Huang J, Liu JZ (2014) Production of shikimic acid from *Escherichia coli* through chemically inducible chromosomal evolution and cofactor metabolic engineering. Microb Cell Fact 13:21
16. Mitsuhashi S, Davis BD (1954) Aromatic biosynthesis. XIII. Conversion of quinic acid to 5-dehydroquinic acid by quinic dehydrogenase. Biochim Biophys Acta 15:268–281
17. Yaniv H, Gilvarg C (1955) Aromatic biosynthesis. XIV. 5-dehydroshikimic reductase. J Biol Chem 213:787–795
18. Adachi O, Tanasupawat S, Yoshihara N, Toyama H, Matsushita M (2003) 3-Dehydroquinate production by oxidative fermentation and further conversion of 3-dehydroquinate to the intermediates in the shikimate pathway. Biosci Biotechnol Biochem 67:2124–2131
19. Adachi O, Ano Y, Toyama H, Matsushita K (2006) High shikimate production from quinate with two enzymatic systems of acetic acid bacteria. Biosci Biotechnol Biochem 70:2579–2582
20. Adachi O, Ano Y, Toyama H, Matsushita K (2006) Purification and properties of NADP-dependent shikimate dehydrogenase from *Gluconobacter oxydans* IFO 3244 and its application to enzymatic shikimate production. Biosci Biotechnol Biochem 70:2786–2789
21. Zhang J, Witholt B, Li Z (2006) Efficient NADPH recycling in enantioselective bioreduction of a ketone with permeabilized cells of a microorganism containing a ketoreductase and a glucose 6-phosphate dehydrogenase. Adv Synth Catal 348:429–433
22. Zhang W, O'Connor K, Wang DIC, Li Z (2009) Bioreduction with efficient recycling of NADPH by coupled permeabilized microorganisms. Appl Environ Microbiol 75(3):687–694

3

Poly-β-hydroxybutyrate production and management of cardboard industry effluent by new *Bacillus sp.* NA10

Anish Kumari Bhuwal[1], Gulab Singh[1], Neeraj Kumar Aggarwal[1*], Varsha Goyal[1] and Anita Yadav[2]

Abstract

Background: In the present study, we aim to utilize the ecological diversity of soil for the isolation and screening for poly β-hydroxybutyrate (PHB)-accumulating bacteria and production of cost-effective bioplastic using cardboard industry effluent.

Results: A total of 120 isolates were isolated from different soil samples and a total of 62 isolates showed positive results with Nile blue A staining, a specific dye for PHB granules and 27 isolates produced PHB using cardboard industry effluent. The selected isolate NA10 was identified as *Bacillus sp.* NA10 by studying its morphological, biochemical, and molecular characteristics. The growth pattern for the microorganism was studied by logistic model and exactly fitted in the model. A maximum cell dry weight (CDW) of 7.8 g l^{-1} with a PHB concentration of 5.202 g l^{-1} was obtained when batch cultivation was conducted at 37°C for 72 h, and the PHB content was up to 66.6% and productivity was 0.072 g l^{-1} h^{-1} in 2.0 L fermentor. Chemical characterization of the extracted PHB was done by H_1NMR, Fourier transform infrared spectroscopy (FTIR), thermal gravimetric analysis (TGA), Gas chromatography–mass spectrometry (GC-MS) analysis to determine the structure, melting point, and molecular mass of the purified PHB. The polymer sheet of extracted polymer was prepared by blending the polymer with starch for packaging applications.

Conclusions: The isolate NA10 can be a good candidate for industrial production of PHB from cardboard industry waste water cost-effectively and ecofriendly.

Keywords: Polyhydroxybutyrate (PHB); Bioplastic; Cardboard industry waste water; *Bacillus* sp.

Background

Today, plastics have become a necessary part of contemporary life due to their durability and resistance to degradation. Worldwide production of petroleum-based synthetic polymer was approximately 270.0 million tons in 2007 [1], and these synthetic polymers are found to be recalcitrant to microbial degradation [2]. Problems related to solid waste management of these petrochemical-derived plastics pose a serious threat to global environment. Therefore, current concern about the environmental fate of polymeric materials have created much interest in the development of biodegradable plastic (bioplastic), such as starch derivatives, polylactic acid, cellulosic polymers and polyhydroxyalkanoates, which

plays an important role. In addition to being biodegradable, they have further advantage of being produced from renewable resources [3]. Among the various biodegradable polymer materials, polyhydroxyalkanoates (PHAs) provide a good fully degradable alternative to petrochemical plastics [4,5]. Polyhydroxybutyrate (PHB) was the first PHA to be discovered and is also the most widely studied and best characterized PHA. It is accumulated as a membrane enclosed inclusion in many bacteria at up to 80% of the dry cell weight and nearly 90% in recombinant *E. coli* [6]. In addition to the easy biodegradability and biocompatibility, it has mechanical properties that are very similar to conventional plastics like polypropylene or polyethylene and have many domestic and commercial applications such as food packing films, biodegradable carriers for medicines and insecticides, disposable cosmetic products, absorbable surgical devices and being immunologically compatible

* Correspondence: Neerajkuk26@rediffmail.com
[1]Department of Microbiology, Kurukshetra University, Kurukshetra, Haryana 136119, India
Full list of author information is available at the end of the article

with human tissue can form microspheres and microcapsules [7]. Recently, PHA has been found useful as a new type of biofuel [8]. Besides all these properties and applications, wider use of PHAs is prevented mainly due to their high production cost compared with the oil-derived plastics [9]. High production cost of PHB production is mainly devoted to the expensive carbon substrates and tedious production procedures [10]. Due to the large impact of the high price of carbon sources on production costs, one of the most important approaches to reduce costs is to use wastes and by-products as raw material for the fermentation process. Novel technologies have been developed to produce PHAs from organic matters in wastewater [12-14], industrial wastes [15,16], municipal waste [17], food wastes [18], and activated sludge of paper and pulp mills [19]. Hence, replacement of non-biodegradable with biodegradable plastic from organic waste can provide multiple benefits to the environment and contribute to sustainable development [20]. Therefore, organic waste from cardboard industry waste water could be a good approach for cost effective production PHA.

Methods

Isolation of PHA-producing bacteria

For the isolation of PHA-producing bacteria, various soil samples were collected from different ecological niches. The samples were stored at room temperature until analysis. In 99 mL sterilized water, 1 g of soil sample was mixed. Then, the sample was serially diluted in sterile distilled water and followed by plating on the carbon-rich nutrient agar medium (beef extract 0.3%, peptone 0.5%, sodium chloride 0.5%, glucose 1%, and agar 2%). For the rapid detection and isolation of PHB-producing bacteria, 0.02% alcoholic solution of Sudan black B was applied to stain bacterial colonies and the plates were kept undisturbed for 30 min. The excess dye was then decanted, and plates were rinsed gently by adding 100% ethanol. Colonies unable to incorporate the Sudan black B appeared white, while PHB producers appeared bluish black [21].

Rapid screening for PHA-producing bacteria

The Sudan black B-positive isolates were further screened by Nile blue A, a more specific stain for PHA by a more rapid and sensitive, viable colony method [22]. This dye at concentrations of only 0.5 µg/mL was directly included in carbon-rich nutrient agar medium, and growth of the cells occurred in the presence of the dye. The PHA-accumulating colonies, after Nile blue A staining, showed bright orange fluorescence on irradiation with UV light and their fluorescence intensity increased with the increase in PHA content of the bacterial cells. The isolates which showed bright orange fluorescence on irradiation with UV light after Nile blue A staining were selected as PHA accumulators. The selected PHA accumulators after Nile blue

A staining were checked for growth and PHA production in both nutrient broth (beef extract 0.3%, peptone 0.5%, sodium chloride 0.5%) and cardboard industry effluent.

Pretreatment of cardboard industry waste water

Untreated cardboard industry effluent was collected from the cardboard industry, Yamunanagar (Haryana), India, and stored at 4°C until used for analysis. The effluent was first filtered through the muslin cloth and then through rough filter paper to remove the undesired suspended solid materials from waste water. After this pretreatment step, cardboard industry waste water was used as quantification and production medium for PHA production by selected bacterial isolate.

Morphological and biochemical characterization-selected bacteria

Microscope Stereo Olympus (America) was used to observe the morphology of bacterial colonies grown on nutrient agar. The growth characteristics such as structure, shape, color, margin, surface characteristics, elevation, cell's arrangement, and Gram staining of the bacterial colonies were observed to characterize the bacterial colonies. Various biochemical tests were performed in selected PHB-producing bacteria, namely, indole production test, methyl red and Voges-Proskauer, citrate utilization test, and H_2S production for their biochemical characterization. The fermentative utilization of various carbohydrates (xylose, mannose, maltose, sucrose, raffinose, dextrose, trehalose, fructose, glucose, ribose, lactose, rhamnose, esculin, inulin, mannitol, arabinose, sorbitol, and melibiose) were also followed for 48 h at 37°C by inoculating the selected isolate separately in the defined medium to which various sugars were added.

Molecular identification

Colony PCR (16S rRNA gene amplification)

Colony polymerase chain reaction (Colony PCR) of the isolate was performed according to Gen Elute™ Bacterial genomic DNA kit (Sigma, St. Louis, MO, USA), and the colonies (approximately 1 mm in diameter) were picked up with a sterilized toothpick and directly transferred to the PCR tubes as DNA templates. The PCR amplification of 16S rRNA was done at 94°C for 4 min, 94°C for 20 s, 52°C for 30 s, 72°C for 2 min, and 72°C for 7 min with hold at 4°C. The universal primer 16sF-5′ AGA GTT TGA TCC TGG CTC AGA 3′ and 16sR-5′ ACG GCT ACC TTG TTA CGA CTT 3′ were used.

Detection of PCR products

PCR-amplified DNA fragments were observed by agarose gel electrophoresis in 1% ± agarose gels (FMC). Ten microliters of each amplified mixture and the molecular mass marker were subjected to agarose gel electrophoresis and ethidium bromide staining and tracked by 0.25% of bromophenol blue. The ampli*ed DNA fragments were visualized by gel documentation box (Genie, Redmond, WA, USA).

Sequencing and analysis of 16S rRNA gene

After that, the PCR product was purified by GenElute™ gel extraction kit (Sigma) method and sequencing of PCR product was done by Sanger sequencing method at 96°C for 1 min, at 96°C for 15 s, at 52°C for 30 s, and at 60°C for 4 min for 25 cycles with a hold at 4°C. The ABI Prism Big Dye Terminator Cycle Sequencing Ready Reaction Kit (Applied Biosystems, Foster City, CA, USA) was used for the sequencing of the PCR product. A combination of universal primers was chosen to sequence the gene sequence. Samples were run on an ABI Prism 3130×1 Genetic Analyzer (Applied Biosystems). The chromatogram was made by Chromas 2.4. software. The obtained sequence was subjected to search for closest possible species by BLAST tools and distance matrix tree tool available at National Centre for Biotechnology Information (NCBI). Phylogenetic tree was constructed by MEGA5.2 phylogenetic tree analysis software.

Optimization of various growth parameters for PHB production in shake flask culture

The optimization for maximum PHA production by selected isolate was carried in 250-ml Erlenmeyer flask using preprocessed cardboard industry waste water as production medium at 100 rpm. Several cultural parameters were evaluated to determine their effect on biomass and PHB production using cardboard industry waste water. The optimized value for each parameter was selected and kept constant for further experiments. Several cultural parameters like production media concentration (25%–100%) time of incubation (24–96 h), temperature of incubation (25°C–55°C), and effect of pH (5.5–8) were evaluated to determine their effect on biomass accumulation and PHA production.

Scale up of poly-β-hydroxybutyrate production and growth kinetics

The production of PHAs was carried out in a 2.0-L fermenter (Minifors CH 4103, Switzerland) with a working volume of 1.25 L using the optimized parameters and production medium. Initially, the production medium (preprocessed undiluted cardboard industry waste water) was added into the fermenter, the pH was adjusted to 7.0, and the medium was sterilized *in situ*. Dissolved oxygen was maintained at 80%–100% air saturation; at the start of the process, 1% (v/v) of overnight culture (at log phase) was inoculated aseptically and the impeller speed was maintained at 100 rpm and temperature at 37°C. The pH was maintained at 7.0 using 1 N NaOH and 1 N HCL. Antifoam (silicone oil) at a concentration of 1:10 (v/v) in water was added after 36th and 54th hour.

The growth curves (DCW vs. time) were prepared to determine the start and end point of exponential phase for NA10 in batch fermentation. Various other growth parameters such as maximum specific growth rate (μ) and doubling time (T_d) were determined according to the method provided by Painter and Marr [24] and Levasseur et al. [25].

$$\mu \, K' = (\ln w_{t2}/w_{t1})/ \, t_2 > t_1; \, t_2 t_1$$

where w_{t2} and w_{t1} are the dry cell weight at the different time points (t_1 and t_2), respectively.

$$T_d = \ln2/\mu_m$$

where T_d is the doubling time and μ_m is the maximum specific growth rate.

Cell dry weight (CDW) and PHB yield coefficient relative to cell dry weight (Yp/x, *g/g*, defined as gram PHB produced per gram dry cell mass produced) [26], PHB concentration (g/L, defined as g PHB measured in 1 L culture), PHB content [g/g, defined as the ratio of PHB concentration (g/L) to dry cell concentration (g/L)], and PHB productivity (g/L/h) [10,11] were measured and calculated per definition during the fermentation process.

All experiments were performed in triplicate to check the reproducibility. The results were analyzed statistically by determining standard deviations values and performing analysis of variance (ANOVA) test.

Extraction and quantitative analysis of PHB

For extraction and quantitative analysis of PHB, culture broth was centrifuged at 8,000 rpm for 15 min after 72 h of incubation at 37°C. The pellet dissolved in 10 mL sodium hypochlorite was incubated at 50°C for 1 h for lyses of cells. The cell extract obtained was centrifuged at 12,000 rpm for 30 min and then washed sequentially with distilled water, acetone, and absolute ethanol. After washing, the pellet was dissolved in 10 mL chloroform (AR grade) and incubated overnight at 50°C [23]. After evaporation at 50°C, 10 mL of sulfuric acid was added to it and placed in water bath for 10 min at 100°C. This converts PHB into crotonic acid, which gives absorbance maximum at 235 nm [24,25]. PHB (Sigma Aldrich) was used as standard for making standard curve.

Polymer analysis

¹H-NMR spectroscopy and FTIR analysis

The identity of individual monomer unit was confirmed by proton nuclear magnetic resonance (¹H-NMR) spectroscopy. ¹H-NMR spectra were acquired by dissolving the polymer in deuterated chloroform (CDCl₃) at a concentration of 10 mg/ml and analyzed on a Bruker Avance

II 500 spectrometer (Madison, WI, USA) at 22°C with 7.4 ms pulse width (30° pulse angle), 1 s pulse repetition, 10,330 Hz spectral width, and 65,536 data points. Tetramethylsilane was used as an internal shift standard. FTIR analysis of the polymer sample was carried out on MB-3000, ABB Fourier transform infrared (FTIR) spectrophotometer in the range 4,000 to 600 cm^{-1}.

TGA

Thermal gravimetric analysis (TGA) was performed using a TGA instrument (Perkin Elmer, Diamond TG/DTA analyzer, USA) calibrated with indium. The temperature was ramped at a heating rate of 10°C/min in nitrogenous environment to a temperature (700°C) well above the degradation temperature of the polymers.

GC-MS analysis

For molecular analysis of purified polymer, a coupled Gas chromatography–mass spectrometry (GC-MS) was performed using a GC-MS-QP 2010 Plus model, with capillary Column-Rtx-5 MS (30 m × 0.25 mm i.d. × 0.25 μm film thickness). The samples were injected (3 μL) in the splitless mode, and the injection temperature was 260°C and column oven temperature was 100°C. The mass spectra obtained were compared with the Nist-08 and Willey-08 mass spectral library.

Results and discussion
Isolation and selection of PHA-producing bacteria

A wide variety of bacteria are known to accumulate PHA granules intracellularly as an energy reserve material. Microbial species from over 90 genera have been reported to accumulate approximately 150 different hydroxyalkanoic acids as polyhydroxyalkanoate polyesters granules [27,28]. These bacteria have been reported from various environments. For the rapid detection and isolation of PHB-producing bacteria, 0.02% alcoholic solution of Sudan black B and Nile blue A, viable colony method [20] was used. The isolation of PHA-producing bacteria was done from various ecological niches. A total of 120 isolates showed black-blue coloration when stained with Sudan black B, a preliminary screening agent for lipophilic compounds, and a total of 62 isolates showed positive result with Nile blue A staining, a specific dye for the presence of PHA granules. Teeka et al. [29] used this method to screen the potential PHA-producing bacteria from soil, and Ramachandra and Abdullah [30] also observed the colonies formed on nutrient-rich medium under ultraviolet light (UV) to screen for the pink fluorescence which indicated the presence of PHA producers. Kitamura and Doi [31] first demonstrated this viable colony method on agar plates; they induced the isolates to accumulate PHA by culturing in E$_2$ medium, containing 2% (w/v) glucose before Nile blue A staining. The PHA-accumulating colonies,

after Nile blue A staining, showed bright orange fluorescence on irradiation with UV light and their fluorescence intensity increased with increase in PHA content of the bacterial cells.

Production of PHB in nutrient broth and cardboard industry waste water

The PHA-positive isolates selected after Nile blue A staining were screened in nutrient broth and best 27 producers were grown in processed cardboard industry waste water in 250-mL Erlenmeyer flasks and were employed to extract PHB after 72 h of incubation under stationary conditions of growth. The PHB from the isolates was extracted by the hypochlorite and chloroform method [21] as described earlier. The isolate NA10 showed maximum PHB production in both nutrient broth and cardboard industry waste water (3.951 g/L) (Table 1) and were selected for further optimization of PHB production.

Organic matter from waste and wastewater have high biological oxygen demand (BOD) and carbon oxygen demand (COD) values and, hence, microorganisms can grow, utilizing the nutrient present in waste water and can convert them into valuable compounds and polymers. Based on this idea, many researchers reported the PHA production from various industrial waste materials. Many researchers have proposed coupling PHA production to biological wastewater treatment [32-34]. Rebah et al. [14] utilized two types of industrial wastewater (starch and slaughterhouse wastewater from a plant located around Quebec region) and a secondary sludge (Quebec municipal waste water treatment plant) for PHB production by cultivating fast-growing Rhizobia but the PHB production was very low up to 10%. In an another study, Mockos et al. [35] reported that selectively enriched pulp mill waste activated sludge can serve as an inoculum for PHA production from methanol-rich pulp mill effluents, and the enriched cultures accumulated nearly 14% PHA on a dry weight basis under nitrogen-limited conditions.

Cardboard industry waste water is typically rich in carbohydrates but poor in fixed nitrogen due to the high C/N ratio. This high carbon-nitrogen ratio favors the growth of PHA-producing bacteria. This waste have high BOD and COD values 680–1,250 mg/L and 3,400–5,780 mg/L and COD/BOD ratio between 3.9–5 [36], which is suitable for microbial growth.

Biochemical and molecular characterization of selected isolate

By using Bergey's manual of determinative bacteriology [37] and by ABIS online, advanced bacterial identification software, the selected isolate was classified up to genus level using the morphological and biochemical characteristics (Tables 2 and 3) and was identified as Bacillus sp. For further characterization, almost complete 16S rRNA

Table 1 List of best PHA accumulating bacteria screened in nutrient broth and cardboard industry effluent

Isolate	Nutrient broth		Cardboard industry effluent	
	PHB (g/L)	Cell dry weight (g/L)	PHB (g/L)	Cell dry weight (g/L)
NA2	2.382 ± 0.012	6.2 ± 0.23	1.73 ± 0.021	4.4 ± 0.12
NA10	3.512 ± 0.011	7.8 ± 0.17	3.951 ± 0.024	8.6 ± 0.13
NA11	3.504 ± 0.019	6.8 ± 0.13	2.604 ± 0.033	8.8 ± 0.16
NA12	4.15 ± 0.014	8.7 ± 0.28	3.387 ± 0.027	9.2 ± 0.23
NA38	2.634 ± 0.017	5.8 ± 0.21	3.128 ± 0.041	7.3 ± 0.17
NA46	2.574 ± 0.021	8.4 ± 0.13	2.822 ± 0.02	8.2 ± 0.33
NA49	1.034 ± 0.033	4.2 ± 0.19	1.87 ± 0.032	4.8 ± 0.13
NA52	3.342 ± 0.014	6.8 ± 0.21	2.373 ± 0.036	4.2 ± 0.18
NA61	1.42 ± 0.031	5 ± 0.23	2.132 ± 0.022	5 ± 0.21
NA67	1.334 ± 0.033	4.4 ± 0.11	2.121 ± 0.017	5 ± 0.31
NA71	2.628 ± 0.016	4.8 ± 0.27	1.732 ± 0.015	3.6 ± 0.42
NA97	3.154 ± 0.009	7.6 ± 0.33	3.132 ± 0.028	10.2 ± 0.28
NA99	1.214 ± 0.018	4.4 ± 0.21	1.73 ± 0.024	4.8 ± 0.26
AN23	2.152 ± 0.032	5.4 ± 0.17	2.143 ± 0.033	8.6 ± 0.22
AN8	4.13 ± 0.033	6.6 ± 0.12	3.658 ± 0.035	7.5 ± 0.33

Growth and production of PHA in nutrient broth and preprocessed cardboard industry waste water after 72 h of incubation at 37°C.

gene sequences were determined. The obtained sequences were aligned and compared with the bacterial sequences available in the GenBank database. The phylogenetic analysis (Figure 1) was done using MEGA 5.2 software by neighbor-joining tree and distance matrix-based nucleotide sequence homology which revealed that isolate NA10 is *Bacillus sp.* Yilmaz et al. [38] isolated 29 strains of the genus *Bacillus* from different soil samples which were taken from grasslands of Ankara, Turkey, and the highest PHB production and productivity percentage was found in *Bacillus brevis* M6 (41.67% *w/v*).

Growth pattern and PHB production in cardboard industry waste water

As such, there is no report available from literature for PHB accumulation by microbial cells using cardboard industry waste water. The growth of *Bacillus* NA10 was recorded in different concentration of cardboard industry waste water (Figure 2), and the maximum growth was noted in undiluted (100% *v/v*) cardboard industry

Table 2 Morphological characteristics of selected isolate NA10

Morphological characters	Results
Colony color	White
Colony texture	Smooth
Gram reaction	+
Cell shape	Rod shaped
Cell arrangement	Isolated colonies
Spore formation	+

waste water after 72 h of incubation at 37°C. Dilution of waste water not only affected the microbial growth but also the PHB accumulation. This may be due to the fact that with increasing dilution of waste water, the carbon and other nutrient got diluted and was not able to support microbial growth [38]. By submerged fermentation, *Bacillus sp.*NA10 accumulated highest PHB (3.952 g/L) from undiluted processed cardboard industry waste water in unoptimized cultural conditions. Quagliano et al. [39] studied the effect of molasses concentration on PHB by increasing the molasses concentration from 1% to 5% (*w/v*), and a maximum of 54% PHA was obtained with 5% molasses concentration with highest Mw 640 kDa. Pozo et al.

Table 3 Biochemical characteristics and carbohydrate tests of selected isolate NA10

Biochemical/carbohydrate test	Results	Carbohydrate test	Results
Citrate utilization test	−	Glucose	+
Indole test	+	Raffinose	+
Methyl red test	−	Xylose	−
V-P Test	−	Sorbitol	−
H_2S production test	+	Lactose	−
Amylase production test	−	Mannose	−
Sucrose	+	Fructose	+
Dextrose	+	Ribose	−
Mannitol	+	Inulin	−
Trehalose	+	Esculin	-
D-Arabinose	−	Mellibiose	−
Maltose	+	Rammanose	−

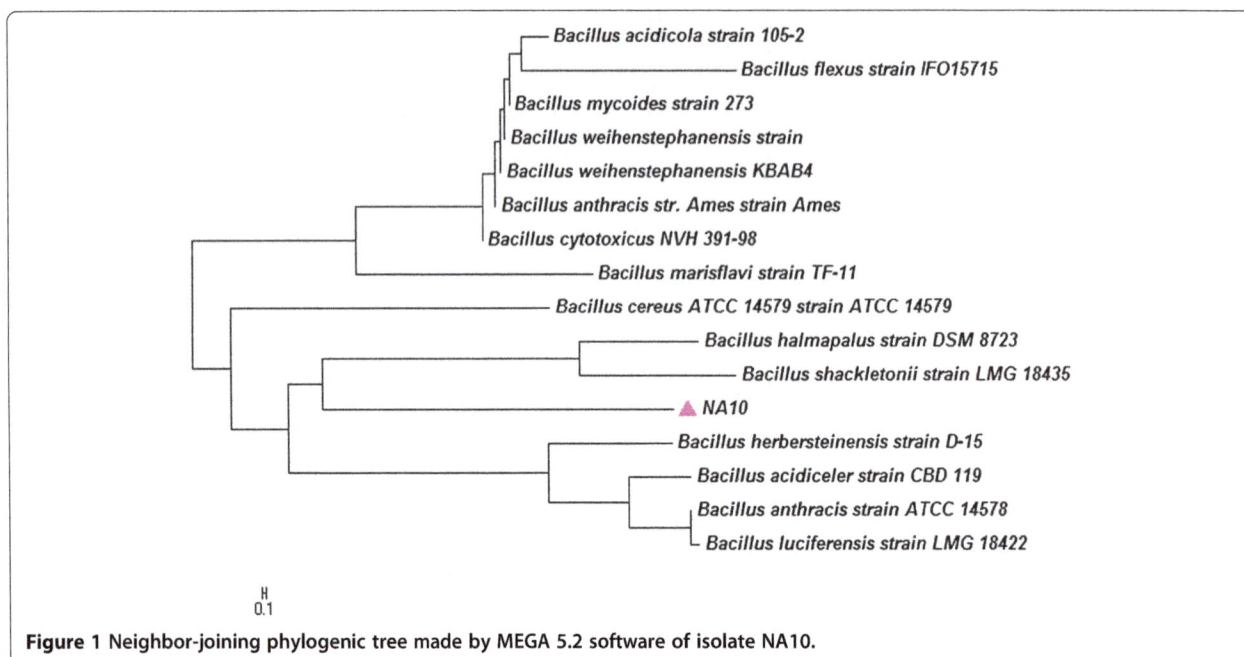

Figure 1 Neighbor-joining phylogenic tree made by MEGA 5.2 software of isolate NA10.

[40] found that NH_4 medium amended with 60% alpechin showed the maximum PHA production and concluded that alpechin was tolerated by *A. chroococcum* strain H23 at high concentrations and that it acts as a substrate to support the growth of the microorganisms.

Optimization of growth parameters
PHB production versus incubation time
The PHB production in cardboard industry waste water was followed for 120 h by submerged fermentation in 250 mL Erlenmeyer flask at an interval of 24 h. The production of PHB increased up to 72 h (4.456 g/L) and, thereafter, got reduced (3.26 g/L after 72 h) (Figure 3a). This reduction in PHB production after 72 h may be due to lack of micronutrients as well as increase in metabolites that might have negative effect on the PHB production. The observation was supported by Yüksekdağ et al. [41] *Alcaligenes eutrophus* [42] and *A. vinelandii* [43].

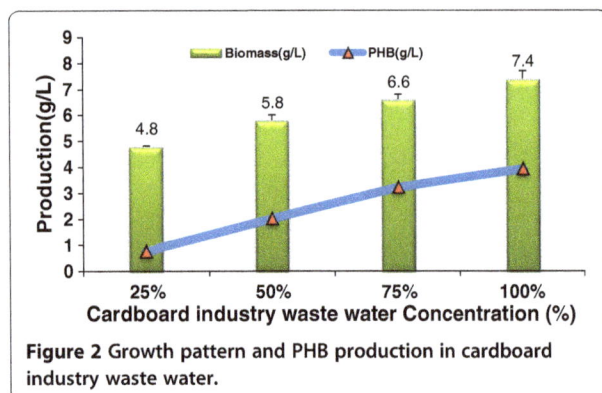

Figure 2 Growth pattern and PHB production in cardboard industry waste water.

Klüttermann et al. [44] also reported that the highest PHB level in *Agrobacterium radiobacter* was achieved after 96 h. After this time, the PHB contents in fermentation broth decreased which might indicate that the bacteria used PHB as a nutrient source due to inadequate nitrogen and carbon sources in the medium.

Effect of incubation temperature on PHB production
The maximum PHB production of 4.274 g/L was recorded at 35°C after 72 h. The increase of temperature beyond 35°C has negative impact on PHB production (Figure 3b). This decrease in PHB production at high temperature is also supported by studies of Grothe et al. [10] and Yüksekdağ et al. [41]. The decrease in PHB production at high temperature could be due to low PHB polymerase enzyme activity [45]. Tamdoğan and Sidal [46] also reported that higher and lower temperatures than 30 C lead to decrease in cell biomass and PHB synthesis by *Bacillus subtilis* ATCC6633.This result coincides with that represented by Aslim et al. [47], Hamieh et al. [48], who reported that optimum incubation temperature for PHB production by *Bacillus subtilis* , *Bacillus pumilis* and *Bacillus thuringiensis* was 37 C.

Effect of pH of the production medium on PHB production
Typically, metabolic process was highly susceptible to even slight changes in pH. Therefore, proper control of pH was critical. The experimental results showed that highest PHB content (4.492 g L^{-1}) was obtained at pH 7.5 by isolate NA10 (Figure 3c). The current observation was in agreement with Sindhu et al. [49] who observed that the maximum PHB was produced (0.01 to 0.5 g/L) at

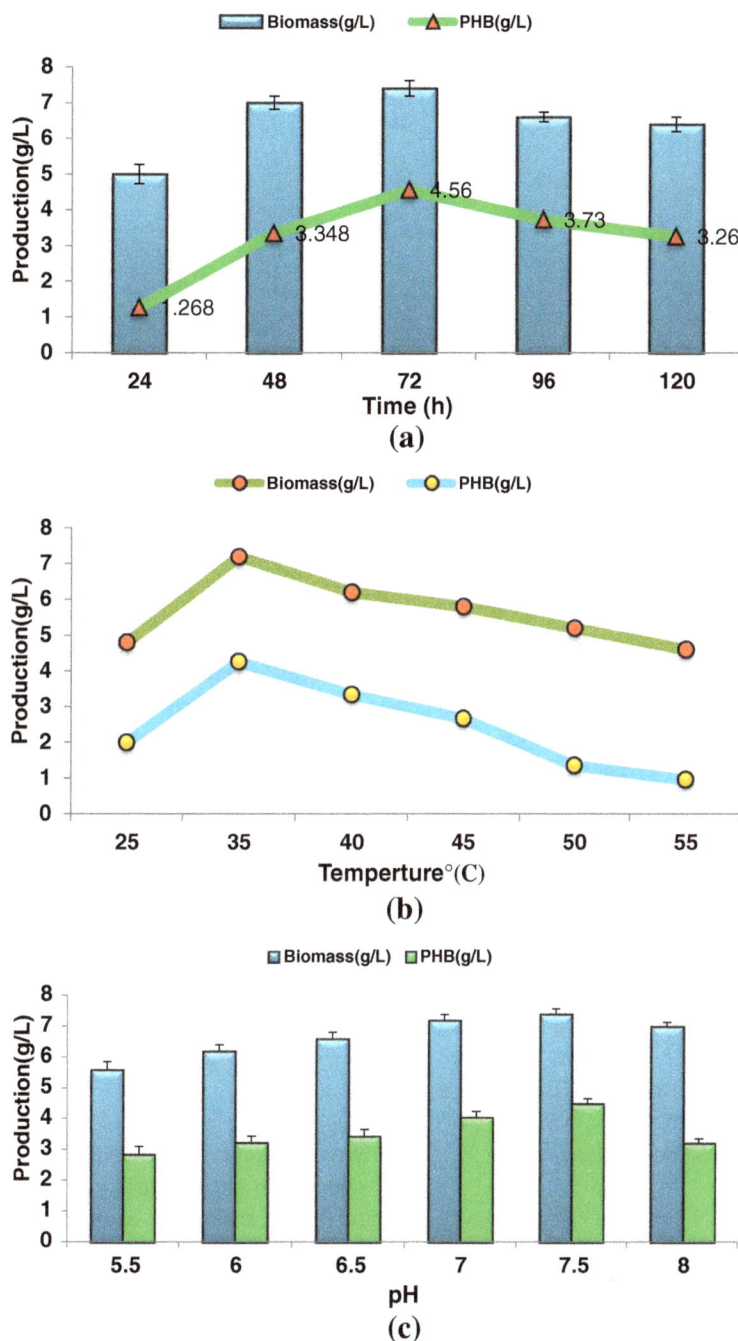

Figure 3 Optimization of various cultural and growth parameters for *Bacillus sp.* NA10. (a) time (b) temperature (c) pH for Bacillus sp. NA10.

pH 7–7.5 by *Bacillus sphaericus* NII 0838 from crude glycerol. Flora et al. [50] revealed that the maximum PHB production (25%) by *B. sphaericus* was at pH range from 6.5–7.5, and the reduction of polymer accumulation at higher pH values is due to the effect on the degradative enzymes of polymer breakdown, so that PHB is utilized at a rate almost equal to the rate of its synthesis. Shivakumar [51] also reported that pH 6.8 to 8.0 was optimum for PHB production by *A. eutrophus*.

Scale up study and growth kinetics

Growth kinetics of the culture was studied in 2.0 L lab scale bioreactor. The profiles of PHB accumulation were analyzed in batch cultivation by growing on the cardboard industry waste water under optimized growth conditions obtained at flask level (Figure 4). The growth pattern for the microorganism was studied by logistic model and exactly fitted as presented in Figure 5. The bacteria NA10 was found to accumulate cell dry weight and specific

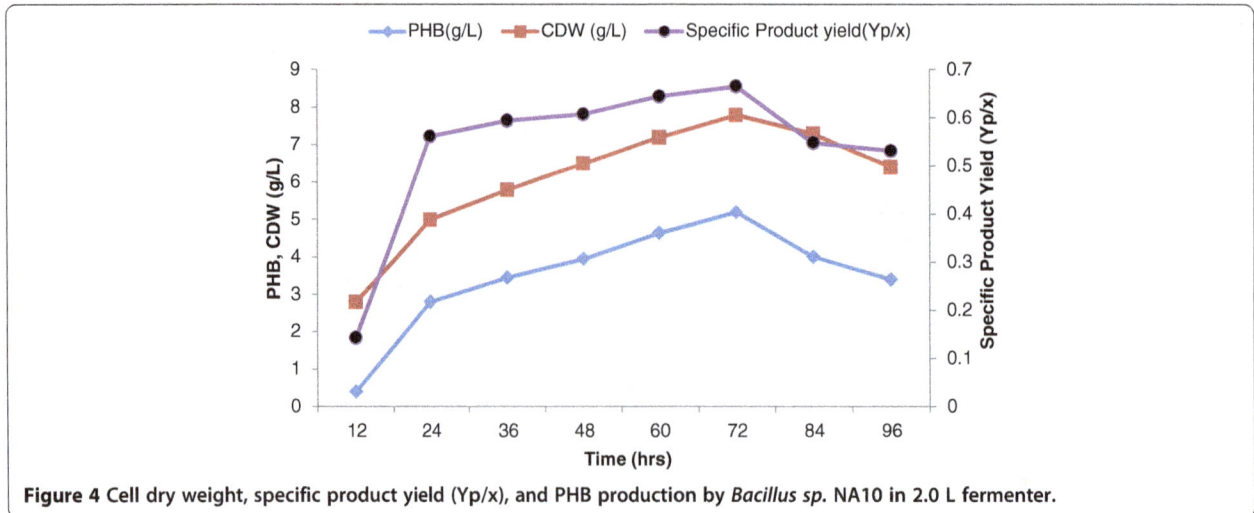

Figure 4 Cell dry weight, specific product yield (Yp/x), and PHB production by *Bacillus sp.* NA10 in 2.0 L fermenter.

product yield after the initial logarithmic growth (24 h) and reached a maximum 7.8 g/l at a rate 0.028 h^{-1} in stationary growth phase at 72 h. Similarly specific product yield (Yp/x) increased by about 50% from initial exponential growth to stationary phase, suggesting that cells accumulated PHB while growing and aging; hence, any strategy that prolongs the stationary phase is likely to further enhance the PHB yield per unit biomass. A maximum CDW of 7.8 g l^{-1} with a PHB concentration of 5.202 g l^{-1} was obtained when batch cultivation was conducted at 37° C after 72 h, and the PHB content was up to 66% and productivity was 0.072 g l^{-1} h^{-1}. Tanamool et al. [52] reported that maximum 0.920 g/l of cell dry mass and 0.034 g/l PHB using sweet sorghum juice by *Ralstonia eutropha* with yield and productivity of 0.037 g PHB/g dry cell and 0.0019 g/(l h), respectively. Similarly, El-Sayed et al. [53] found specific growth rate of 0.055 h^{-1} on sucrose media while a specific growth rate of 0.053 h^{-1} was obtained with *R. eutropha* on glucose medium.

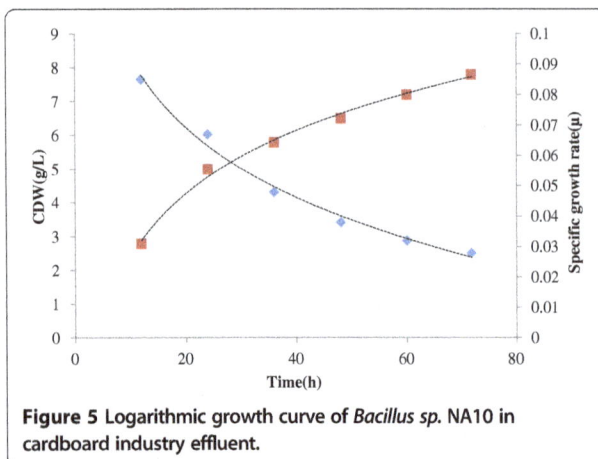

Figure 5 Logarithmic growth curve of *Bacillus sp.* NA10 in cardboard industry effluent.

Polymer analysis
^{1}H-NMR spectroscopy

The structures of polyesters were investigated by ^{1}H NMR (Figure 6a). The ^{1}H NMR spectra of the PHAs extracted from *Bacillus sp.* NA10 shows a doublet at 1.26 ppm which is attributed to the methyl group coupled to one proton, a doublet of quadruplet at 2.580 ppm which is attributed to a methylene group adjacent to an asymmetric carbon atom bearing a single proton, and a multiplet at 5.25 ppm characteristic of the methyne group. Two other signals are also observed, the broad one at 1.04 ppm which is due to water, and another one at 7.27 ppm which is attributed to chloroform. From these results, it can be concluded that *Bacillus sp.* NA10 cells grown with cardboard industry effluent as the carbon source produce PHA exclusively in the form of PHB.

FTIR

Polymer extracted from NA10 was used for recording IR spectra in the range 4,000 to 600 cm^{-1}. IR spectra (Figure 6b) showed two intense absorption band at 1,720 and 1,273 cm^{-1} which are specific for C = O and C–O stretching vibrations, respectively. The absorption bands at 2,932 and 2,962 cm^{-1} are due to C-H stretching vibrations of methyl and methylene groups. These prominent absorption bands confirm the structure of poly-β-hydroxybutyrate.

TGA

TGA curves were obtained in the temperature range of 30°C to 700°C for PHB. TGA results of *Bacillus sp.* NA10 showed that the T_m is 182.34°C, and the enthalpy of PHA fusion is 83.62 J/g. The result showed similarity with the data obtained from standard PHB (174.29°C and 86.49 J/g) and from other works [54,55]. The PHA gave a rapid thermal degradation between 250°C and 290°C while

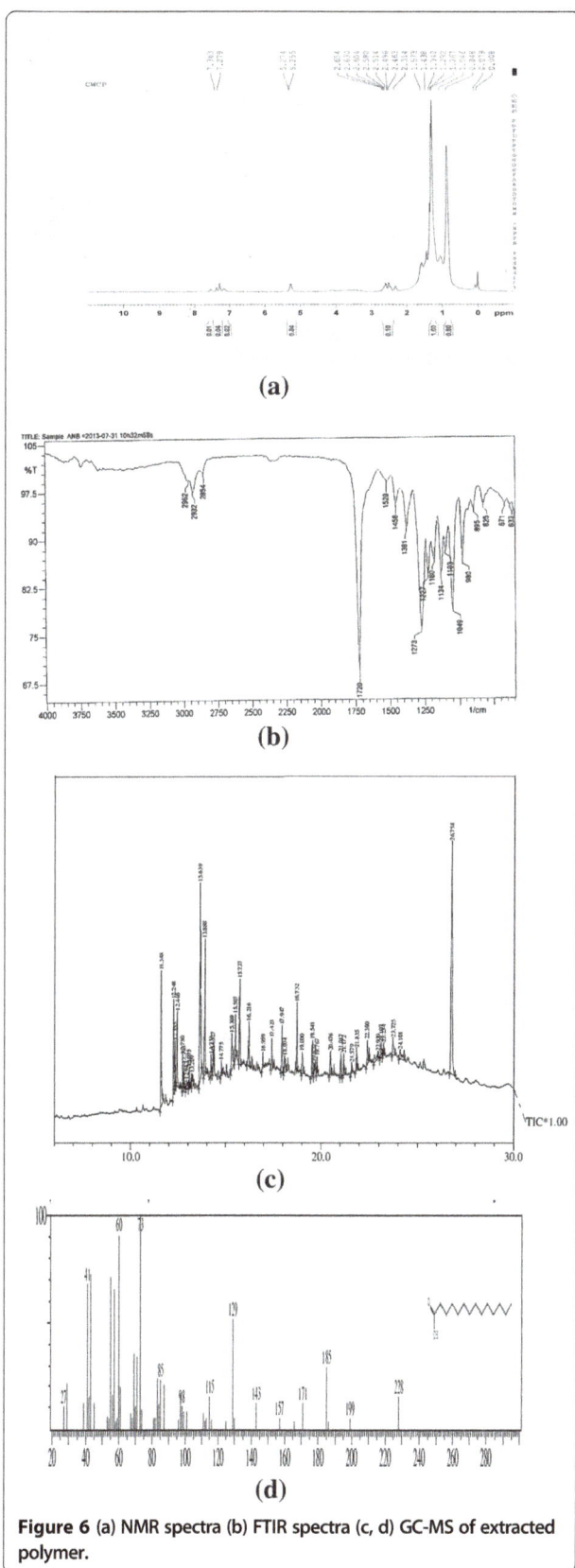

Figure 6 (a) NMR spectra (b) FTIR spectra (c, d) GC-MS of extracted polymer.

the standard PHB represented at between 250°C and 295°C. It is indicated that biopolymeric material obtained can possibly be further used in a large-scale processing of bioplastic [56,57].

GC-MS analysis of extracted PHB

GC-MS analysis helps in elucidating the structure of components. The key compounds of concern were identified based on their retention peak. PHA from *Bacillus sp.* NA10 cultured in processed cardboard industry waste significantly contained hexadecanoic acid, methyl ester (56%), and tetradecanoic acid (32%). These compounds signify that the monomer chains were of biodegradable polyester family [8]. Figure 6c,d shows that a common molecular fragment of the 3HB ion chromatogram of the PHB was produced. A predominant peak corresponding to the tetramer of 3HB (hexadecanoic acid) was noted at 13.639 to 13.88 min, respectively, in GC-purified product from *Bacillus sp.* NA10. The retention times and ion fragment patterns of the peaks at 11.6, 12.24, and 15.73 min were identical to those of the dimer methyl esters of 3HV and 3HBV, respectively, but in low percentage up to 32% and 11%, respectively. The similar results of GC-MS were observed by He et al. [58] with 3-hydroxydecanoate (HD or C10) 63% and 3-hydroxyoctanoate (HO or C8) 21% with other medium chain length (mcl) monomers. From the data obtained by GC-MS, the molecular weight of PHB obtained from isolate *Bacillus sp.* NA10 is 242 kDa while, the commercial PHB have a molecular weight of 275 kDa. Galego et al. [59] extracted the PHB with a molecular weight of about 177 kDa. This lowering in molecular weight is due to the presence of other polyhydroxyalkanoic acids e.g. 3HV and 3HBV up to a low percentage (32%) with major polymer 3HB.

Figure 7 PHB-starch blended sheet.

Preparation of polymer blended (PHB-starch) sheet

The PHB produced was found to be brittle and breaks easily. Blending PHB with other polymers is an economic way to improve its mechanical properties. The PHB blend sheet (Figure 7) was prepared by mixing extracted polymer with soluble starch following the conventional solvent cast technique. Starch is used for blending because it has few advantage of being biodegradable, biocompatible, cheap, and also readily available while all other biodegradable polymer such as polyethylene glycol and polylactic acid are costly. Starch-blended sheet shows better mechanical and thermal properties making it more reliable for packaging industry. As PHB and starch was not completely miscible, the blend showed insoluble particle aggregation on the surface. Identical observations were reported by Choi et al. [60] in which they prepared PHB sheet with EPB blends. The blends showed a higher thermal stability compared to PHB sheet. Parra et al. [61] reported PHB blend preparation with polyethyleneglycol (PEG) with improved properties.

Conclusions

In this study, inexpensive cardboard industry waste water was tried as a carbon source to produce PHB. Different bacterial strains were isolated from soil and screened for polyhydroxybutyrate production using cardboard manufacturing industry waste water as a carbon source. The bacterial isolate *Bacillus sp.* NA10 can be regarded as potential strain for conversion of cardboard industry waste water into PHB. The selected isolate efficiently utilized cardboard industry waste water as sole carbon source for growth and PHB biosynthesis, accumulating PHB up to 66.6% of the cell dry mass with $0.072 \ \mathrm{g \ l^{-1} \ h^{-1}}$ productivity. Currently, this bacterial strain is further studied to increase the productivity of PHB by studying the effect of various carbon and nitrogen supplementation in cardboard industry waste water.

Competing interests
The authors declare that they have no competing interests.

Authors' contributions
AKB carried out all the experimental work such as isolation, screening, fermentation studies, chemical, and statistical analysis and drafted the manuscript. GS (Gulab Singh) collected the substrate and helped in fermentor running and chemical and statistical analysis. NKA guided all the research work and participated in the design of the study. VG and AY helped in isolation and screening and provided the moral support in the whole study. All authors read and approved the final manuscript.

Acknowledgements
The authors express their sincere gratitude to the Rana Cardboard industry, Yamuna Nagar, for providing the untreated waste water. The authors are very thankful to the Department of Chemistry, Kurukshetra University, Kurukshetra, for providing the necessary facilities for NMR, FTIR, and TGA analysis of the polymer.

Author details
[1]Department of Microbiology, Kurukshetra University, Kurukshetra, Haryana 136119, India. [2]Department of Biotechnology, Kurukshetra University, Kurukshetra 136119, India.

References
1. Lazarevic D, Aoustin E, Buclet N, Brandt N (2010) Plastic waste management in the context of a European recycling society: comparing results and uncertainties in a life cycle perspective. Resour Conserv Recycling vol 55:246–259
2. Flechter A (1993) Plastics from bacteria and for bacteria: PHA as natural, biodegradable polyesters. Springer, New York, pp 77–93
3. Nath A, Dixit M, Bandiya A (2008) Enhanced PHB production and scale up studies using cheese whey in fed batch culture of *Methylobacterium* sp. ZP 24. Bioresource Technol 99:5749–5755
4. Anderson AJ, Dawes EA (1990) Occurrence, metabolism, metabolic role, and industrial uses of bacterial polyhydroxyalkanoates. Microbiol Rev 54:450–472
5. Yu J, Stahl H (2008) Microbial utilization and biopolyester synthesis of bagasse hydrolysates. Bioresource Technol 99:8042–8048
6. Lee IY, Chang HN, Park YH (1995) A simple method for recovery of microbial poly-3-hydroxybutyrate by alkaline solution treatment. J Microbiol Biotechnol 5(4):238–240
7. Gao X, Chen JC, Wu Q, Chen GQ (2011) Polyhydroxyalkanoates as a source of chemicals, polymers and biofuels. Curr Opin Biotechnol 22:768–774
8. Choi JI, Lee SY (1997) Process analysis and economic evaluation for poly (3-hydroxybutyrate) production by fermentation. Bioprocess Eng 17:335–342
9. Yamane T, Chen XF, Ueda S (1996) Growth associated production of poly(3-hydroxyvalerate) from *n*-pentanol by a methylotrophic bacterium, *Paracoccus denitificans*. Appl Environ Microbiol 62:380–384
10. Grothe E, Moo-Young M, Chisti Y (1999) Fermentation optimization for the production of poly (ß-hydroxybutyric acid) microbial thermoplastic. Enzyme Microb Technol 25:132–141
11. Wen Q, Chen Z, Tian T, Chen W (2010) Effects of phosphorus and nitrogen limitation on PHA production in activated sludge. J Environ Scien 22 (10):1602–1607
12. Ceyhan N, Ozdemir G (2011) Polyhydroxybutyrate (PHB) production from domestic wastewater using *Enterobacter aerogenes* 12Bi strain. Afr J Microbiol Res 5(6):690–702
13. Wong HH, Lee SY (1998) Poly-(3-hydroxybutyrate) production from whey by high- density cultivation of recombinant *Escherichia coli*. Appl Microbiol Biotechnol 50:30–33
14. Rebah FB, Prévost D, Tyagi RD, Belbahri L (2009) Poly-β-hydroxybutyrate production by fast-growing *Rhizobia* cultivated in sludge and in industrial wastewater. Appl Biochem Biotech 158(1):155–163
15. Reddy VST, Thirumala M, Mahmood KS (2009) Production of P HB and P (3HB-co-3HV) biopolymers of *Bacillus megaterium* strain OU303A isolated from municipal sewage sludge. W J Microbiol Biotechnol 25:391–397
16. Koller MB, C hiellini E, Fernandes EG, Horvat P, K utschera C, Hesse P, Braunegg G (2008) Polyhydroxyalkanoate production from whey by *Pseudomonas hydrogenovora*. Bioresour Technol 99(11):4854–4863
17. Bengtsson S, Werker A, Welander T (2008) Production of polyhydroxyalkanoates by glycogen accumulating organisms treating a paper mill wastewater. Water Scien Technol 58:323–330
18. Du G, Yu J (2002) Green technology for conversion of food scraps to biodegradable thermoplastic polyhydroxyalkanoates. Environ Sci Technol 36:5511–5516
19. Juan ML, Gonzalez LW, Walker GC (1998) A novel screening method for isolating exopolysaccharide-deficient mutants. Appl Environ Microbiol 64:4600–4602
20. Spiekermann P, Rehm BH, Kalscheuer R, Baumeister D, Steinbüchel A (1999) A sensitive, viable-colony staining method using Nile red for direct screening of bacteria that accumulate polyhydroxyalkanoic acids and other lipid storage compounds. Arch Microbiol 171(2):73–80
21. Singh G, Mittal A, Kumari A, Goel V, Aggarwal NK, Yadav A (2011) Optimization of poly-β-hydroxybutyrate production from *Bacillus species*. Eur J Biol Scien 3(4):112–116
22. Law J, Slepecky RA (1961) Assay of poly-hydroxybutyric acid. J Bacteriol 82:52–55

23. Lee YU, Yoo YJ (1991). Kinetics for the growth of Alcaligenes eutrophus and the biosynthesis of poly-β-hydoxybutyrate. Korean J Appl. Microbiol. Biotechnol 19:186–92

24. Painter PR, Marr AG (1963) Mathematics of microbial populations. Ann Rev Microbiol 22:219–221

25. Levasseur M, Thompson PA, Harrison Paul J (1993) Physiological acclimation of marine phytoplankton to different nitrogen sources. J Phycol 29(5):587–595

26. Wang F, Lee SY (1997) Poly(3-hydroxybutyrate) production with high productivity and high polymer content by a fed-batch culture of Alcaligenes latus under nitrogen limitation. Appl Environ Microbiol 63(9):3703–3706

27. Du G, Yu J, Chen J, Lun S (2001) Continuous production of poly-3-hydroxybutyrate by Ralstonia eutropha in a two stage culture system. J Biotechnol 88:59–65

28. Steinbüchel A (2001) Perspectives for biotechnological production and utilization of biopolymers: metabolic engineering of polyhydroxyalkanoats biosynthesis pathways as a successful example. Macromol Bioscien 1:1–24

29. Teeka J, Cheng I, Xuehang T, Alissara R, Takaya H, Koichi Y, Masahiko S (2010) Screening of PHA-producing bacteria using biodiesel-derived waste glycerol as a sole carbon source. J Water Environ Technol 8:371–381

30. Ramachandran H, Abdullah AA (2010) Isolation of PHA-producing bacteria from Malaysian Environment. Proceedings of the 7th IMT-GT UNINET and The 3rd International PSUUNS Conferences on Bioscience 178–179

31. Kitamura S, Doi Y (1994) Staining method of poly (3-hydroxyalkanoic acid) producing bacteria by Nile Blue. Biotechnol Tech 8:345–350

32. Salehizadeh H, Van Loosdrecht MCM (2004) Production of polyhydroxyalkanoates by mixed culture: recent trends and biotechnological importance. Biotechnol Adv 22:261–279

33. Dias JM, Lemos PC, Serafim LS, Oliveira EC, Albuquerque M, Ramos M, Oliveira R, Reis MA (2006) Recent advances in polyhydroxyalkanoate production by mixed aerobic cultures: from the substrate to the final product. Macromol Biosci 6:885–906

34. Ganzeveld KJ, Van Hagen A, Van Agteren MH, Koning DW, Uiterkamp AMJS (1999) Upgrading of organic waste: production of the copolymer poly-3-hydroxybutyrate-co-valerate by Ralstonia eutrophus with organic waste as sole carbon source. J C lean Prod 7:413–420

35. Mockos G R, Loge FJ, Smith WA, and Thompson DN (2008). Selective enrichment of a methanol- utilizing consortium using pulp & paper mill waste streams. http://www.ncbi.nlm.nih.gov/pubmed/18418753" \o "Applied biochemistry and biotechnology. Appl Biochem Biotechnol 148(1–3):211–26.

36. Rao MN, Datta AK (2007) Waste water treatment. 702 Co. Pvt. Ltd, New Delhi. pp 203–208

37. Holt John G, Krieg NR, Peter S, Staley HA, Williams Satnley T, James T (2009) Bergey's manual of determinative bacteriology, 9th edn. Lippincott Williams & Wilkins, Baltimore, United States

38. Mirac Y, Haluk S, Yavuz B (2004) Determination of poly-β-hydroxybutyrate (PHB) production by some Bacillus spp. World J Microbiol Biotechnol 21:565–566

39. Quagliano JCF, Amarilla G, Fernandes DM, Miyazaki SS (2001) Effect of simple and complex carbon sources, low temperature culture and complex carbon feeding policies on poly-3-hydroxybutyric acid (PHB) content and molecular weight (Mw) from Azotobacter chroococcum 6B. World J Microbiol Biotechnol 17:9–14

40. Pozo C, Martı'nez-Toledo MV, Rodelas B, Gonza'lez-Lo'pez J (2002) Effects of culture conditions on the production of polyhydroxyalkanoates by Azotobacter chroococcum H23 in media containing a high concentration of alpechín (wastewater from olive oil mills) as primary carbon source. J Biotechnol 97:125–131

41. Yüksekdağ ZN, Aslim B, Beyatli Y, Mercan N (2004) Effect of carbon and nitrogen sources and incubation times on polybeta-hydroxybutyrate (PHB) synthesis by Bacillus subtilis 25 and Bacillus megaterium 12. African J Biotechnol 3(1):63–66

42. Khanna S, Srivastava A (2005) Recent advances in microbial polyhydroxyalkanoates. Process Biochem 40:607–619

43. Page WJ, Manchak J, Rudy B (1992) Formation of poly (hydroxybutyrate-co-hydroxyvalerate) by Azotobacter vinelandii UWD. Appl Environ Microbiol 58:28–66

44. Klu'ttermann K, Tauchert H, Kleber HP (2002) Synthesis of poly-beta-hydroxybutyrate by Agrobacterium radiobacter after growth on D-Carnitine. Acta Biotechnol 22:261–269

45. Singh G, Mittal A, Kumari A, Goyal V, Yadav A, Aggarwal NK (2013) Cost effective production of poly-β-hydroxybutyrate by Bacillus subtilis NG 05 using sugar industry waste water. J Polym Environ 21:441–44

46. Tamdoğan N, Sidal U (2011) Investigation of poly-β-Hydroxybutyrate (PHB) production by Bacillus subtilis ATCC 6633 under different conditions. Kafkas Univ Vet Fak Derg 17(Suppl A):173–176

47. Aslim B, Yüksekdağ ZN, Beyatli Y (2001) Determination of growth quantities of certain Bacillus species isolated from soil. Turk Electr J Biotechnol (Sp. Issue):24–30

48. Hamieh A, Olama Z, Holail H (2013) Microbial production of polyhydroxybutyrate, a biodegradable plastic using agro-industrial waste products. G Adv Res J Microbiol 2(3):054–064

49. Sindhu R, Ammu B, Parameswaran B, Deepthi SK, Ramachandran KB, Soccol CR, Pandey A (2011) Improving its thermal properties by blending with other polymers. Brazilian J Microbiol 54(4):783–794

50. Flora G, Bhatt K, Tuteja U (2010) Optimization of culture conditions for poly-β-hydroxybutyrate production from isolated Bacillus species. J Cell Tissue Res 10:2235–2242

51. Shivakumar S (2009) Optimization of process parameters for maximum poly-β-hydroxybutyrate production by Bacillus thuringiensis IAM 12077. Pol J Microbiol 58(2):149–154

52. Tanamool V, Danvirutai P, Thanonkeo P, Imai T, Kaewkannetra P (2009) Production of poly-β-hydroxybutyric acid (PHB) from sweet sorghum juice by Alcaligenes eutrophus TISTR 1095 and Alcaligenes latus ATCC 29714 via batch fermentation. The 3th International Conference on Fermentation Technology for Value Added Agroculture Products 1–6

53. El-Sayed AA, Abdelhady HM, Abdel Hafez AM, Khodair TA (2009) Batch production of polyhydroxybutyrate (PHB) by Ralstonia eutropha and Alcaligenes latus using bioreactor different culture strategies. J Appl Sci Res 5(5):556–564

54. Khanna S, Srivastava A (2006) Optimization of nutrient feed concentration and addition time for production of poly(β-hydroxybutyrate). Enzyme Microbiol Technol 39:1145–1151

55. Tanamool V, Imai T, Danvirutai P, Kaewkannetra P (2011) Biosynthesis of polyhydroxyalkanoate (PHA) by Hydrogenophaga sp. isolated from soil environment during batch fermentation. J Life Sci 5:1003–1012

56. Yezza A, Halasz A, Levadoux W, Hawari J (2007) Production of poly-β-hydroxybutyrate (PHB) by Alcaligenes latus from maple sap. Apply Microbiol Biotechnol 77:269–274

57. Pachekoski WM, Agnelli JAM, Belem LP (2009) Thermal, mechanical and morphological properties of poly (hydroxybutyrate) and polypropylene blends after processing. Material Res 12:159–164

58. He W, Tian W, Zhang G, Chen GQ, Zhang Z (1998) Production of novel polyhydroxyalkanoates by Pseudomonas stutzeri 1317 from glucose and soybean oil. FEMS Microbiol Lett 169:45–49

59. Galego N, Rozsa C, Sánchez R, Fung J, Vázquez A, Tomás JS (2000) Characterization and application of poly (β-hydroxyalkanoates) family as composite biomaterials. Polym Test 19:485–492

60. Choi JY, Lee JK, You Y, Park WH (2003) Epoxidized polybutadiene as a thermal stabilizer for poly(3-hydroxybutyrate). II. Thermal stabilization of poly (3-hydroxybutyrate) by epoxidized polybutadiene. Fiber Polym 4:195–198

61. Parra D, Rosa F,D, Rezende SP, Ponce J, Luga⁻o AB (2011) Biodegradation of irradiated poly-3-hydroxybutyrate (PHB) films blended with poly (ethyleneglycol). J Polym Environ 19:918–925

4

Recombinant D-galactose dehydrogenase partitioning in aqueous two-phase systems: effect of pH and concentration of PEG and ammonium sulfate

Anvarsadat Kianmehr[1,2]*, Maryam Pooraskari[3], Batoul Mousavikoodehi[4] and Seyede Samaneh Mostafavi[5]

Abstract

Background: D-Galactose dehydrogenase (GalDH; EC 1.1.1.48) belongs to the family of oxidoreductases that catalyzes the reaction of β-D-galactopyranose in the presence of NAD^+ to D-galacto-1,5-lactone and NADH. The enzyme has been used in diagnostic kits to neonatal screen for galactosemia diseases. This article reports the partitioning optimization of recombinant *Pseudomonas fluorescens* GalDH in aqueous two-phase systems (ATPS).

Methods: Preliminary two-phase experiments exhibited that the polyethylene glycol (PEG) concentration, pH value, and concentration of salt had a significant influence on the partitioning efficiency of recombinant enzyme. According to these data, response surface methodology (RSM) with a central composite rotatable design (CCRD) was performed to condition optimization.

Results: The optimal partition conditions were found using the 14.33% PEG-4000 and 11.79% ammonium sulfate with pH 7.48 at 25°C. Yield, purity, recovery, and specific activity were achieved 92.8%, 58.9, 268.75%, and 373.9 U/mg, respectively. PEG and ammonium sulfate concentration as well as pH indicated to have a significant effect on GalDH partitioning. Enzyme activity assay and sodium dodecyl sulfate-polyacrylamide gel electrophoresis (SDS-PAGE) analysis demonstrated the suitability of predicted optimal ATPS as well. The K_m and molecular weight values for the purified GalDH were 0.32 mM and 34 kDa, respectively.

Conclusions: Ultimately, our data showed the feasibility of using ATPS for partitioning and recovery of recombinant GalDH enzyme.

Keywords: Aqueous two-phase systems (ATPS); D-Galactose dehydrogenase (GalDH); Response surface methodology (RSM); Partition; *Pseudomonas fluorescens*

Background

Liquid-liquid extraction using aqueous two-phase systems (ATPS) has been applied for recovery and purification of many industrial enzymes [1,2]. When two aqueous solutions of certain incompatible substances, such polyethylene glycol (PEG) and dextran or PEG and salt, are mixed above a critical concentration, two-phase separation occurs. Separation techniques based on two-phase partitioning have proved to be suitable tools for recovery of biomolecules. Compared with the traditional techniques, ATPS have the advantages such as ensuring high values of the purification parameters, preserving the targeted biomolecules, yielding separation performance, and ease to scale-up. Successful applications of ATPS for downstream processing of proteins on industrial scales have been demonstrated [3,4].

D-Galactose dehydrogenase (GalDH; D-galactose: NAD^+ oxidoreductase; EC 1.1.1.48) belongs to the family of oxidoreductases that catalyzes the dehydrogenation reaction of β-D-galactopyranose in the presence of NAD^+ to D-galacto-1,5-lactone and NADH. The kinetic mechanism of Bi-Bi has been determined for this enzyme, with the

* Correspondence: kiabiotpro@yahoo.com
[1]Genetic and Metabolism Research Group, Department of Biochemistry, Pasteur Institute of Iran, 13164 Tehran, Iran
[2]Department of Medical Biotechnology, School of Advanced Medical Sciences, Tabriz University of Medical Sciences, 13164 Tabriz, Iran
Full list of author information is available at the end of the article

NAD$^+$ binding first to the enzyme. The substrates of GalDH are D-galactose and NAD$^+$, whereas its products are D-galactono-1,4-lactone, NADH, and H$^+$ [5]. GalDH has been identified in plants (e.g., green peas and *Arabidopsis thaliana*), algae (e.g. *Iridophycus flaccidum*), bacteria, and mammals. However, GalDH from *Pseudomonas fluorescens* bacterium is the best investigated enzyme, as its recombinant form has been produced in *Escherichia coli* [6,7]. GalDH is a significant tool for the measurement of β-D-galactose, α-D-galactose, and lactose as well. The enzyme has been used in diagnostic kits to screen blood serum of neonates for galactosemia diseases [8]. Galactosemia is an inborn metabolic disorder that without strict dietary control results in mental retardation, microcephaly, and seizures. Newborn screening using GalDH is a simple method which has proved sensitive, reliable, rapid, and cheap compared to other methodologies [9]. This enzyme has been purified by conventional methods including ammonium sulfate precipitation followed by chromatography which are usually time-consuming and expensive [6,8]. Owing to the commercial importance of GalDH, developing the efficient and scalable alternative methods for downstream processing is of great interest. In this work, we aimed to use ATPS technology for partitioning of *P. fluorescens* GalDH. The best partition conditions are generally achieved by systematic variation of different parameters such as temperature, pH, size and concentration, and type of polymer and salt. However, despite the apparent simplicity, partition of compounds is very complex due to the several factors involved. In fact, the classical optimization approach varying the level of one parameter at a time, while holding the rest of the variables constant, is generally time-consuming [10]. For these reasons, mathematical modeling has been utilized to identify parameters mainly those that affect the partition of proteins in ATPS [11,12]. An effective statistical technique is the response surface methodology (RSM) which is a useful statistical tool for studying of systems where several independent variables influence the responses [13]. In recent years, the use of RSM in performing biological process has gained importance. The main advantage of RSM is the reduced number of tests needed to calculate multiple factors and their interactions [14]. In this communication, the RSM was applied to identify the suitable operating conditions for partitioning of recombinant *P. fluorescens* GalDH in ATPS.

Methods
Materials
Polyethylene glycols with different molecular weights were purchased from Merck (Darmstadt, Germany). D-Galactose and NAD$^+$ were obtained from Sigma-Aldrich (St. Louis, MO, USA) and utilized in enzyme activity assay. The salts and all other chemicals were of analytical grade.

P. fluorescens strain which produces GalDH enzyme has been isolated form a soil sample by Anvarsadat Kianmehr.

Construction of expression plasmid for *P. fluorescens* GalDH
Primers for polymerase chain reaction (PCR) amplification were designed based on the available nucleotide sequence of GalDH of the *P. fluorescens* genome using DNASIS MAX software (DNASIS version 3.0, Hitachi Software Engineering Co., Ltd., Tokyo, Japan). The *gdh* gene was amplified from the genomic DNA with specific primers GDHFw (5′-T*GGATCC*ATGCAACCGATTCGT CTCG-3′) and GDHRev (5′-GCG*AAGCTT* TTAATCG TAGAACGGC-3′), which contained the restriction sites for *Bam*HI and *Hin*dIII, respectively. PCR amplification was performed under condition: preincubation at 95°C for 1 min and then 30 cycles of 95°C for 1 min, 61°C for 1 min and 72°C for 2 min. The PCR reaction product was cut with *Bam*HI and *Hin*dIII and then ligated into the pET-28a (+) expression vector. The construct bearing the *gdh* gene was named pET28aGDH and transformed into *E. coli* BL-21 (DE3).

Cell cultivation and production of recombinant GalDH
A recombinant strain of *E. coli* BL21 (DE3) was grown overnight in Luria-Bertani (LB) medium containing 40 µg/mL of kanamycine at 37°C and 150 rpm. When cell density reached an OD600 of 0.8, GalDH enzyme was expressed by the addition of 0.7 mM sterile isopropyl-β-D-thiogalactopyranoside (IPTG). After 5 h of induction at 30°C, cells were harvested and stored at −20°C for further use. Pelleted *E. coli* cells were suspended in lysis buffer (50 mM Tris-HCl, 50 mM NaCl, 1 mM EDTA, pH 8.0), mechanically broken by sonication using a pulse sequence of 15 s on and 10 s off and clarified by centrifugation at 4,000 rpm at 4°C for 1 h. The supernatant was employed as a crude enzyme in partition experiments [6].

ATPS preparation
ATPS were prepared in 15-mL graduated tubes by mixing the appropriate amounts of PEG-4000, (NH$_4$)$_2$SO$_4$, and enzyme solution. A final weight of 10-g system was obtained by adding a sufficient amount of 0.1 M potassium phosphate buffer (pH 8.0). Systems were agitated for 1 h at room temperature and then centrifuged at 3,000 rpm at 25°C for 40 min to speed up the phase separation. The volumes of the phases were determined, and the samples from the two phases were carefully tested for enzyme assay and total protein concentration. To avoid interference of the phase components, samples were analyzed against blanks containing the same compositions, but without enzyme [6]. In this work, all partition experiments were done at 25°C.

Analytical techniques

Enzyme activity was determined by monitoring the reduction of NAD^+ at 340 nm. Mixture assay contained 10 mM D-galactose, 100 mM Tris-HCl buffer (pH 8.6), 2.5 mM NAD^+, and the enzyme solution in a total volume of 1 mL. The change of absorbance at 340 nm was measured and corrected for blank values not including D-galactose. One unit of GalDH activity (U) is defined as the amount of enzyme catalyzing the formation of 1 μmol NADH per minute under the assay conditions [8]. The total protein concentration was determined by a Bio-Rad protein assay kit with bovine serum albumin (BSA) as a standard [15]. The purity of recombinant enzyme in ATPS was analyzed by a 12% sodium dodecyl sulfate-polyacrylamide gel electrophoresis (SDS-PAGE). Samples were diluted in a sample loading buffer and heated at 100°C for 5 min prior to being loaded into electrophoretic gel. After separation, the gel was stained with Coomassie Brilliant Blue R-250 and then destained by diffusion in a solution containing 40% (v/v) methanol and 10% (v/v) acetic acid [16]. The kinetic parameters of the final purified enzyme were calculated from the secondary plots of intercepts versus reciprocal concentrations of the other substrate.

Determination of partition parameters

To evaluate the partition performance of GalDH, different parameters were defined [10]. These include the partition coefficient (K_E or K_P), which is calculated as the ratio of the enzyme activity or protein concentration in the top phase divided by the correspondent value in the bottom phase. Specific activity (SA), which is defined as the enzyme activity (U/ml) in the phase sample divided by the total protein concentration (mg/ml) and is expressed in U/mg of protein.

The recovery ($R\%$) is the ratio of the enzyme activity in the top phase (A_t) to the initial activity added to the system (A_{ori}). Purification factor (PF) is calculated as the specific activity in the top phase (SA_t) divided by the initial specific activity in the original sample (SA_{ori}). Yield ($Y\%$) is determined as

$$Y(\%) = 100 V_t K / (V_t K + V_b) \qquad (1)$$

where V_t and V_b are the top and bottom phase volumes, respectively.

Design of experiments and statistical analysis

A three-factor central composite face-centered design (CCFD) was used to optimize recombinant GalDH partitioning using Design-Expert software (version 8.0.4, State-Ease, Inc., USA). The selected variables were PEG-4000 concentration (X_1), $(NH_4)_2SO_4$ concentration (X_2), and pH (X_3). For each of the three variables, high (coded value +1) and low (coded value −1) points were chosen

Table 1 Factors and value levels used in the central composite design

Factors	Factor code	Inferior level[a] (−1)	Center point[a] (0)	Superior level[a] (+1)
PEG (%, w/w)	X_1	13	14	15
Salt (%, w/w)	X_2	11	12	13
pH	X_3	7	7.5	8

[a]Level 0 represents the central level of each factor in the intended ranges to evaluate the background variability in the process; levels −1 and +1 are the factorial points, i.e., the high and low levels of each factor in the intended range.

on the basis of preliminary test about their effects on GalDH partition. The level and ranges chosen for the variables are shown in Table 1. A complete CCFD experiment design allows estimation of a full quadratic model for each response. These kind of designs are easy to build because they are based on two-level factorials that have been augmented with a center point and $2k$ (k is the number of studied variables). Their general description is Number of experiments = $2^{k-p} + 2k + cp$, where k is the number of studied variables, p is the fractionalization element (full design, $p = 0$), and cp is the number of central points. All the 20 experiment ($2^3 + 2.3 + 6$) points which included six replications at the center point are described in Table 2. The experimental data obtained from

Table 2 Experimental results from the central composite design

Assay	Factor level			Responses			
	X_1	X_2	pH (X_3)	R (%)	Y (%)	PF	SA
1	1	−1	−1	220	72	47	185.317
2	1	−1	1	230	70.58	48.3	307.811
3	0	0	0	256.25	96.09	61	359.216
4	1	1	−1	195	62.83	43	121.153
5	0	−1.68	0	200	60	42.61	268.246
6	0	0	−1.68	195	60.43	53	278.261
7	−1	1	+1	163	59.62	30	296.022
8	0	0	0	268.75	92.80	58.9	373.913
9	−1	1	−1	181.25	42.64	35	129.486
10	1.68	0	0	259	62.83	39.81	155.593
11	−1.68	0	0	188	54.54	38	92.5069
12	1	1	+1	209	56.04	35.02	135.932
13	0	0	0	270	92.30	55	283.94
14	−1	−1	+1	205	65.75	36	103.347
15	0	1.68	0	175	62.68	26.25	89.3498
16	0	0	0	262.5	92.64	56.64	324.638
17	0	0	1.68	198	65.03	42.73	82.8794
18	0	0	0	270	96.35	61.76	332.703
19	−1	−1	−1	220	57.28	28.98	74.5062
20	0	0	0	268	79.41	63	333.025

the design were analyzed by the response surface regression procedure using the following second-order polynomial equation:

$$Y_i = b_0 + \sum b_i x_i + \sum b_{ii} x_i^2 + \sum b_{ij} x_i x_j \qquad (2)$$

where Y_i is the predicted response; b_0, b_i, b_{ii}, and b_{ij} are regression coefficient for the intercept, first-order model coefficients, and the linear mode coefficient for the interaction between variables i and j, respectively; and x_i's are the coded independent variables. Analysis of variance (ANOVA) was used to estimate the statistical significance of the full quadratic models. The suitability of the proposed model was evaluated by Fisher's statistical test (F test) by testing for significance between sources if variation in experimental data, i.e., the significance of the regression (SOR), the lack of fit (LOF), model p value, and the coefficient of determination (R^2), results. The F value is defined as the ratio of the mean square of regression (MR_R) to the error (MR_e), representing the significance of each controlled variable on the tested model. The regression equations were also summated to the F test to determine the coefficient R^2. The fitted polynomial equation was expressed as three-dimensional surface plots to visualize the relationship between the responses (dependent variables) and the experimental levels of each factor (independent variables) employed in the design. Parameters with less than 95% significance ($p > 0.05$) were removed, and the experimental data was refitted to only the significant ($p < 0.05$) factors to obtain the final reduced model. The combination of different optimized parameters, which gave maximum response, i.e., maximum recovery of favorite enzyme in PEG phase, was tested experimentally to confirm the validity of the model [17,18].

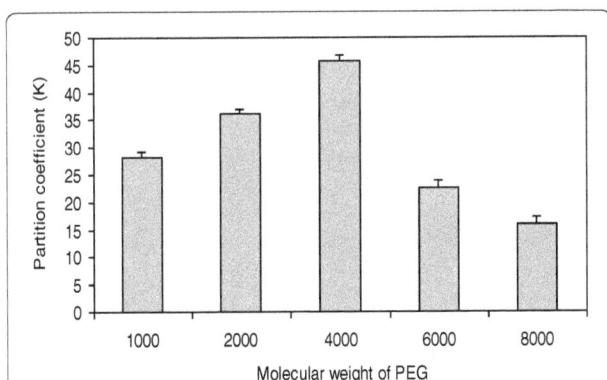

Figure 1 Influences of PEG MW on partition coefficient of GalDH in ATPS containing 14% (*w/w*) PEG-4000 and 12% (*w/w*) (NH$_4$)$_2$SO$_4$ at pH 7.0.

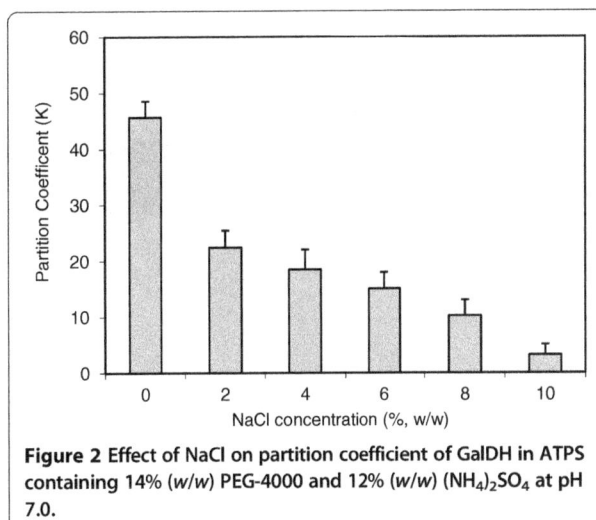

Figure 2 Effect of NaCl on partition coefficient of GalDH in ATPS containing 14% (*w/w*) PEG-4000 and 12% (*w/w*) (NH$_4$)$_2$SO$_4$ at pH 7.0.

Results and discussion

Optimization of GalDH partition process

In order to elucidate the main factors that will be included in recombinant *P. fluorescens* GalDH partition in ATPS, a series of preliminary studies were performed. This was done in a system composed of 14% (*w/w*) PEG-4000 and 12% (*w/w*) (NH$_4$)$_2$SO$_4$ ATPS at pH 7.0 and 25°C. The influence of PEG MW was investigated using four different polymer molecular weights. As shown in Figure 1, GalDH showed high affinity for the top phase. The optimal system was attained using PEG-4000, which suggested that the decreasing polymer MW until 4,000 daltons was favorable for enzyme partitioning. In contrast, the increase of PEG MW from 4,000 to 8,000 daltons resulted in less available space of GalDH in the upper phase, which led to the decrease of partition coefficient. This behavior was in agreement with an exclusion effect owing to the diminution of the free volume available in the top phase [19]. In accordance with the above, during examination within the framework of this research polymer, MW was not changed. To study the effect of neutral salt on the partition features of GalDH, the addition of 0% to 10% (*w/w*) NaCl in 14% (*w/w*) PEG-4000/12% (*w/w*) (NH$_4$)$_2$SO$_4$ system was examined. Based on the obtained data (Figure 2), the

Table 3 Results of ANOVA for the influence of variables on SA in GalDH partitioning using PEG-4000/(NH$_4$)$_2$SO$_4$ ATPS

Source	df	SS	MS	F value	p value
Model	9	163,600	18,172.92	3.23	0.0409
Residual	10	56,264.39	5,626.44		
Lack of fit	5	51,441.15	10,288.23	10.67	0.0106
Pure error	5	4,823.25	964.65		
Total	19	219,800	964.65		

df, degree of freedom; SS, sum of squares; MS, mean squares.

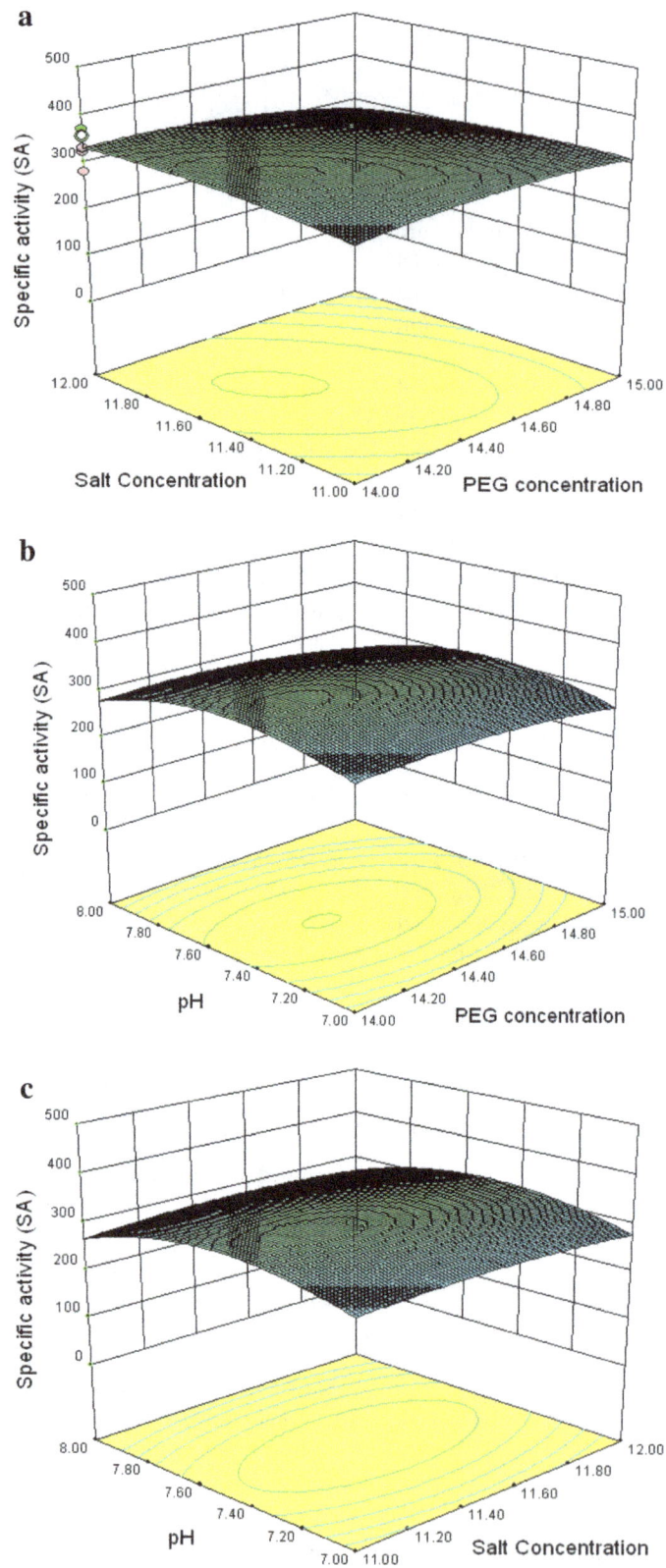

Figure 3 Response surface plot of SA. As a function of **(a)** PEG-4000 and $(NH_4)_2SO_4$ concentrations, **(b)** pH and PEG-4000 concentrations, and **(c)** pH and $(NH_4)_2SO_4$ concentrations.

highest K (45.67%) was achieved at no addition of NaCl. This result suggested that NaCl addition was not required in further modeling. The addition of salt at high concentrations leads to aggregation followed by protein precipitation, because a large amount of water molecules are strongly bound to the salts. As a consequence, the interactions among proteins become more powerful than between protein and water. Similar results for the influences of PEG MW and NaCl addition have been reported for other enzymes of oxidoreductases such as phenylalanine dehydrogenase (PheDH) [1] and proline dehydrogenase (ProDH) [20]. Meanwhile, pH range applied during this work was chosen according to the pI of target enzyme (pI = 4.86). The optimal pH range for this pI in two-phase partitioning is usually between 7.0 and 8.0 [4]. This was a typical behavior for enzymes which have negative charges [21]. The pH parameter was an important factor in process optimization, so it was included in the experimental conditions. Briefly, pH, polymer concentration, and salt concentration, which were the most important variables affecting enzyme partition were chosen for optimization in the remaining steps of this process. The optimum conditions for recombinant GalDH recovery through partitioning in PEG-ammonium sulfate ATPS were achieved by defining the experimental space around the conditions selected from preliminary tests using an experimental design methodology. The design variables and their ranges are determined as follows: PEG concentration (13% to 15%, w/w), pH (7.0 to 8.0), and salt concentration (11% to 13%, w/w). Experiments according to the design matrix of variables in Table 1 were carried out, and the relevant results are shown in Table 2. PEG and ammonium sulfate concentration as well as pH exhibited to have a significant effect on GalDH partitioning. The ANOVA was employed for the determination of significant variables and all their possible linear and quadratic interactions on the response variables. Table 3 lists the significant parameters and statistical test results of the models. The model determination coefficient, R^2, was calculated to be 0.74 for SA, which indicates a good response between prediction and experimental data. The F values were calculated by the mean square (MS) of the model to the mean square of the residual error term. The model F value (10.67) and probability $>F$ value (0.0106) for SA of recombinant GalDH indicated that the models for selected response were significant. The variables found to be statistically non-significant ($p > 0.05$) were removed, and the experimental data was refitted to only the significant ($p < 0.05$) factors to achieve the appropriate model. Larger values of F and smaller values of p showed that the variables would be significant ($p < 0.05$). The achieved data confirmed that the factors, pH, PEG concentration, and salt concentration, significantly affected enzyme partitioning. The response surface plots provide a method to visualize

the relationship between responses and experimental levels of each variables and the type of interactions between two test variables [22]. Through the response surface plots, the interactions of variables and the optimum level of each variable for maximum response can be well understood. The three-dimensional graphs were generated, in which the effect of three defined factors are shown. Figure 3 depicts the response surface plots of SA against different pH values, salt concentration, and PEG concentration when one of the three variables is fixed. The individual optimum region led to a SA value (373.9 U/mg) which was achieved in the ATPS system of 14.33 % (w/w) PEG-4000 concentration, 11.79% (w/w) $(NH_4)_2SO_4$ concentration, and pH 7.48. This enhancement of SA was likely the consequence of elimination of inhibitors from the optimal system which increases the enzyme activity. The highest estimated PF, Y, and R at this condition were 58.9, 92.8%, and 268.75%, respectively. As observed, the response surface plots obtained for the enzyme SA exhibited a slight increase in SA when the concentration of

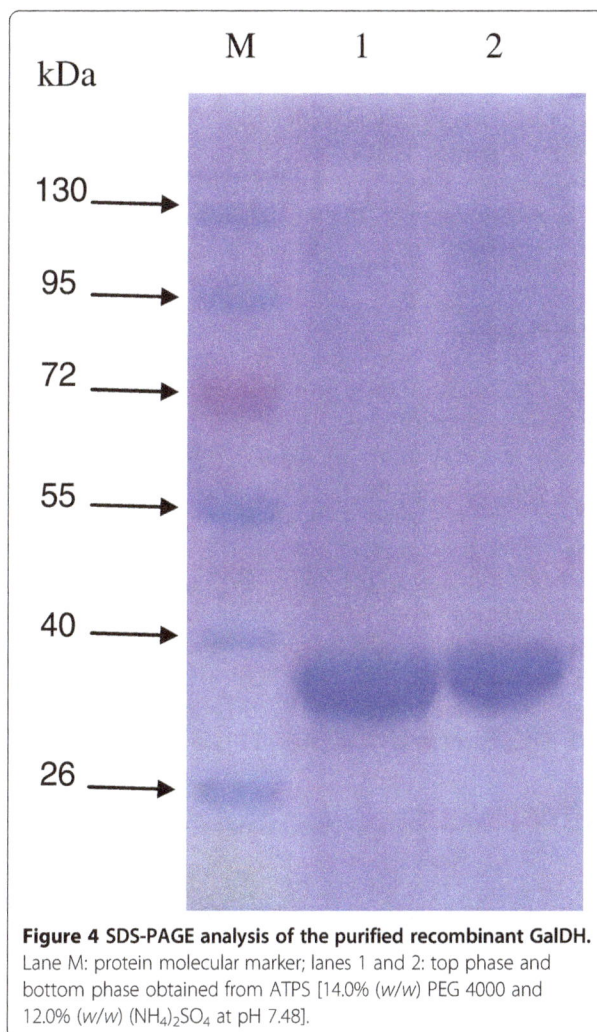

Figure 4 SDS-PAGE analysis of the purified recombinant GalDH.
Lane M: protein molecular marker; lanes 1 and 2: top phase and bottom phase obtained from ATPS [14.0% (w/w) PEG 4000 and 12.0% (w/w) $(NH_4)_2SO_4$ at pH 7.48].

PEG was increased (Figure 3a,b) and a slight decline of SA as the concentration of salt was decreased (Figure 3a,c). Also, the increase of pH value resulted to reduction in enzyme SA (Figure 3b,c). The highest SA was obtained for an intermediate value of pH (7.48). This implies that there are other factors related with the characteristics of the desired enzyme (physico-chemical properties) besides the ATPS components that affect the partitioning behavior [4]. Collectively, it can be concluded that pH, salt concentration, and PEG concentration had a significant influence on the separation and recovery. Meanwhile, the highest values of the responses are achieved near the centers of the graph, indicating that the center points selected for this work were appropriate.

SDS-PAGE and kinetic analysis of recombinant GalDH

The partitioning of recombinant enzyme in ATPS was evaluated by SDS-PAGE analysis (Figure 4). Purified GalDH was found in the PEG-rich phase and appeared as a single protein band on Coomassie Brilliant Blue stained SDS-PAGE gel. The achieved data also demonstrated the applicability of the studied system for GalDH recovery. The subunit MW of recombinant GalDH was estimated to be about 34 kDa, which was similar to the reported value from *P. fluorescens* [5,6]. The K_m for the purified recombinant GalDH was measured to be 0.32 mM. This result showed that the obtained enzyme with ATPS had considerable affinity for its substrate and, therefore, was suitable for the diagnostic applications.

Conclusions

The work presented here showed the potential application of ATPS for partitioning and recovery of recombinant GalDH in a single step. The RSM combined to a proper factorial experimental design proved to be a powerful tool in designing and modeling the best two-phase condition for enzyme partitioning. It was concluded that the ATPS consisting 14.33% (*w/w*) PEG-4000 and 11.79% (*w/w*) (NH$_4$)$_2$SO$_4$, pH 7.48 at 25°C were the most optimum system to perform GalDH partition. Under these experimental conditions, the response values for PF, *R*, *Y*, and SA were 58.9, 268.75%, 92.8%, and 373.9 U/mg, respectively, and these results were also confirmed by the evaluation of activity assay and purity of final product.

Competing interests
The authors declare that they have no competing interests.

Authors' contributions
AK carried out the design of experiments with RSM and purification and characterization of the target enzyme. MP did the general molecular biology methodology in this research. BM participated in the microbiology sections and also the paper writing. SS participated in the coordination of the manuscript as well as doing the statistical analysis. All authors read and approved the final manuscript.

Acknowledgements
We would like to express our thanks to Dr. Hamid Shahbaz Mohammadi for revising the text.

Author details
[1]Genetic and Metabolism Research Group, Department of Biochemistry, Pasteur Institute of Iran, 13164 Tehran, Iran. [2]Department of Medical Biotechnology, School of Advanced Medical Sciences, Tabriz University of Medical Sciences, 13164 Tabriz, Iran. [3]Department of Cell and Molecular Biology, Islamic Azad University, East Tehran Branch, 13164 Tehran, Iran. [4]Department of Microbiology, Islamic Azad University, North Tehran Branch, 13164 Tehran, Iran. [5]Department of Cell and Molecular Biology, Faculty of Science, Islamic Azad University, Sanandaj Branch, 13164 Sanandaj, Iran.

References
1. Shahbaz Mohammadi H, Omidinia E (2007) Purification of recombinant phenylalanine dehydrogenase by partitioning in aqueous two-phase systems. J Chromatogr B 854:273–278
2. Hatti-Kaul R (1999) Methods in biotechnology: aqueous two-phase systems: methods and protocols. Humana Press Inc., Totowa
3. Albertsson PA (1986) Partition of cell particles and macromolecules, 3rd edn. Wiley, New York
4. Shahbaz Mohammadi H, Omidinia E (2011) The Features of Partitioning Behavior of Recombinant Amino acid Dehydrogenase in Aqueous Two-phase Systems. In: Ehliers TP, Wilhelm JK (eds) Polymer phase behavior. Nova Science, New York, pp 235–264
5. Sperka S, Zehelein E, Fiedler S, Fischer S, Sommer R, Buckel P (1989) Complete nucleotide sequence of *Pseudomonas fluorescens* D-galactose dehydrogenase gene. Nucleic Acids Res 17:5402–5402
6. Buckel P, Zehelein E (1981) Expression of *Pseudomonas fluorescens* D-galactose dehydrogenase in *E. coli*. Gene 16:149–159
7. Prachayasittikul V, Ljung S, Isaraankura-Na-Ayudhya C, Bulow L (2006) NAD (H) recycling activity of an engineered bifunctional enzyme galactose dehydrogenase/lactate dehydrogenase. Int J Biol Sci 2:10–16
8. Mazitsos BA, Rigden DJ, Tsoungas PG, Clonis YD (2002) Galactosyl-mimodye ligands for *Pseudomonas fluorescens* β-galactose dehydrogenase. Eur J Biochem 269:5391–5405
9. Fughmura Y, Kawamura M, Naruse H (1981) A new method of blood galactose estimation for mass screening of galactosomia. Tohoku J Exp Med 133:371–378
10. Teles de Faria J, Coelho Sampaio F, Converti A, Lopes Passos FM, Paula Radrigues Minim V, Antonio Minim L (2009) Use of response surface methodology to evaluate the extraction of *Debaryomyces hansenii* xylose reductase by aqueous two-phase system. J Chromatogr B 877:3031–3037
11. Dembczynski R, Białas W, Jankowski T (2013) Partitioning of lysozyme in aqueous two-phase systems containing ethylene oxide-propylene oxide copolymer and potassium phosphates. Food Bioprod Process 91:292–302
12. Garai D, Kumar V (2013) Aqueous two phase extraction of alkaline fungal xylanase in PEG/phosphate system: optimization by Box–Behnken design approach. Biocatal Agric Biotechnol 2:125–131
13. Zhu W, Song J, Ouyang F, Bi J (2005) Application of response surface methodology to the modeling of α-amylase purification by aqueous two-phase systems. J Biotechnol 118:157–165
14. Singh P, Singh Shera S, Banik J, Mohan Banik R (2013) Optimization of cultural conditions using response surface methodology versus artificial neural network and modeling of L-glutaminase production by *Bacillus cereus* MTCC 1305. Bioresour Technol 137:261–269
15. Bradford MM (1976) Rapid and sensitive method for the quantification of microgram quantities of protein utilizing the principles of protein-dye binding. Anal Biochem 72:248–254
16. Sambrook J, Fritsch EF, Maniatis T (1994) Molecular cloning: a laboratory manual, 2nd edn. Cold Spring Harbor Laboratory press, Cold Spring Harbor, pp 1847–1857
17. Treier K, Lester P, Hubbuch J (2012) Application of genetic algorithms and response surface analysis for the optimization of batch chromatographic systems. Biochem Eng J 63:66–75

18. Xu Q, Shen Y, Wang H, Zhang N, Xu S, Zhang L (2013) Application of response surface methodology to optimize extraction of flavonoids from fructus sophorae. Food Chem 138:2122–2129

19. Yücekan I, Onal S (2011) Partitioning of invertase from tomato in poly (ethylene glycol)/sodium sulfate aqueous two-phase systems. Process Biochem 40:226–232

20. Shahbaz Mohammadi H, Omidinia E (2013) Process integration for the recovery and purification of recombinant *Pseudomonas fluorescens* proline dehydrogenase using aqueous two-phase systems. J Chromatogr B 929:11–17

21. Yan-Min L, Yan-Zaho Y, Xi-Dan Z, Chuan-Bo X (2010) Bovine serum albumin partitioning in polyethylene glycol (PEG)/potassium citrate aqueous two-phase systems. Food Bioprod Process 88:40–46

22. Kammoun R, Chouyekh H, Abid H, Naili B, Bejar S (2009) Purification of CBS 819.72 α-amylase by aqueous two-phase systems: modeling using response surface methodology. Biochem Eng J 46:306–312

5

Effect of fruit pulp supplementation on rapid and enhanced ethanol production in very high gravity (VHG) fermentation

Veeranjaneya Reddy Lebaka[1*], Hwa-Won Ryu[2] and Young-Jung Wee[3]

Abstract

Background: The energy crisis and climate change necessitate studying and discovering of new processes involved in the production of alternative and renewable energy sources. Very high gravity (VHG) fermentation is one such process improvement aimed at increasing both the rate of fermentation and ethanol concentration. The technology involves preparation and fermentation of media containing 300 g or more of dissolved solids per liter to get a high amount of ethanol.

Findings: *Saccharomyces cerevisiae* was inoculated to the very high gravity medium containing 30% to 40% *w/v* glucose with and without supplementation of three selected fruit pulps (mango, banana, and sapota). The fermentation experiments were carried out in batch mode. The effect of supplementation of 4% fruit pulp/puree on the metabolic behavior and viability of yeast was studied. Significant increase in ethanol yields up to 83.1% and dramatic decrease in glycerol up to 35% and trehalose production up to 100% were observed in the presence of fruit pulp. The fermentation rate was increased, and time to produce maximum ethanol was decreased from 5 to 3 days with increased viable cell count. The physical and chemical factors of fruit pulps may aid in reducing the osmotic stress of high gravity fermentation as well as enhanced ethanol yield.

Conclusions: It was found that fruit pulp supplementation not only reduced fermentation time but also enhanced ethanol production by better utilization of sugar. Production of high ethanol concentration by the supplementation of cheap materials in VHG sugar fermentation will eliminate the expensive steps in the conventional process and save time.

Keywords: High gravity fermentation; Osmotic stress; Ethanol; Fruit pulp supplementation

Background

The energy crisis and climate change necessitate studying and discovering of new processes involved in the production of alternative and renewable energy sources. Bioethanol is regarded as a promising alternative energy source, which is both renewable and environmentally friendly [1,2]. The commonly used ethanol producer in industries is *Saccharomyces cerevisiae* and the initial sugar concentration will not exceed 20% and the conventional ethanol production process needs high energy, high cost and low productivity [3]. Very high gravity (VHG) fermentation is one such process improvement aimed at increasing both the rate of fermentation and ethanol concentration. The technology involves preparation and fermentation of media containing 300 g or more of dissolved solids per liter [4]. VHG fermentation influences the five basic fermentation assets: (1) plant and equipment, (2) raw materials, (3) utilities and consumables, (4) personnel, and (5) money [5]. High gravity fermentation is an accepted method to produce more ethanol in existing fermenters and distil houses (affecting item 1), and uses less cooling equipment and produces less effluent (affecting item 3), resulting in higher yield (affecting item 2) and less staff work (affecting item 4); all these properties decrease the money investment for ethanol production. Another advantage is an increase in opportunities for harvest of high protein spent yeast [4,6].

* Correspondence: lvereddy@yahoo.com
[1]Department of Microbiology, Yogi Vemana University, Kadapa, Andhra Pradesh 516003, India
Full list of author information is available at the end of the article

However, the high sugar content of the very high gravity fermentation medium causes an increase in the osmotic pressure, which has a pessimistic effect on yeast cells, and the fermentations are rarely fast and complete. The ethanol produced by the yeast also poses negative effects on yeast metabolism like enzyme inhibition and membrane solubility and needs some protectants to counteract these effects at the end of the fermentation process [3]. S. cerevisiae can ferment an increased amount of sugars in the medium when all required nutrients are provided in adequate amounts [6]. Specific nutrients, such as nitrogen, trace elements, or vitamins, are required to obtain rapid fermentation and high ethanol levels, which are desirable to minimize capital costs and distillation energy. On a laboratory scale, media are often supplemented with peptone, yeast extract, amino acids, and vitamins [6-8]. However, such addition is not feasible in industrial fermentation processes due to the associated high costs. Thus, it is necessary to exploit inexpensive nutrient sources to supply all nutritional requirements for yeast growth and fermentation. Many investigators studied the effect of inexpensive substances like soy flour, oils and fatty acids, fungal mycelia, and fruit pulp [9-12] on the improvement of ethanol production. In our laboratory, we have tested finger millet and horse gram powder supplementation in VHG fermentation and successfully improved the ethanol production [13,14].

In view of the above, we have screened 15 different commercially available fruits to determine their effect on very high gravity fermentation in terms of fermentation rate and enhancement of ethanol yield. In this paper, we presented the results of tropical fruit pulps mango (*Mangifera indica*), banana (*Musa paradisiaca*), and sapota (*Achras sapota*), which show a significant effect in enhancing the ethanol production during the screening process.

Materials and methods
Organism and cultural conditions
Yeast strain S. cerevisiae 3215 was used in all the experiments. The yeast strain was obtained from National Collection of Industrial Microorganisms (NCIM, Pune, India). The culture was maintained on MPYD (malt extract 3 g/L, peptone 5 g/L, yeast extract 3 g/L, and dextrose 2 g/L) agar (1.5 g/L) slants at 4°C. The inoculum was prepared by inoculating the slant culture into 25 mL of the sterile MPYD liquid medium taken in a 100-mL flask and growing it on a rotary shaker (100 rpm) for 48 h. The above produced yeast culture (5%, 1×10^6 cells/mL) was used as inoculum to initiate the fermentation.

Fruit pulp preparation for supplementation
The fruits selected for the supplementation, mango (*M. indiaca*), banana (*M. praradisiaca*), and sapota (*A. sapota*),

were purchased from the local market of Kadapa, India. The fruits were peeled off, and the pulp was separated from stones in the case of mango and sapota and prepared as puree with a macerator. The prepared puree will have good suspension in the fermentation medium.

Fermentation
The very high gravity fermentation medium composition is the same as the abovementioned MPYD medium with high glucose concentration (300 to 400 g/L); 4% fruit pulp/puree was supplemented to 300 and 400 g/L sugar medium to evaluate the potential effect of fruit pulps in enhancing the ethanol production, and the medium without supplementation of fruit pulp was treated as control. Fermentations were conducted at 30°C in 250-mL Erlenmeyer flasks with 100 mL of fermentation medium. The initial pH was adjusted to 5.5. The progress of fermentation was monitored by periodical sample analysis. The fermentation was stopped after 5 days, and samples were kept at –4°C until the analysis.

Analytical estimations
Sugar concentration was estimated using the Shaffer and Somogyi method [15] as follows: Reducing sugars were estimated using the idometric method of Shaffer and Somogyi (1933). Sugars containing a free sugar syrup group undergo enolization when placed in an alkaline solution. Enediol forms of sugars are highly reactive to acids. The reduced copper was quantified by idometric titration using starch as an indicator. The 1 L reagent contains sodium carbonate (25 g), Rochelle salt (25 g), copper sulfate (75 mL from 100 g/L solution), sodium bicarbonate (20 g), potassium iodide (5 g), and potassium iodate (3.567 g). A 5 mL solution containing 0.5 to 2.5 mg dextrose units was pipetted into test tubes and 5 mL of reagent was added; then, the solution with the added reagent was mixed well by stirring. Tubes capped with bulbs were placed in a boiling water bath for 15 min and cooled under running water. Next, 2 mL of idodine-oxalate titrated with 0.005 N sodium thiosulfate was added using starch as indicator. Ethanol was determined with the help of gas chromatography [16]. The fermented samples were centrifuged at 5,000 rpm for 10 min. The supernatant was used for ethanol analysis. An Agilent Systems Gas Chromatograph with Flame Ionization Detector (GC-FID) Model 6890 Plus instrument (Agilent Technologies Inc., Santa Clara, CA, USA) was used, and conditions were as follows: 5% Carbowax 20M glass column (6 ft (2 m), 2-mm inner diameter (ID), 1/4 mm). Nitrogen was used as a carrier gas with a flow of 20 mL/min, and the eluted compounds were detected using a flame ionization detector (FID). For this, the fuel gas was hydrogen with a flow rate of 40 mL/min, and the oxidant was air with a flow rate of 40 mL/min; *n*-propanol was used as internal

standard. Glycerol on diluted samples was estimated using Boehringer kits (Boehringer Mannheim (Roche, Basel Switzerland); enzymatic test (340 nm) 3 × 11 determinations (code number 10148270035)). Trehalose was estimated in the supernatant by the anthrone method as described previously [17]. The pellet obtained after the trehalose extraction was used for the protein estimation.

Cell viability

Cellular viability was determined by the methylene blue staining technique [18]. A 100 mL sterile solution of methylene blue (3.3 mM in 68 mM sodium citrate) was mixed with 100 mL of a yeast suspension diluted to reach an OD of 0.4 to 0.7 at 620 nm. This mixture was shaken, and after a 5-min incubation, it was placed in a Thomas counting chamber. The number of stained (inactive cells) and unstained (active cells) were counted in five different fields with total of at least 200 to 300 cells.

Statistical analysis

All the experiments were carried out three times (triplicate), and the mean value with standard deviation and significant (P) was determined. SPSS version 11.0 was used for analysis of variance.

Results

The present study provides potential observation of fruit pulps as supplements in small quantity during fermentation stimulating the rate of alcohol production and final alcohol concentration in very high gravity fermentation. Two sets of batch fermentation experiments with two levels of sugar concentrations, 300 and 400 g/L (30% and 40%), were carried out with and without fruit pulp supplementation, in order to evaluate the effect of fruit pulps. The unsupplemented batch fermentation experiments yielded only 9% (w/v) of alcohol in 300 g/L, and a good amount of residual sugars was left

and incompletely fermented by *S. cerevisiae*. However, the 4% fruit pulp supplementation led to a significant increase in ethanol production, and the final concentration reached 14.5% (w/v) in a shorter time (72 h) with a productivity of 2.1 g/h/L (Table 1). In the three fruit pulps selected, mango supplementation gave the highest yields of ethanol when compared to banana and sapota. In the fruit pulp-supplemented medium, 10% (v/v) of ethanol production was achieved in just 48 h after the inoculation. Besides the high fermentation rate in the supplemented medium, it also decreased the duration of fermentation from 5 to 3 days.

An attempt was made to increase ethanol production up to 18% to 20% (w/v) as in the case of sake fermentation, by increasing the sugar concentration from 300 to 400 g/L with supplementation of fruit pulp. In the 400 g/L sugar fermentation with 4% fruit pulp supplementation, the ethanol concentration was 12.5% (Table 1). The sugars were utilized maximally up to 300 g/L. In the three fruit pulps selected, mango supplementation gave the highest yields of ethanol when compared to banana and sapota. The ethanol production after 5 days in the control experiments was only 7.5%.

Effect of fruit pulp supplementation on cell viability

After 30 to 35 h of fermentation in both supplemented and control media, cell growth rate was decreased. After 40 to 50 h, the growth ceased, but glucose fermentation continued slowly until the number of the viable cell count decreased and became very low. The viability percentage of yeast cells in the supplemented medium was greater than that in the control medium. The supplementation of fruit pulp led to an increase in the rate of fermentation and ethanol yield through the extended growth phase of cells (Figure 1). In the three fruit pulps selected, mango supplementation gave a higher cell viability than banana and sapota supplementation.

Table 1 Effect of fruit pulp supplementation on ethanol production in 30 and 40% sugar fermentation

	Serial number	Supplement	Alcohol concentration (*w/v*)					
			24 h	% IMP	48 h	% IMP	72 h	% IMP
30% Sugar	1	Control	2.5 ± 0.3	-	5.5 ± 0.3	-	9.0 ± 0.7	-
	2	Mango	4.5 ± 0.5	90 ± 4.5	10 ± 0.8	90.5 ± 8.0	14.5 ± 1.2	80.5 ± 10
	3	Banana	4.0 ± 0.5	80 ± 4.0	9.0 ± 1.0	81.5 ± 9.0	13.2 ± 1.0	73.1 ± 8.0
	4	Chiku	4.0 ± 0.4	80 ± 4.2	7.5 ± 0.6	68 ± 6.2	12 ± 0.8	66.5 ± 4.8
40% Sugar	1	Control	2.0 ± 0.3	-	4.5 ± 0.5	-	7.5 ± 0.6	-
	2	Mango	4.0 ± 0.2	100 ± 2.0	8.4 ± 0.7	93 ± 7.0	12.5 ± 1.0	83.3 ± 10
	3	Banana	3.3 ± 0.3	82.5 ± 2.5	8.0 ± 0.6	87.6 ± 6.2	11 ± 0.8	73.3 ± 7.0
	4	Chiku	3.2 ± 0.2	77.5 ± 2.0	7.2 ± 0.8	80 ± 6.8	10 ± 1.0	66.5 ± 8.1

% IMP, percentage of improvement.

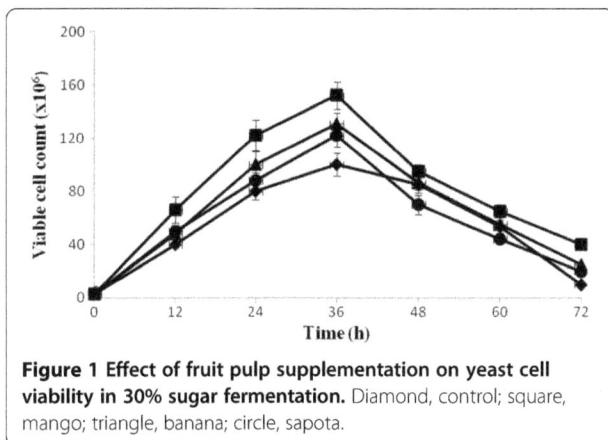

Figure 1 Effect of fruit pulp supplementation on yeast cell viability in 30% sugar fermentation. Diamond, control; square, mango; triangle, banana; circle, sapota.

Effect of fruit pulp supplementation on glycerol production and trehalose

Concentrations of glycerol, one of the stress indicators and releasers, were decreased in the fruit pulp-supplemented experiments from 954 to 620 mg/L in the 300 g/L fermentation and from 1,266 to 823 mg/L in the 400 g/L fermentation (Table 2). In the three fruit pulps selected, mango supplementation gave less glycerol when compared to banana and sapota supplementation. Trehalose is a disaccharide which is typically produced by yeast when it experiences stress conditions. In the present study, trehalose concentration was in low in the fruit pulp-supplemented experiments when compared to the unsupplemented control experiments (Table 3). In the control 30% (w/v) fermentation experiments, the trehalose concentration was 40 mg/g yeast cells, and in the fruit pulp-supplemented experiments, it was 21 mg/g yeast cells. In the 40% sugar (w/v) control fermentation experiments, the trehalose concentration was 52 mg/g yeast cells, while in the fruit pulp supplementation with aeration experiments, it was 38 mg/g yeast cells. In the three fruit pulps selected, mango supplementation decreased the trehalose to low levels when compared to banana and sapota supplementation.

Discussion

The present study provides potential observation of fruit pulps as supplements in small quantity during fermentation

Table 2 Effect of fruit pulp supplementation on glycerol production in 30 and 40% sugar fermentation

Serial number	30% Sugar		40% Sugar	
	Supplement	Glycerol (mg/L)	Supplement	Glycerol (mg/L)
1	Control	954 ± 62	Control	1,266 ± 75
2	Mango	620 ± 35	Mango	823 ± 54
3	Banana	757 ± 58	Banana	938 ± 68
4	Chiku	826 ± 73	Chiku	1,040 ± 47

stimulating the rate of alcohol production and final alcohol concentration in very high gravity fermentation. In unsupplemented controls of 30% glucose fermentation experiments, compared with the supplemented medium, the sugar was not utilized completely. It is evident that at the end of fermentation, yeast requires certain nutrients that aid tolerance to the high concentrations of alcohol it forms. Nearly 70% to 75% of the volume of the final ethanol concentration was formed within 48 h of fermentation, and almost all the final concentration of ethanol was formed in 60 h; the remaining 1% or 2% (v/v) took some time for its secretion out of the cell. In addition to nutrients, fruit pulps also contain good amounts of polyphenols (all flavones, stilbenes, flavonones, isoflavones, catechins, chalcones, tannins, and anthocyanidins), which are frequently attributed to antioxidant, metal ion-chelating, and/or free radical scavenging activity [19]. This may help in keeping the yeast cells viable for longer duration and producing such high concentrations of ethanol in 48 h. The supplemented medium had higher viable cell count than the control medium. There was a dramatic drop in cell count from 10×10^7 to 3×10^7 in the control medium with increase in ethanol concentration from 5% to 9% (v/v). But in the supplemented medium, the cell viability went up even up to 12% (v/v) ethanol. This indicates that the threshold concentration of ethanol to yeast inhibition is 9% (v/v). In all cases, the cell viability increased even at high ethanol concentration (12% v/v) in the fruit pulp-supplemented medium compared with the control medium. Alfenore et al. [6] made a similar observation in fed-batch fermentation by vitamin feeding strategy that enhanced the final ethanol up to 19% (v/v) in 45 h.

The important byproduct formed during ethanol fermentation is glycerol. Commonly, its production is high in high gravity fermentation. Glycerol is the well-known compatible solute in *S. cerevisiae*. Osmophilic yeasts accumulate glycerol to compensate for high osmotic pressure [20,21]. In the present study, the formation of glycerol was found to be high at the growth/logarithmic phase. After cessation of cell growth, glycerol was not present much in the media. The percentage of glycerol in the supplemented media was low when compared with that in the control medium. These results confirmed the previous reports that the growth rate of yeast cells is reduced irreversibly in proportion to an increase in external osmolarity [21]. Another important reserve carbohydrate and stress protectant for the yeast is trehalose. Trehalose is also considered as one of the most effective saccharines in preventing phase transition in the lipid bilayer and thereby protecting membranes against damages, and considering the relation of intracellular trehalose concentration with the cellular resistance to osmotic stress, trehalose was supposed to act as an osmoprotectant under osmotic stress [20]. In the supplemented medium, trehalose concentration was

Table 3 Effect of fruit pulp supplementation on trehalose accumulation in 30 and 40% sugar fermentation

Serial number	30% Sugar		40% Sugar	
	Supplement	Trehalose (mg/g yeast cells)	Supplement	Trehalose (mg/g yeast cells)
1	Control	40 ± 3.4	Control	52 ± 2.5
2	Mango	21 ± 2.8	Mango	38 ± 3.2
3	Banana	29 ± 3.3	Banana	43 ± 4.6
4	Chiku	34 ± 2.6	Chiku	49 ± 3.8

decreased at the end of the fermentation which shows that the cells are not under stress when compared to the control. This could explain the fact that stress induced the genes involved in trehalose synthesis and those involved in degradation, and why the genes responded in a similar pattern in osmotic and oxidative stress [22]. It has been reported that the production pattern of protein synthesis is changed dramatically by osmotic and heat stress, and also depriving amino acids or proteins inhibits translation initiation through the phosphorylation pathway [23,24].

Higher amount of ethanol in the 400 g/L sugar medium was not obtained probably due to the initial high glucose concentration that strongly inhibited fermentation. Even in such high osmolarity, the supplemented media yield 12.5% (v/v) ethanol with a productivity of 1.73 g/h/L. It is likely that the supplementation of fruit pulp may add sugars, thereby contributing to the increased osmotic pressure. The old yeast cells ferment slowly when compared with actively growing yeast cells. It is possible to produce high ethanol concentrations by extending the growth phase of yeast to longer periods as in the case of beer production. It is expected that fruit pulp supplementation would overcome nutritional deficiencies of yeast and allow them to stay longer in the growth phase and that antioxidants protected the yeast cells from osmotic stress and aeration allowed yeast to produce membrane lipids to be sustained at higher alcohol concentrations. During VHG ethanol fermentation, maintaining the redox potential at a constant level is essential, as yeast requires a small amount of oxygen to facilitate the synthesis of sterols and unsaturated fatty acids, which serve as the building blocks for constructing cell membranes [25].

Conclusions

It is concluded that fruit pulp supplementation enhanced the rate and yield of ethanol production in a very high gravity medium. It is observed that the selected fruit pulps were not much effective in the 400 g/L sugar fermentation when compared to the 300 g/L sugar fermentation. The decrease in both glycerol and trehalose concentrations by the supplementation would suggest that the fruit pulp constituents might be involved in lowering the osmotic stress induced by high sugar at the beginning of fermentation and high ethanol stress at the end of the fermentation. The increased ethanol production by the fruit pulp supplementation is a significant finding that could also be applied to an industrial fermentation of ethanol utilizing molasses and other raw materials as substrates. This may reduce the cost of ethanol production in developing countries like India. The nature of active principles from fruit pulps and their mechanism that aids in tolerating high osmotic stress and enhance ethanol production rate are being investigated by the authors.

Abbreviations
% IMP: percentage of improvement; %: percent; g/l: gram per liter; mg/l: milligram per liter; mM: millimolar; MPYD: malt extract, peptone, yeast extract, and dextrose (medium); v/v: volume per volume; VHG: very high gravity; w/v: weight per volume.

Competing interests
The authors declare that they have no competing interests.

Authors' contributions
LV participated in the design of the study, carried out the fermentations, analyzed the results, and wrote the manuscript. YJ participated in the experimental procedure and the GC and result analysis. HW conceived the study and participated in analyzing the results and correcting the manuscript. All authors read and approved the final manuscript.

Acknowledgements
The author would like to acknowledge the Council of Scientific and Industrial Research, Government of India and Department of Science and Technology, Government of India for the financial support given in the form of research projects entitled 'Studies on Rapid and Enhanced Production of Ethanol through Very High Gravity (VHG) Fermentation' (Ref No: 38 (1310)/11/EMR-II) and 'Biotechnological production of Acetone-Butanol-Ethanol (ABE) from agricultural biomass using solventogenic bacteria' (Ref No: SR/FT/LS-79/2009).

Author details
[1]Department of Microbiology, Yogi Vemana University, Kadapa, Andhra Pradesh 516003, India. [2]School of Biological Sciences and Technology, Chonnam National University, Gwangju 500-757, Korea. [3]Department of Food Science and Technology, College of Natural Resources, Yeungnam University, Gyeongbuk 712-749, Korea.

References
1. Cardona C, Sa´nchez O (2007) Fuel ethanol production: process design trends and integration opportunities. Bioresour Technol 98:2415–2457
2. Reddy LVA (2013). Potential bioresources as future sources of biofuels production: an overview. V. K. Gupta and M. G. Tuohy (eds.), Biofuel technologies, Springer-Verlag Berlin Heidelberg doi:10.1007/978-3-642-34519-7_9, 2013

3. Pereira FB, Guimarães PMR, Teixeira JA, Domingues L (2010) Optimization of low-cost medium for very high gravity ethanol fermentations by *Saccharomyces cerevisiae* using statistical experimental designs. Bioresour Technol 101:7856–7863

4. Thomas KC, Hynes SH, Jones AM, Ingledew WM (1993) Production of fuel alcohol from wheat by VHG technology. Appl Biochem Biotechnol 43:211–226

5. Reeve P (1998) Sweat your fermentation assets. Brewer 12:212–215

6. Bafrncova P, Smogrovicova D, Salvikova I, Patkova J, Domeny Z (1999) Improvement of very high gravity ethanol fermentation by media supplementation using *Saccharomyces cerevisiae*. Biotechnol Lett 21:337–341

7. Casey GP, Magnus CA, Ingledew WM (1984) High-gravity brewing: effects of nutrition on yeast composition, fermentative ability, and alcohol production. Appl Environ Microbiol 48:639–646

8. Alfenore S, Molina-Jouve C, Guillouet SE, Uribelarrea JL, Goma G, Benbadis L (2002) Improving ethanol production and viability of *Saccharomyces cerevisiae* by vitamin feeding strategy during fed batch process. Appl Microbiol Biotechnol 60:67–72

9. Damoano D, Wang SS (1985) Improvements in ethanol concentration and fermentor ethanol productivity in yeast fermentations using whole soy flour in batch and continuous recycle systems. Biotechnol Lett 71:35–140

10. Deepak S, Visvanathan L (1984) Effects of oils and fatty acids on the tolerance of distillers yeast to alcohol and temperature. Enzyme Microb Technol 6:78–80

11. Patil SG, Patil BG (1989) Chitin supplement speeds up the ethanol production in cane molasses fermentation. Enzyme Microb Technol 11:38–43

12. Patil SG, Patil BG, Gokhale VD, Bastawde KB, Puntambekar S, Ranjekar PK (2000) Process for the production of alcohol. US Patent no: 6016699.

13. Reddy LVA, Reddy OVS (2005) Improvement of ethanol production in very high gravity fermentation by horse gram (*Dolichos biflorus*) flour supplementation. Lett Appl Microbiol 41:440–445

14. Reddy LVA, Reddy OVS (2006) Rapid and enhanced production of ethanol in very high gravity (VHG) sugar fermentation by *Saccharomyces cerevisiae*: role of finger millet (*Eleusinae coracana* L.) flour. Process Biochem 41:726–729

15. Shaffer PA, Somogyi M (1933) Copper iodometric reagents for sugar determination. J Biol Chem 100:695–713

16. Antony JC (1984) Malt beverages and malt brewing materials: gas chromatographic determination of ethanol in beer. J Assoc Off Annal Chem 67:192–193

17. Aranda JS, Salgado E, Taillandier P (2004) Trehalose accumulation in *Saccharomyces cerevisiae* cells: experimental data and structured modeling. Biochem Eng J 17:129–140

18. Postgate JP (1967) Viable counts and viability. In: Norris JR, Ribbons DW (eds) Methods in microbiology, vol. 1. Academic Press, New York

19. Ferguson LR (2001) Role of plant polyphenols in genomic stability. Mutat Res 475:89–111

20. Li LL, Ye YR, Pan L, Zhu Y, Zheng S, Lin Y (2009) The induction of trehalose and glycerol in *Saccharomyces cerevisiae* in response to various stresses. BBRC 387:778–783

21. Klipp E, Nordlander B, Krüger R, Gennemark P, Hohmann S (2005) Integrative model of the response of yeast to osmotic shock. Nat Biotechnol 23:975–982

22. Da Costa M, Da Silva C, Mariani D, Fernandes P, Pereira M, Panek A, Eleutherio E (2008) The role of trehalose and its transporter in protection against reactive oxygen species. Biochem Biophys Acta 1780:1408–1411

23. Siderius M, Van Wuytswinkel O, Reijenga K, Kelders M, Mager W (2000) The control of intracellular glycerol in *Saccharomyces cerevisiae* influences osmotic stress response and resistance to increased temperature. Mol Microbiol 36:1381–1390

24. Uesono Y, Tohe A (2002) Transient inhibition of translation initiation by osmotic stress. J Biol Chem 277:13848–13855

25. Lin YH, Chien WS, Duan KJ, Chang PR (2011) Effect of aeration timing and interval during very-high-gravity ethanol fermentation. Process Biochem 46:1025–1028

6

In vitro conversion of glycerol to lactate with thermophilic enzymes

Chalisa Jaturapaktrarak[1], Suchada Chanprateep Napathorn[1*], Maria Cheng[2], Kenji Okano[2], Hisao Ohtake[2] and Kohsuke Honda[2*]

Abstract

Background: *In vitro* reconstitution of an artificial metabolic pathway has emerged as an alternative approach to conventional *in vivo* fermentation-based bioproduction. Particularly, employment of thermophilic and hyperthermophilic enzymes enables us a simple preparation of highly stable and selective biocatalytic modules and the construction of *in vitro* metabolic pathways with an excellent operational stability. In this study, we designed and constructed an artificial *in vitro* metabolic pathway consisting of nine (hyper)thermophilic enzymes and applied it to the conversion of glycerol to lactate. We also assessed the compatibility of the *in vitro* bioconversion system with methanol, which is a major impurity in crude glycerol released from biodiesel production processes.

Results: The *in vitro* artificial pathway was designed to balance the intrapathway consumption and regeneration of energy and redox cofactors. All enzymes involved in the *in vitro* pathway exhibited an acceptable level of stability at high temperature (60°C), and their stability was not markedly affected by the co-existing of up to 100 mM methanol. The one-pot conversion of glycerol to lactate through the *in vitro* pathway could be achieved in an almost stoichiometric manner, and 14.7 mM lactate could be produced in 7 h. Furthermore, the *in vitro* bioconversion system exerted almost identical performance in the presence of methanol.

Conclusions: Many thermophilic enzymes exhibit higher stability not only at high temperatures but also in the presence of denaturants such as detergents and organic solvents than their mesophilic counterparts. In this study, compatibilities of thermophilic enzymes with methanol were demonstrated, indicating the potential applicability of *in vitro* bioconversion systems with thermophilic enzymes in the conversion of crude glycerol to value-added chemicals.

Keywords: *in vitro* metabolic engineering; thermophilic enzymes; glycerol; methanol

Background

Integration of diverse biocatalytic modules to construct an advanced microbial cell factory has emerged as a powerful approach for the production of industrially important metabolites [1]. Bioprospecting efforts for exploring novel biocatalytic molecules with unique properties have inspired the design and construction of a wider variety of artificial metabolic pathways [2]. However, installation of an artificially engineered metabolic pathway in living organisms often leads to a competition with natural metabolic pathways for intermediates

and cofactors, resulting in insufficient yield of desired metabolites. A possible solution to this problem is to avoid the use of living microorganisms and to construct an *in vitro* artificial metabolic pathway in which only a limited number of enzymes are involved. Until now, a variety of *in vitro* synthetic pathways have been designed and constructed for the production of alcohols [3,4], organic acids [5,6], carbohydrates [7], hydrogen [8,9], bioplastic [10], and even electricity [11]. Particularly, employment of enzymes derived from thermophiles and hyperthermophiles enables the simple preparation of catalytic modules with excellent selectivity and thermal stability [5,12]. Furthermore, although the detailed mechanisms remain to be clarified, many thermophilic enzymes have also been reported to display higher tolerance towards denaturants such as detergents and

* Correspondence: Suchada.Cha@chula.ac.th; honda@bio.eng.osaka-u.ac.jp
[1]Department of Microbiology, Faculty of Science, Chulalongkorn University, Phayathai Road, Patumwan, Bangkok 10330, Thailand
[2]Department of Biotechnology, Graduate School of Engineering, Osaka University, 2-1 Yamadaoka, Suita, Osaka 565-0871, Japan

organic solvents than their mesophilic counterparts [13,14], and activities of some thermophilic enzymes are even improved with organic solvents [15]. These excellent stabilities of thermophilic enzymes allow great flexibility in the operational conditions of *in vitro* bioconversion systems.

Concerns about the global warming and depletion of fossil fuel reserves have led to the rapid increase of biodiesel production. Generally, 10 kg of crude glycerol, which is the primary byproduct of the biodiesel industry, is released for every 100 kg of biodiesel and the growing production of biodiesel has resulted in a worldwide surplus of crude glycerol [16]. Although many studies have been conducted to use crude glycerol as a starting material for the fermentation-based production of industrially valuable chemicals, these attempts often suffer from the inhibitory effects of impurities contained in crude glycerol on the growth and biocatalytic activity of living organisms [17,18]. Particularly, methanol, which is the most abundant impurity in crude glycerol, accounts for up to 70% (*w/w*) of a raw glycerol obtained through a biodiesel production process [19]. In this study, we focused on the high operational stabilities of thermophilic enzymes and employed them as modules to construct an *in vitro* synthetic pathway for the conversion of glycerol to lactate, which is one of the most important and versatile biomass-derived chemical [20], in the presence of methanol.

Methods

Materials

Glycerol and methanol were purchased from Wako Pure Chemical Industries Ltd. (Osaka, Japan). Other intermediates of the synthetic pathway, including glycerol-3-phosphate (G3P), dihydroxyacetone phosphate (DHAP), glyceraldehyde-3-phosphate (GAP), 3-phosphoglycerate (3-PG), 2-phosphoglycerate (2-PG), phosphoenolpyruvate (PEP), and pyruvate were obtained from Sigma-Aldrich Japan (Tokyo, Japan). NAD^+, NADH, ADP, and ATP were products of Oriental Yeast Co. Ltd. (Osaka, Japan). 2-(4-Iodophenyl)-3-(4-nitrophenyl)-5-(2,4-disulfophenyl)-2H-tetrazolium, monosodium salt (WST-1), and 1-methoxy-5-methylphenazinium methylsulfate (1-methoxy PMS) were purchased from Dojindo Laboratories (Kumamoto, Japan). All other reagents were commercially available and of analytical grade.

Microorganisms and plasmid

Escherichia coli JM109 was used for general cloning purpose. *E. coli* Rosetta 2 (DE3) was used for gene expression. Recombinant *E. coli* was aerobically cultivated at 37°C in Luria-Bertani medium supplemented with 100 μg ml^{-1} ampicillin and 34 μg ml^{-1} chloramphenicol. Gene expression was induced by the addition of 0.2 mM

isopropyl β-D-1-thiogalactopyranoside at the late log phase. The expression vector encoding the glycerol kinase of *Thermococcus kodakarensis* (GK$_{Tk}$, gi| 3986088) was donated by Dr. Y. Koga, Osaka University [21]. Sources of expression vectors for triose phosphate isomerase (TIM$_{Tt}$, gi| 3169211), enolase (ENO$_{Tt}$, gi| 55979971), pyruvate kinase (PK$_{Tt}$, gi| 55979972), lactate dehydrogenase (LDH$_{Tt}$, gi| 55981082) of *Thermus thermophilus*, non-phosphorylating GAP dehydrogenase of *T. kodakarensis* (GAPN$_{Tk}$, gi|57640640), and cofactor-independent phosphoglycerate mutase of *Pyrococcus horikoshii* (iPGM$_{Ph}$, gi| 14589995) were described previously [5]. The expression vector for G3P dehydrogenase of *T. thermophilus* (G3PDH$_{Tt}$, gi|55981709) was obtained from the Riken *T. thermophilus* HB8 expression plasmid set [22]. Gene encoding NADH oxidase of *Thermococcus profundus* (NOX$_{Tp}$, gi|187453160) was cloned and expressed in *E. coli* as described elsewhere [12].

Analytical method

Lactate was quantified by high-performance liquid chromatography (HPLC) equipped with two tandemly connected ion exclusion columns (Shim-pack SPR-H, 250 mm × 7.8 mm, Shimadzu Corp., Kyoto, Japan). The columns were eluted at 50°C using 4 mM *p*-toluenesulfonic acid as a mobile phase at a flow rate of 0.2 ml min^{-1}. The eluent was mixed with a pH-buffered solution (16 mM Bis-Tris, 4 mM *p*-toluenesulfonic acid, and 0.1 mM EDTA) supplied at a flow rate of 0.2 ml min^{-1} and then analyzed for lactate using a conductivity detector (CDD-20A, Shimadzu Corp.). Methanol concentration was quantified by an enzymatic assay using the alcohol dehydrogenase (Sigma-Aldrich Japan) and the horseradish peroxidase (Sigma-Aldrich Japan) according to the protocol provided by the manufacturer.

Enzyme assay

E. coli cells were collected by centrifugation, resuspended in 50 mM HEPES-NaOH (pH 7), and then disrupted by a UD-201 ultrasonicator (Kubota Corp., Osaka, Japan). After the removal of cell debris by centrifugation, the cell-free extract was incubated at 70°C for 30 min. The heat-precipitated proteins were removed by centrifugation, and the resulting supernatant was used as an enzyme solution. One unit of an enzyme was defined as the amount consuming 1 μmol of the substrate per min under the below-mentioned standard assay conditions. Protein concentration was measured with the Bio-Rad protein assay kit (Bio-Rad Laboratories Inc., Hercules, CA, USA) using bovine serum albumin as the standard.

Enzyme activities were spectrophotometrically determined at 60°C by monitoring consumption or generation of NADH at 340 nm. When necessary, NADH generation was coupled with the reduction of WST-1 and detected at 438 nm. GK_{Tk} activity was determined by coupling with $G3PDH_{Tt}$. The standard assay mixture for GK_{Tk} was composed of 50 mM HEPES-NaOH (pH 7), 0.2 mM glycerol, 0.2 mM ATP, 1 mM NAD^+, 5 mM $MgCl_2$, 0.5 mM $MnCl_2$, 0.15 mM WST-1, 6 μM 1-methoxy PMS, an excess amount of $G3PDH_{Tt}$, and an appropriate amount of GK_{Tk}. The mixture without glycerol was pre-incubated at 60°C for 2 min, and the reaction was initiated by the addition of the substrate. Enzyme reaction was monitored through the reduction of WST-1 to the corresponding formazan dye at 438 nm using a UV-2450 spectrophotometer (Shimadzu Corp.). $G3PDH_{Tt}$ assay was performed in the same manner except that 0.2 mM G3P was used as a substrate. TIM_{Tt} was assayed in a mixture containing 50 mM HEPES-NaOH (pH 7), 0.2 mM DHAP, 1 mM NAD^+, 5 mM $MgCl_2$, 0.5 mM $MnCl_2$, 1 mM glucose-1-phosphate (G1P), an excess amount of $GAPN_{Tk}$, and an appropriate amount of the enzyme. After a pre-incubation at 60°C for 2 min, the substrate was added to the mixture and the reduction of NAD^+ was monitored at 340 nm. For the determination of $GAPN_{Tk}$ activity, 0.2 mM GAP was used instead of DHAP. Similarly, $iPGM_{Ph}$ activity was assessed by coupling with ENO_{Tt}, PK_{Tt}, and LDH_{Tt}. The enzyme was assayed in a mixture containing 50 mM HEPES-NaOH (pH 7), 0.2 mM 3-PG, 0.2 mM ADP, 0.2 mM NADH, 5 mM $MgCl_2$, 0.5 mM $MnCl_2$, and excess amounts of ENO_{Tt}, PK_{Tt}, and LDH_{Tt}. The reaction rate was determined by monitoring the concomitant decrease of NADH at 340 nm. Assays for ENO, PK, and LDH were performed in the same mixture using 0.2 mM each of 2-PG, PEP, and pyruvate, respectively. NOX_{Tp} activity was determined by monitoring the oxidation of NADH under an air atmosphere. A reaction mixture comprising 50 mM HEPES-NaOH (pH 7.0), 5 mM $MgCl_2$, 0.5 mM $MnCl_2$, 0.02 mM flavin adenine dinucleotide (FAD), and 0.2 mM NADH was preincubated at 60°C for 2 min and then the reaction was initiated by adding an appropriate amount of enzyme.

Lactate production

The reaction mixture (4 ml) was composed of 50 mM HEPES-NaOH (pH 7), 0.2 mM glycerol, 1 mM NAD^+, 0.2 mM NADH, 0.2 mM ATP, 0.2 mM ADP, 0.02 mM FAD, 0.5 mM FBP, 1 mM G1P, 5 mM $MgCl_2$, and 0.5 mM $MnCl_2$. Enzymes were added to the reaction mixture to give the following final concentrations: 0.04 U ml^{-1} GK_{Tk}, 0.18 U ml^{-1} $G3PDH_{Tt}$, 0.04 U ml^{-1} TIM_{Tt}, 0.1 U ml^{-1} $GAPN_{Tk}$, 0.09 U ml^{-1} $iPGM_{Ph}$, 0.07 U ml^{-1} ENO_{Tt}, 0.09 U ml^{-1} PK_{Tt}, 0.08 U ml^{-1} LDH_{Tt}, and 0.04 U ml^{-1} NOX_{Tp}. The mixture was put in a 10-ml cylindrical vessel and kept at 60°C with stirring. Glycerol (160 mM) solution was continuously supplied to the mixture at a flow rate of 1 μl min^{-1} (0.04 μmol glycerol ml^{-1} min^{-1}) using a Shimadzu LC-20 AD solvent delivery unit. Alternatively, a model solution of crude glycerol, which consisted of 160 mM glycerol and 770 mM methanol, was used as a substrate and fed to the reaction mixture in the same manner. NAD^+ was put in the substrate solution at 4 mM and supplied into the reaction mixture with the substrate to complement the thermal degradation (0.001 μmol NAD^+ ml^{-1} min^{-1}). Aliquots (50 μl) of the reaction mixture were sampled at every 1-h intervals, diluted fourfold with distilled water. The sample was ultrafiltrated using Amicon 3 K (Merk Milipore, Billerica, MA, USA) and then analyzed by HPLC.

Results and discussion

Design of the *in vitro* synthetic pathway

Figure 1 illustrates the newly designed synthetic pathway for the conversion of glycerol to lactate. To construct an *in vitro* synthetic pathway, it is vital to prevent the depletion of energy and redox cofactors (ATP/ADP, and NAD^+/NADH) by balancing their intrapathway consumption and regeneration. In a previous study, we constructed an ATP/ADP-balanced chimeric Embden-Meyerhof (EM) pathway by swapping the enzyme couple of GAP dehydrogenase and phosphoglycerate kinase in the bacterial/eukaryotic EM pathway with the non-phosphorylating GAP dehydrogenase ($GAPN_{Tk}$) involved in the modified EM pathway of a hyperthermophilic archaeon, *Thermococcus kodakarensis* [5]. Similarly, we employed the GAPN-mediated non-ATP-forming dehydrogenation of GAP to 3-PG for balancing the consumption and regeneration of ATP and ADP through the glycerol converting pathway. On the other hand, the conversion of one molecule of glycerol to lactate through the designed pathway was accompanied by the generation of one molecule of NADH. To re-oxidize the cofactor and to maintain the redox balance of the whole pathway, a hyperthermophilic NADH oxidase was integrated into the pathway. NADH oxidases catalyze the reduction of O_2 using NAD(P)H as a reductant and can be divided into two groups: those catalyzing two-electron reduction of O_2 to H_2O_2 and those catalyzing four-electron reduction of O_2 to H_2O. In this study, we employed the NADH oxidase from *T. profundus* (NOX_{Tp}), which preferably catalyzes four-electron reduction of O_2 [23], to eliminate the inhibitory effects of H_2O_2 on enzymes. The chemical equation of the overall reaction through the synthetic pathway can be shown as follows:

$$HOCH_2CHOHCH_2OH + 1/2\,O_2$$
$$= CH_3CHOHCOO^- + H_2O + H^+$$

The standard Gibbs energy change ($\Delta G°$) of the reaction was calculated to be −256.4 kJ/mol.

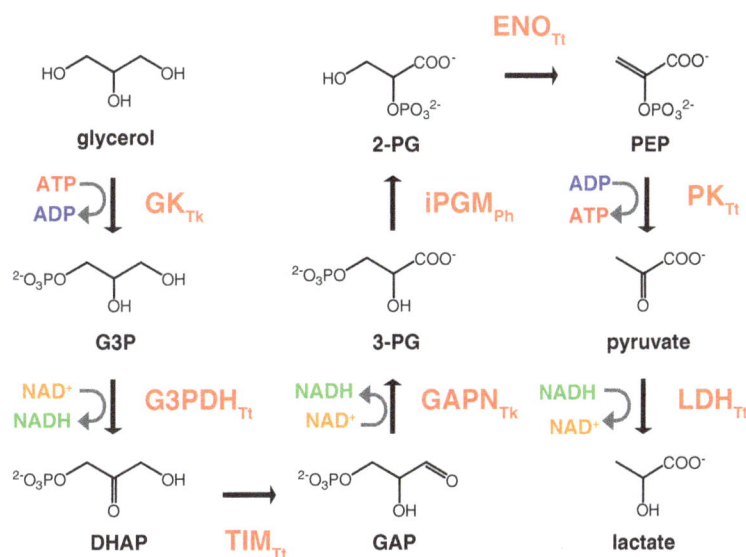

Figure 1 Schematic illustration of the *in vitro* synthetic pathway constructed in this study.

Enzyme stability

Crude extracts of recombinant *E. coli* cells were heat-treated at 70°C for 30 min to denature indigenous proteins and then used in following studies. SDS-PAGE analysis of the crude extract revealed that most of host-derived proteins were removed by the heat precipitation (Additional file 1: Figure S1). The enzyme stability was assessed by measuring the remaining activity of enzymes after the incubation at 60°C (Figure 2). Most enzymes could retain more than 80% of their initial activity for 8 h, except that PK_{Tt} lost 35% of the activity after the incubation for the same time period. We also investigated the effect of methanol, which is the primary impurity contained in crude glycerol, on the enzyme stability. Although residual activities of TIM_{Tt}, $iPGM_{Ph}$, ENO_{Tt}, and LDH_{Tt} were moderately lower than those in the absence of methanol (16% to 32% decrease), the destabilization profile of other enzymes were not significantly affected by at least up to 100 mM of methanol.

Optimization of reaction conditions

The lactate production rate through the synthetic pathway was determined at different pH and temperatures by incubating 0.1 U ml^{-1} each of GK_{Tt}, $G3PDH_{Tt}$, TIM_{Tt}, $GAPN_{Tk}$, $iPGM_{Ph}$, ENO_{Tt}, PK_{Tt}, LDH_{Tt}, and NOX_{Tp} with 10 mM glycerol and appropriate concentrations of cofactors and metal ions (Figure 3). Glucose-1-phosphate (G1P) was put in the reaction mixture as an activator for $GAPN_{Tk}$ [24]. Although LDH_{Tt} is allosterically inhibited by NAD$^+$ [5], lactate dehydrogenases can generally be activated by fructose-1,6-bisphosphate (FBP). In fact, we found that LDH_{Tt} activity in the presence of 1 mM NAD$^+$ could be recovered to the similar

level to that under the standard assay conditions by the addition of 0.5 mM FBP (Table 1). When the reaction was carried out at 60°C in different buffers, the highest lactate production rate of 0.036 μmol min^{-1} ml^{-1} was observed in HEPES-NaOH (pH 7.0) (Figure 3A). The reaction was then performed in this buffer at 50°C, 60°C, and 70°C (Figure 3B). Although no significant difference was observed in production rates at 60°C and 70°C ($P > 0.1$, Student's *t*-test), the reaction temperature of 60°C was employed for further studies to mitigate the thermal inactivation of the enzymes and the decomposition of thermo-labile intermediates and cofactors.

Lactate production

Unlike highly branched metabolic pathways in living organisms, *in vitro* synthetic pathways, in which only a limited number of enzyme reactions are sequentially aligned, appear to be less sensitive to the imbalance in enzyme concentrations. Although the existence of a rate-limiting enzyme leads to the accumulation of the specific intermediate, it is eventually converted by downstream enzymes without being routed into the co-existing pathway. However, the accumulation of chemically labile intermediates will result in their spontaneous degradation and decrease in the overall yield of product. We previously demonstrated that the flux through an *in vitro* metabolic pathway can be spectrophotometrically determined by dividing the whole pathway into some partial pathways, in each of which the NAD(H)-dependent enzymes are assigned to be the last step and by monitoring the concomitant consumption or production of NAD(P)H through the partial pathways [5]. This real-time monitoring technique enables us to identify

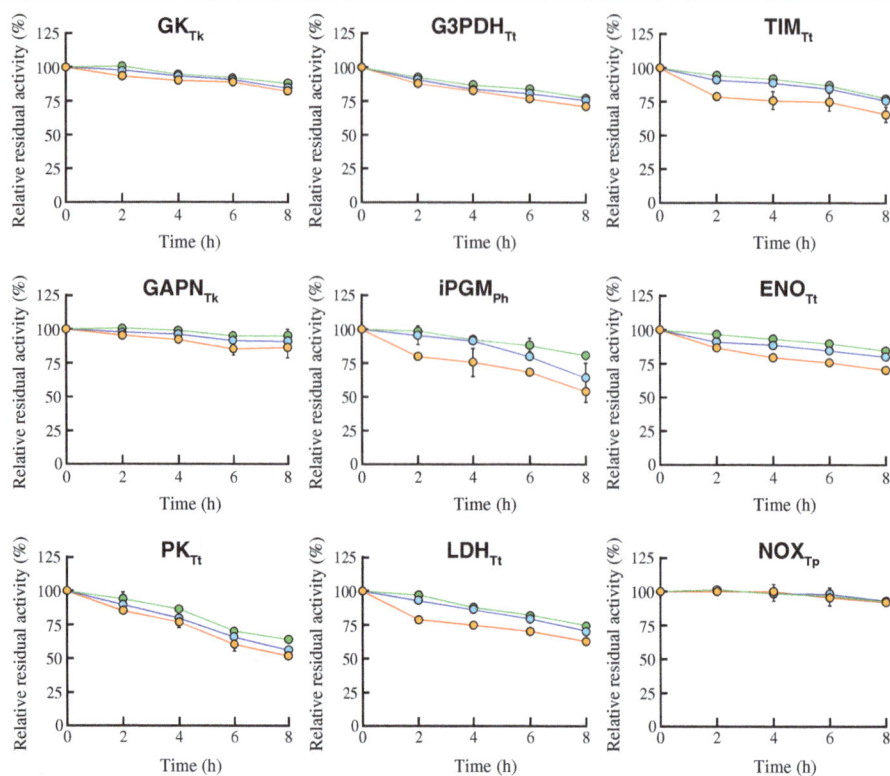

Figure 2 Enzyme stability. Enzyme solutions were incubated at 60°C for indicated time periods and residual activities were determined under the standard assay conditions (green circle). Enzyme stabilities were also assessed at 60°C in the presence of 50 (blue circle) and 100 mM methanol (orange circle).

rate-limiting enzymes in an *in vitro* pathway by increasing the concentration of each enzyme, one by one. The optimum concentrations of enzymes to achieve a desired flux can be experimentally determined by modulating the concentrations of the rate-limiting enzymes [4,5]. Accordingly, we divided the glycerol converting pathway in three parts, namely from glycerol to DHAP, from DHAP to 3-PG, and from 3-PG to lactate, and then adjusted the enzyme concentrations in each partial pathway separately. As a result, the optimum enzyme concentrations to achieve a lactate production rate of 0.04 μmol ml^{-1} min^{-1} were determined as follows: 0.04 U ml^{-1} GK$_{Tt}$, 0.18 U ml^{-1} G3PDH$_{Tt}$, 0.04 U ml^{-1} TIM$_{Tt}$, 0.1 U ml^{-1} GAPN$_{Tk}$, 0.09 U ml^{-1} iPGM$_{Ph}$, 0.07 U ml^{-1} ENO$_{Tt}$, 0.09 U ml^{-1} PK$_{Tt}$, and 0.08 U ml^{-1} LDH$_{Tt}$. Accordingly, NOX$_{Tp}$ was put in the reaction mixture to give a NADH re-oxidizing rate of 0.04 μmol ml^{-1} min^{-1}. A glycerol solution (160 mM) was continuously supplied to the mixture at a rate of 1 μl min^{-1}, which is identical to the experimentally determined lactate production rate through the synthetic pathway, to maintain the pool size of the substrate and a constant flux through the pathway. Although the synthetic pathway was designed to achieve the balanced reduction and oxidation of NAD$^+$ and NADH, the thermal decomposition of the cofactors

was not negligible (Additional file 2: Figure S2). Owing to this fact, NAD$^+$ was also continuously supplied to the reaction mixture at a rate identical to that of its thermal decomposition (0.001 μmol min^{-1} ml^{-1}). The lactate production rate could be remained almost constant at the expected level (0.04 μmol min^{-1} ml^{-1}) for the initial 5 h (Figure 4). Following this, 11.5 mM lactate could be produced with an overall molar conversion yield of 95.5%. Decrease in the production rate became significant after the initial 5 h, and the conversion yield dropped down to 88% at 7 h. This appeared partly due to the dilution of the reaction mixture (10.5% increase in the total volume at 7 h) caused by the continuous feeding of the substrate solution as well as the loss of enzymes by the sampling (8.8% decrease in the total concentration at 7 h). Decrease in the production rate might also result from the thermal inactivation of PK$_{Tt}$ (Figure 2). Substitution of PK$_{Tt}$ with another pyruvate kinase derived from hyperthermophiles with higher optimum growth temperature than *T. thermophilius* may be a possible means for improving the operational stability of the *in vitro* bioconversion system.

The final lactate concentration after the reaction for 7 h was 14.7 mM. The overall turnover number of ATP/ADP was calculated to be 36.8, while that of NAD$^+$/NADH was

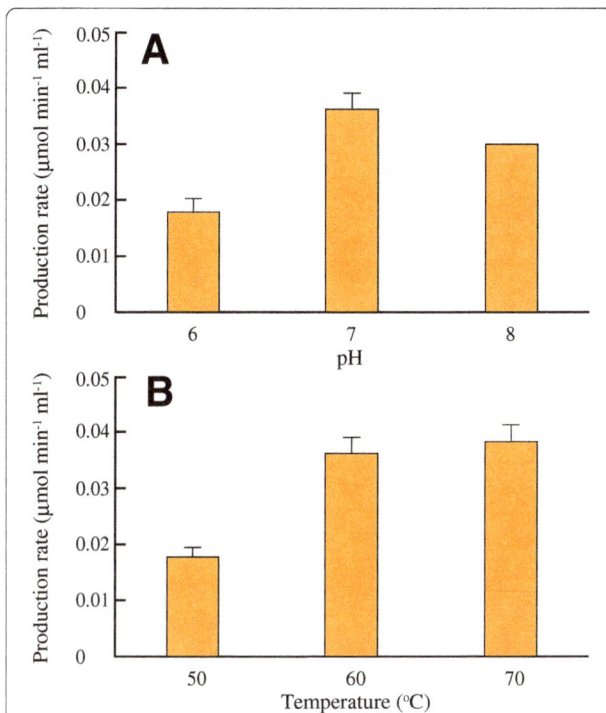

Figure 3 Effects of pH (A) and temperature (B) on the lactate production through the *in vitro* synthetic pathway. Enzymes (0.1 U ml^{-1} each) were incubated in a mixture of 10 mM glycerol, 1 mM NAD$^+$, 0.2 mM NADH, 0.2 mM ATP, 0.2 mM ADP, 0.02 mM FAD, 0.5 mM FBP, 1 mM G1P, 5 mM MgCl$_2$, 0.5 mM MnCl$_2$, and an appropriate buffer. Reaction was performed at indicated pH and temperature for 30 min and then terminated by removing the enzymes with ultrafiltration. Fifty millimolar of MES-NaOH (pH 6) and HEPES-NaOH (pH 7 and 8) were used to adjust pH.

17.5. Enantiomeric purity of the product was determined using a BF-5 biosensor equipped with a D-lactate quantification unit (Oji Scientific Instruments, Amagasaki, Japan). Concentration of D-lactate in the reaction mixture was under the detection limit (approximately 0.05 mM), indicating that glycerol was enantio-specifically converted to L-lactate.

Compatibility of the *in vitro* bioconversion system with methanol

The chemical composition of crude glycerol is highly varied with the types of catalysts and feedstocks used for

Table 1 Effect of NAD$^+$ and FBP on the activity of LDH$_{Tt}$

FBP (mM)	NAD$^+$ (mM)			
	0	0.2	0.5	1
0	192 ± 4.2	2.51 ± 0.10	1.08 ± 0.03	0.31 ± 0.02
0.5	1,150 ± 15	980 ± 3.5	396 ± 6.2	230 ± 2.9

Enzyme assays were performed with and without 0.5 mM FBP in the reaction mixture containing indicated concentrations of NAD$^+$. Results were expressed as specific enzyme activities, which were assessed using a heat-treated cell-free extract and normalized by the protein concentration of the corresponding non-heated cell-free extract (U mg^{-1} protein).

biodiesel production processes [18]. Hansen et al. analyzed the chemical composition of 11 types of crude glycerol obtained from 7 Australian biodiesel manufacturers and revealed that the glycerol content in the crude glycerol varied in the range of 38% to 96% and up to 16.1% of methanol was contained as an impurity [25]. Moreover, Asad-ur-Rehman et al. reported that a raw glycerol obtained during the biodiesel preparation from sunflower oil contained 50% methanol, which is more abundant than the glycerol content (30%) [19]. In order to assess the compatibility of the *in vitro* system with crude glycerol, a model solution of crude glycerol consisting of 30% (*w/v*) glycerol and 50% (*w/v*) methanol was prepared and used as a substrate for lactate production. The model solution was diluted by distilled water to give a final glycerol concentration of 160 mM (thereby the final methanol concentration was 770 mM) and supplied into the reaction mixture in the same manner as the lactate production with pure glycerol. The time profile of the lactate production with the model solution was almost identical to that with pure glycerol (Figure 4). After the reaction for 7 h, methanol concentration in the reaction mixture reached 47.8 mM, which was significantly lower than the calculated concentration of 80.5 mM probably due to volatilization. These results were in reasonable agreement with our observation that stabilities of most enzymes involved in the synthetic pathway were not markedly affected by up to 100 mM of methanol and demonstrated the potential applicability of *in vitro* bioconversion systems with

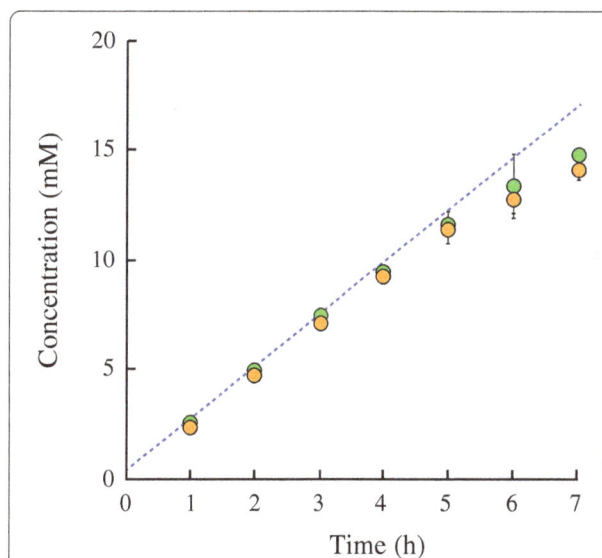

Figure 4 Lactate production through the *in vitro* synthetic pathway. Production assays were performed using pure glycerol (green circle) and a mixture of glycerol and methanol (orange circle). Total concentration of glycerol fed into the reaction mixture was indicated by a dotted line.

thermophilic enzymes for the conversion of crude glycerol to value-added chemicals.

Conclusions

In this study, we constructed an artificial *in vitro* metabolic pathway for the conversion of glycerol to lactate. The *in vitro* pathway consisted of nine thermophilic and hyperthermophilic enzymes and designed to balance the intrapathway consumption and regeneration of cofactors. The one-pot conversion of glycerol to lactate through the *in vitro* pathway could be achieved in an almost stoichiometric manner. Although the final product concentration obtained in this study was modest, the overall yield and the production rate of lactate were comparable to those in conventional fermentation processes [20]. Besides, the *in vitro* bioconversion system can be operated in a simple buffer solution and thus would markedly simplify downstream processes including product recovery and purification. Furthermore, the *in vitro* bioconversion system exerted almost identical performance in the presence of methanol, demonstrating the potential of thermophilic-enzyme-based *in vitro* metabolic engineering approaches in the utilization of crude glycerol as a starting material for the production of value-added chemicals. On the other hand, although their contents are generally less abundant than those of methanol, crude glycerol contains many other impurities, including fatty acids, soap, and salts. Compatibility tests of thermophilic enzymes with these impurities and the implementation of the *in vitro* bioconversion with a real crude glycerol would be needed in future studies.

Additional files

Additional file 1: Figure S1. SDS-PAGE analysis of the crude extracts of recombinant *E. coli* cells. Crude extracts were prepared from approximately 1 mg (wet weight) of the cells overproducing indicated thermophilic enzymes and separated on 12% acrylamide gels before (−) and after (+) heat treatment at 70°C for 30 min. Dual Xtra prestained protein standard (Bio-Rad Laboratories Inc.) was used as a protein marker (lanes indicated by M).

Additional file 2: Figure S2. Thermal decomposition of NAD⁺ and NADH. A mixture of NAD⁺ (1 mM, indicated by blue circle) and NADH (0.2 mM, green circle) in 50 mM HEPES-NaOH (pH 7.0) was incubated at 60°C. After the incubation for indicated time periods, residual concentrations of the cofactors were determined by HPLC as described elsewhere (Morimoto et al. 2014). Orange circles indicate the total concentration of NAD⁺ and NADH. Data represent the averages of triplicate assays.

Abbreviations

GK_{Tk}: glycerol kinase from *Thermococcus kodakarensis*; $G3PDH_{Tt}$: glycerol-3-phosphate dehydrogenase from *Thermus thermophilus*; TIM_{Tt}: triose phosphate isomerase from *Thermus thermophilus*; $GAPN_{Tk}$: non-phosphorylating glyceraldehyde-3-phosphate dehydrogenase from *Thermococcus kodakarensis*; $iPGM_{Ph}$: cofactor-independent phosphoglycerate mutase from *Pyrococcus horikoshii*; ENO_{Tt}: enolase from *Thermus thermophilus*; PK_{Tt}: pyruvate kinase from *Thermus thermophilus*; LDH_{Tt}: lactate dehydrogenase from *Thermus thermophilus*; NOX_{Tp}: NADH oxidase from *Thermococcus profundus*; G3P: glycerol-3-phosphate; DHAP: dihydroxyacetone phosphate; GAP: glyceraldehyde-3-phosphate; 3-PG: 3-phosphoglycerate; 2-PG: 2-phosphoglycerate; PEP: phosphoenolpyruvate;

G1P: glucose-1-phosphate; FBP: fructose-1,6-bisphosphate; FAD: flavin adenine dinucleotide; WST-1: 2-(4-iodophenyl)-3-(4-nitrophenyl)-5-(2,4-disulfophenyl)-2H-tetrazolium, monosodium salt; 1-methoxy PMS: 1-methoxy-5-methylphenazinium methylsulfate; HPLC: high-performance liquid chromatography.

Competing interests
The authors declare that they have no competing interests.

Authors' contribution
CJ performed the experiments and wrote the manuscript. SCN supervised the work and wrote the manuscript. MC co-performed the experiments, particularly on the enzyme stability. KO and HO contributed general advice and resource support, as well as edited the manuscript. KH designed all the experiments and wrote the manuscript. All authors read and approved the final manuscript.

Acknowledgements
This work was supported in part by the Japan Science and Technology Agency, PRESTO/CREST program. This work was also partly supported by the Japan Society for the Promotion of Science, KAKENHI Grant (26450088). CJ was supported by Japanese Funds-in-Trust, UNESCO Biotechnology School in Asia program and Graduate School Thesis Grant, Chulalongkorn University.

References

1. Rabinovitch-Deere CA, Oliver JWK, Rodriguez GM, Atsumi S (2013) Synthetic biology and metabolic engineering approaches to produce biofuels. Chem Rev 113:4611–4632
2. Bond-Watts BB, Bellerose RJ, Chang MCY (2011) Enzyme mechanism as a kinetic control element for designing synthetic biofuel pathways. Nat Chem Biol 7:222–227
3. Guterl JK, Garbe D, Carsten J, Steffler F, Sommer B, Reiße S, Philipp A, Haack M, Rühmann B, Koltermann A, Kettling U, Thomas Brück T, Sieber V (2012) Cell-free metabolic engineering: production of chemicals by minimized reaction cascades. ChemSusChem 5:2165–2172
4. Krutsakorn B, Honda K, Ye X, Imagawa T, Bei X, Okano K, Ohtake H (2013) *In vitro* production of *n*-butanol from glucose. Metab Eng 20:84–91
5. Ye X, Honda K, Sakai T, Okano K, Omasa T, Hirota R, Kuroda A, Ohtake H (2012) Synthetic metabolic engineering - a novel, simple technology for designing a chimeric metabolic pathway. Microb Cell Fact 11:120
6. Ye X, Honda K, Morimoto Y, Okano K, Ohtake H (2013) Direct conversion of glucose to malate by synthetic metabolic engineering. J Biotechnol 164:34–40
7. You C, Chen H, Myung S, Sathitsuksanoh N, Ma H, Zhang XZ, Li J, Zhang YHP (2013) Enzymatic transformation of nonfood biomass to starch. Proc Natl Acad Sci U S A 110:7182–7187
8. Woodward J, Orr M, Cordaray K, Greenbaum E (2000) Enzymatic production of biohydrogen. Nature 405:1014–1015
9. Zhang YHP, Evans BR, Mielenz JR, Hopkins RC, Adams MWW (2007) High-yield hydrogen production from starch and water by a synthetic enzymatic pathway. PLoS ONE 2:e456
10. Opgenorth PH, Korman TP, Bowie JU (2014) A synthetic biochemistry molecular purge valve module that maintains redox balance. Nat Commun 5:4113
11. Zhu Z, Tam TK, Sun F, You C, Zhang YHP (2014) A high-energy-density sugar biobattery based on a synthetic enzymatic pathway. Nat Commun 5:3026
12. Ninn PH, Honda K, Sakai T, Okano K, Ohtake H (2014) Assembly and multiple gene expression of thermophilic enzymes in *Escherichia coli* for *in vitro* metabolic engineering. Biotechnol Bioeng (in press)
13. Owusu RK, Cowan DA (1989) Correlation between microbial protein thermostability and resistance to denaturation in aqueous: organic solvent two-phase systems. Enzyme Microb Technol 11:568–574
14. Atomi H (2005) Recent progress towards the application of hyperthermophiles and their enzymes. Curr Opin Chem Biol 9:166–173
15. Pennacchio A, Pucci B, Secundo F, La Cara F, Rossi M, Raia CA (2008) Purification and characterization of a novel recombinant highly enantioselective short-chain NAD(H)-dependent alcohol dehydrogenase from *Thermus thermophilus*. Appl Environ Microbiol 74:3949–3958

16. Nguyen AQ, Kim YG, Kim SB, Kim CJ (2013) Improved tolerance of recombinant *Escherichia coli* to the toxicity of crude glycerol by overexpressing trehalose biosynthetic genes (*otsBA*) for the production of β-carotene. Bioresour Technol 143:531–537

17. Venkataramanan KP, Boatman JJ, Kurniawan Y, Taconi KA, Bothun GD, Scholz C (2012) Impact of impurities in biodiesel-derived crude glycerol on the fermentation by *Clostridium pasteurianum* ATCC 6013. Appl Microbiol Biotechnol 93:1325–1335

18. Yang F, Hanna MA, Sun R (2012) Value-added uses for crude glycerol - a byproduct of biodiesel production. Biotechnol Biofuels 5:13

19. Asad-ur-Rehman SWRG, Nomura N, Sato S, Matsumura M (2008) Pretreatment and utilization of raw glycerol from sunflower oil biodiesel for growth and 1,3-propanediol production by *Clostridium butyricum*. J Chem Technol Biotechnol 83:1072–1080

20. Okano K, Tanaka T, Ogino C, Fukuda H, Kondo A (2010) Biotechnological production of enantiomeric pure lactic acid from renewable resources: recent achievements, perspectives, and limits. Appl Microbiol Biotechnol 85:413–423

21. Koga Y, Haruki M, Morikawa M, Kanaya S (2001) Stability of chimeras of hyperthermophilic and mesophilic glycerol kinases constructed by DNA shuffling. J Biosci Bioeng 91:551–556

22. Yokoyama S, Matsuo Y, Hirota H, Kigawa T, Shirouzu M, Kuroda Y, Kurumizaka H, Kawaguchi S, Ito Y, Shibata T, Kainosho M, Nishimura Y, Inoue Y, Kuramitsu S (2000) Structural genomics projects in Japan. Nat Struct Biol 7:943–945

23. Jia B, Park S-C, Lee S, Pham BP, Yu R, Le TL, Han SW, Yang J-K, Choi M-S, Baumeister W, Cheong G-W (2008) Hexameric ring structure of a thermophilic archaeon NADH oxidase that produces predominantly H$_2$O. FEBS J 275:5355–5366

24. Matsubara K, Yokooji Y, Atomi H, Imanaka T (2011) Biochemical and genetic characterization of the three metabolic routes in *Thermococcus kodakarensis* linking glyceraldehyde 3-phosphate and 3-phosphoglycerate. Mol Microbiol 81:1300–1312

25. Hansen CF, Hernandez A, Mullan BP, Moore K, Trezona-Murray M, King RH, Pluske JR (2009) A chemical analysis of samples of crude glycerol from the production of biodiesel in Australia, and the effects of feeding crude glycerol to growing-finishing pigs on performance, plasma metabolites and meat quality at slaughter. Anim Prod Sci 49(1):54–161

7

Virtual screening for angiotensin I-converting enzyme inhibitory peptides from *Phascolosoma esculenta*

Yalan Liu[1], Lujia Zhang[1], Mingrong Guo[1], Hongxi Wu[2], Jingli Xie[1*] and Dongzhi Wei[1]

Abstract

Background: Many short peptides have proved to exhibit potential anti-hypertensive activity through the inhibition of the Angiotensin I-converting enzyme (ACE) activity and the regulation of blood pressure. However, the traditional experimental screening method for ACE inhibitory peptides is time consuming and costly, accompanied with the limitations as incomplete hydrolysis and peptides loss during purification process. Virtual methods with the aid of computer can break such bottle-neck of experimental work. In this study, an attempt was made to establish a library of di- and tri-peptides derived from proteins of *Phascolosoma esculenta*, a kind of seafood, through BIOPEP (http://www.uwm.edu.pl/biochemia/index.php/pl/biopep), and to screen highly active ACE inhibitory peptides by molecular docking with the help of LibDock module of Discovery Studio 3.5 software.

Results: Two hundred and eighty four (284) di- and tri-peptides, derived from *P. esculenta* proteins after a virtual hydrolysis with pepsin, trypsin and a mixture of pepsin and trypsin, were predicted to possess ACE inhibitory activity, among which there are 99 ACE inhibitory peptides with estimated IC_{50} less than 50 μM. Nine peptides were synthesized for the comparison between the estimated and the experimentally determined IC_{50}. The results indicated that errors between the estimated and measured $\log(1/IC_{50})$ are all less than 1.0 unit.

Conclusions: Virtual method for peptide library construction and ACE inhibitory peptides screening efficiently demonstrated that *P. esculenta* proteins are prospect resource for food-origin ACE inhibitory peptide.

Keywords: Virtual screening; Angiotensin I-converting enzyme (ACE); ACE inhibitory peptide; *Phascolosoma esculenta*

Background

Hypertension is a worldwide health problem, the prevalence of which have affected up to 30% of the adult population according to the World Health Organization. Hypertension carries a high-risk factor for arteriosclerosis, myocardial infarction, and end-stage renal disease [1,2]. It is predicted that by 2025, about 20% of the world population will suffer from hypertension [3].

Although the cause of hypertension currently cannot be well determined, it is understood that the renin-angiotensin system regulates an organism's water, electrolytes, and blood, and the angiotensin I-converting enzyme (ACE) (peptidyldipeptide hydrolase, EC 3.4.15.1) plays an

important role in regulating the blood pressure [4]. ACE is a hypertension-responsible glycoprotein distributed in vascular endothelial, absorptive epithelial, and male germinal cells [5,6]. ACE cleaves the carboxyl terminal His-Leu dipeptide from inactive decapeptide angiotensin I to active angiotensin II, a powerful vasoconstrictor which can trigger hypertension [7-10]. ACE also influences the kallikrein-kinin system by promoting the degradation and inactivation of bradykinin, which can lead to reduction of hypertension. Therefore, excessive activity of ACE leads to hypertension. Molecules which can inhibit the activity of ACE are considered useful drugs for hypertension management [11]. Currently, synthetic ACE inhibitors, such as captopril, enalapril and lisinopril, are available on the market [12]; however, they tend to have side effects [13].

Since the discovery of the first anti-hypertensive peptide in snake venom [14], more attention has been paid to natural sources, especially peptides. Peptides derived

* Correspondence: jlxie@ecust.edu.cn
[1]State Key Laboratory of Bioreactor Engineering, Department of Food Science and Engineering, East China University of Science and Technology, 130# Meilong Rd., P.O. Box 283, Shanghai 200237, People's Republic of China
Full list of author information is available at the end of the article

from cheese whey [15], fermented milk [16], mushroom [17], soy bean [18,19], corn gluten [20], insect protein [21], peanut flour [22], and egg [23] have been proven to inhibit the activity of ACE. However, few studies were reported about their side effects [24,25]. Nutritionists claim that peptides found in food are safer than 'traditional' drugs, and they are promising synthetic drug substitutes [26].

Among the ACE inhibitory peptides, shorter ones (di- and tri-peptides) usually have significant advantages over longer ones. They easily pass through blood circulation system [27,28] and then reach action sites faster without being hydrolyzed by digestive enzymes during the gastrointestinal digestion [29,30]. For these reasons, the present study focused on di- and tri-peptides.

The discovery of ACE inhibitory peptides with potential anti-hypertensive effect is mostly based on experiments, which require amounts of labors and funds. Besides, the possible active peptides can not be totally harvested due to the incomplete hydrolysis and peptides loss in the purification by the experimental protocols. Recently, as the computation simulation technology for drug design and discovery of molecular interaction are booming, the virtual screening or *in silico* experiment may replace the traditionally experimental screening of anti-hypertensive peptides to some extent. Computational approaches, which are based on computational evaluation of interactions between receptor and ligand, are proved feasible for virtual screening [31]. Molecular docking is a powerful and a widely used tool in molecular simulation, which is approximated to a lock-and-key process. The docking protocol is to 'dock' a ligand into an active site of a receptor; then, the interactions between them were 'scored' to assess the potential bioactivity of candidate compounds. The most advantage of docking is its high-throughput screening in short time with little cost [32].

In this study, an attempt was made to investigate the ACE inhibitory activity of di- and tri-peptides derived from *Phascolosoma esculenta*, a marine deposit-feeding benthonic invertebrates, also a traditional seafood with over 70% protein (dry weight) in Southeast China [33,34]. Database of di- and tri-peptides derived from *P. esculenta* were established, and their ACE inhibitory activities were predicted by virtual hydrolysis and screening method. Finally, di- and tri-peptides which have obvious ACE inhibitory activity were synthesized for verifying the validity of such virtual strategy.

Methods

Materials

There are 22 proteins of *P. esculenta* with the protein messages including entry name and sequence obtained

from UniProt (http://www.uniprot.org/) (Table 1). They were used as original materials for database of di- and tri-peptides. With the help of BIOPEP (http://www.uwm.edu.pl/biochemia/index.php/pl/biopep), the 22 proteins were virtually hydrolyzed with pepsin, trypsin, and a mixture of pepsin and trypsin.

Molecular docking experiments

LibDock, a module of Discovery Studio 3.5 software (DS3.5, Accelrys, San Diego, CA, USA), was used for molecular docking experiments. Scoring results (LibDock score) about ligand-receptor combination were used as the final criterion to estimate the ACE inhibitory activity of ligands. Based on a previous study [35], the corresponding relationship between LibDock score and IC_{50} was

$$\text{LibDock score} = 10.063 \log(1/IC_{50}) + 68.08 \,,$$

where IC_{50} is 50% inhibitory concentration (in μM) towards ACE. According to the LibDock score, ACE inhibitory activity of ligands could be estimated.

Table 1 Properties of 22 *P. esculenta* proteins in UniProt (http://www.uniprot.org/)

Name	Number of amino acids	MW (kDa)	PI
D2J0B2	226	25.2	6.0
C3PUI4	378	43.2	9.1
D2J288	726	83.6	4.9
C3PUI8	304	34.6	8.5
C3PUI2	267	30.3	6.9
B6CQR3	658	71.6	5.1
A5A2J9	135	15.4	6.5
C3PUI5	121	13.9	6.7
B3TCX0	84	9.3	4.7
C3PUJ0	571	63.3	9.2
C3PUI9	231	25.7	5.9
B6CPA3	120	13.63	5.8
A5A2K2	220	24.4	5.8
C3PUH9	519	57.2	6.2
C3PUI7	450	50.2	9.2
B3TFG2	174	20.2	5.1
C3PUI1	54	6.5	10.8
C3PUI3	157	17.7	9.5
A3EX91	137	14.8	4.7
C3PUJ1	323	36.1	9.1
C3PUI0	231	25.9	4.8
C3PUI6	94	10.5	9.2

MW molecular weight, *PI* isoelectric point.

ACE was used as receptor in docking simulation, whose crystal structures was available in the Protein Data Bank (PDB) (http://www.pdb.org), from where the three-dimensional structure of ACE was imported [PDB:1O8A]. Before the docking procedure, water molecules were removed and zinc ions were retained. The 284 di- and tri-peptides derived from *P. esculenta* were used as ligands, of which structures and energies were generated with Chem-BioDraw software [36] and minimized with the CHARMM program [37], respectively. Parameters used in the docking process are shown in Table 2.

Synthesis of peptides

Five tri-peptides (GYF, WAL, AYF, GLR, and ILK) and four di-peptides (FK, QF, EL, and HK) generated through *in silico* hydrolysis of *P. esculenta* protein, with purity of 95%, were synthesized by GL Biochem Co. Ltd. (Shanghai, China) for IC_{50} testing.

Measurement of ACE inhibitory activity

The ACE inhibitory activity was measured according to the method of Cushman and Cheung [38] with slight

Table 2 Parameters for molecular docking experiments performed with the LibDock of DS3.5

Parameter name	Parameter value
Docking sphere	10 Å
Input site sphere	
x	48.65
y	82.55
z	54.04
Number of HPTPot	100
Docking tolerance	0.25
Docking preference	User specified
Max hits to save	10
Max number of hits	100
Minimum LibDock score	100
Final score cutoff	0.5
Max BFGS steps	50
Rigid optimization	False
Max conformation hits	30
Max start conformations	1,000
Steric fraction	0.10
Final cluster radius	0.5
Apolar SASA cutoff	15.0
Polar SASA cutoff	5.0
Surface grid steps	18
Conformation method	Best
Minimization algorithm	Do not minimize
Parallel processing	True

modifications. Ten milligram of the sample was dissolved in 1 mL distilled water and then diluted to seven different concentrations for ACE inhibitory measurements. Fifteen microliters of the sample solution in certain concentration (Seven different concentrations) were needed, which were determined by the pre-experiment about ACE inhibition ratio. The whole principle is that the concentration which ACE inhibition ratio reaches 50% is included within the concentration range. The concentrations for GYF, FK, WAL, QF, and AYF are 10, 20, 30, 40, 50, 60, and 70 μg/mL, and for EL, GLR, HK, and ILK are 20, 40, 60, 80, 100, 120, and 140 μg/mL, respectively) and 15 μL substrate hippuryl-L-histidyl-L-leucine (HHL) (8.3 mM Hip-His-Leu in 50 mM sodium borate buffer containing 0.5 M NaCl at pH 8.3) were mixed together and then pre-incubated at 37°C for 5 min. The reaction was initiated by adding 5 μL of ACE solution (310 mU/mL) and incubated for 60 min at the same temperature. The reaction was terminated by the addition of 1.0 M HCl (200 μL). Ten microliters of the reaction solution was injected directly onto a Thermo BDS-C18 column (3.0 mm × 250 mm, 5 μm, Thermo Scientific Co. Ltd., Waltham, MA, USA). The mobile phase consisted of 10% acetonitrile and 90% water with 0.1% trifluoroacetic acid (TFA). The flow rate was 0.7 mL/min and the absorbance was monitored at 228 nm. All determination was carried out at least in triplicate. The inhibition activity was calculated using the following equation:

$$\text{ACE inhibition } (\%) = [1-(A_{\text{inhibitor}}/A_{\text{control}})] \times 100 \,,$$

where $A_{\text{inhibitor}}$ is the absorbance with ACE, HHL, and sample, and A_{control} is the absorbance of hippuric acid (HA) with ACE and HHL without the sample. Dose-dependent ACE inhibition was investigated using at least five different concentrations of peptides. The concentration of peptides that inhibited ACE activity by 50% (IC_{50}) was calculated using a non-linear regression from a plot of ACE inhibition versus sample concentrations.

Study on structural-active relationship of ACE inhibitory peptides

The chemical properties of C-terminal and N-terminal amino acids of 99 peptides with estimated IC_{50} less than 50 μM were summarized to deduce the structural-active relationship of ACE inhibitory peptides.

Results and discussion

Pool of di- and tri-peptides derived from *P. esculenta* proteins

Pepsin, trypsin, and the mixture of pepsin and trypsin were used to virtually hydrolyze the 22 proteins from *P. esculenta*, with the help of BIOPEP (http://www.uwm.edu. pl/biochemia/index.php/pl/biopep). In total, 2,667 peptides were virtually produced, and among them, 1,084 were di-

and tri-peptides, which accounted for about 40.6% (Figure 1). After excluding the repeated ones, there were 1,017 non-repeated peptides, among which 284 were di- and tri-peptides. The sequences and the frequencies of these 284 short peptides are shown in Table 3. These 284 peptides were used as the ligands for docking experiment with ACE.

Estimated IC_{50} distribution of ACE inhibitory di- and tri-peptides

The estimated ACE inhibitory IC_{50} of the 284 di- and tri-peptides derived from *P. esculenta* proteins were obtained according to LibDock scores, which were summarized in Figure 2. Ninety-nine (99) peptides had an estimated IC_{50} less than 50 μM (34.9% of 284 peptides), 100 peptides had an estimated IC_{50} between 50 and 100 μM (35.2% of 284 peptides), and 37 peptides had an estimated IC_{50} less than 500 μM, accounting for 13.0%. Most reported ACE inhibitory peptides with IC_{50} less than 100 μM showed potent *in vivo* anti-hypertensive activity [39]. Therefore, *P. esculenta* is a prospective anti-hypertensive peptide-containing resource since more than two thirds di- and tri-peptides theoretically possess obvious ACE inhibitory activity. The sequences and estimated IC_{50} of di- and tri-peptides with estimated IC_{50} less than 50 μM are shown in Table 4.

Short peptides were usually used for predicting potent ACE inhibitory activity. Pripp docked 58 di-peptides into protein target using the Molegro Virtual Docker version 1.1.1 software and found significant relationship between docking results and experimental IC_{50} values [32]. Several tri-peptides consisting of I or L and positive charged amino acids and aromatic amino acids were synthesized, and their ACE inhibitory activities were measured to clarify the amino acid sequence for inhibition of ACE [40]. Larger peptides, for instance, the sequence length more than 5, were also focused in some work [41]; however, such works

were reported with lower R^2 (coefficient of variation) because of the complexity in the modeling due to the bigger peptide [42,43].

Confirmation of virtual screening method

In order to confirm the validity of virtual screening method of the present work, nine peptides were synthesized for IC_{50} testing. The sequences of these peptides were obtained from *P. esculenta* protein through virtual hydrolysis and screening by docking experiments. The estimated log(1/IC_{50}) and measured log(1/IC_{50}) of the nine peptides were compared (Table 5). The error between the estimated log (1/IC_{50}) and measured log(1/IC_{50}) is less than 1.0 unit. Desirable limit for model is that the error between estimated log(1/IC_{50}) and measured log(1/IC_{50}) is less than 1.5 units [30]. A reported quantitative structure-activity relationship (QSAR) model was constructed on 168 di-peptides and 140 tri-peptides collected from literatures, and the model verification was made on seven reported di-peptides and tri-peptides (not included in 168 di-peptides and 140 tri-peptides), of which the error was between 0.07 and 1.39 [29]. On the ground of such criterion, the present model is efficient and credible.

Previous studies suggested that the structural-active relationship of ACE inhibitory peptides largely depended on their amino acid composition, sequence, and configuration, though the full mechanism of interaction between peptides and ACE is not established so far [44,45]. For the short peptides as di- and tri-peptides, the amino acid composition and configuration are more significant. The di- and tri-peptides which have an estimated IC_{50} within 50 μM were used to study the structural-active relationship of these ACE inhibitors.

There are four kinds of C-terminal residues for 99 sequences (Figure 3) due to the cutting specificity of pepsin

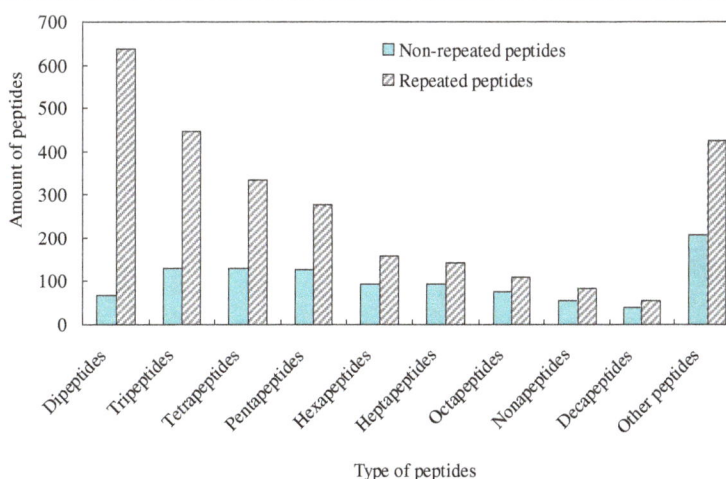

Figure 1 Length distribution of peptides derived from *P. esculenta* proteins by virtual hydrolysis. The enzymes used in virtual hydrolysis are pepsin, trypsin, and a mixture of pepsin and trypsin.

Table 3 Sequence and frequency of di- and tri-peptides derived from *P. esculenta* proteins by virtual hydrolysis

Peptide	Frequency	Peptide	Frequency	Peptide	Frequency	Peptide	Frequency
ADL	1	GYF	2	NDK	2	STL	2
AEF	2	HAQ	2	NF	10	SVF	2
AEK	1	HER	1	NGK	1	SVL	4
AEL	2	HF	9	NIL	2	SWK	2
AF	19	HGL	2	NK	2	SWL	2
AGF	2	HIK	2	NL	13	SYL	1
AIF	4	HK	4	NLK	1	TAL	2
AK	2	HKF	1	NMR	1	TDK	1
AL	35	HL	17	NPF	6	TDR	1
ALR	1	HSL	2	NR	2	TEF	2
AMF	2	HTK	1	NRF	2	TF	7
APF	2	HTL	1	NSF	2	TGF	2
AQF	2	IAR	2	NTL	3	TGL	4
AR	2	ICL	4	NVL	5	TIL	2
ASK	1	IF	20	NVR	2	TK	4
AVK	2	IGR	1	NWL	2	TL	31
AVR	2	IHR	2	PCK	1	TLK	1
AYA	2	IIL	4	PDL	2	TML	2
AYF	3	IK	7	PF	19	TNR	1
CF	2	IL	46	PGF	2	TPF	1
CK	1	ILK	1	PIL	2	TSL	4
CL	9	IMF	2	PK	7	TTK	1
CVF	2	IMK	2	PKL	1	TVK	1
DF	4	IPK	2	PL	30	TVL	2
DK	16	IPL	5	PNK	2	TVR	1
DL	11	IQK	2	PPL	2	TWK	1
DMF	2	IR	4	PRL	1	TYF	2
DNR	1	IRF	1	PSF	2	VAL	6
DPK	1	ISF	2	PSK	1	VDL	2
DR	4	ISL	2	PSL	2	VEK	3
DSK	2	ISR	1	PTL	4	VER	2
DSL	4	ITK	1	PTR	2	VF	9
DWL	1	ITL	2	PVK	1	VGF	6
EAF	2	IVL	1	PVL	2	VGL	3
EAR	1	IWL	4	QAL	2	VIR	3
EDK	2	KAL	1	QDF	2	VK	2
EEF	1	KDL	2	QEL	2	VKL	1
EEL	2	KF	2	QF	5	VL	14
EER	1	KGF	1	QGL	2	VMK	2
EF	2	KIL	1	QIR	1	VML	2
EGL	1	KL	4	QK	2	VNL	4
EIF	1	KPL	1	QL	7	VPK	1
EIL	1	KRF	1	QR	1	VPL	4
EK	7	KSL	1	QYK	1	VSF	2

Table 3 Sequence and frequency of di- and tri-peptides derived from *P. esculenta* proteins by virtual hydrolysis *(Continued)*

EL	16	KTL	1	RAL	1	VSL	3
EML	1	KVF	1	REL	1	VVF	2
ENK	2	LDK	1	RL	3	VVK	2
ENL	6	LEK	1	RPF	1	WAF	2
ER	2	LK	2	RSF	2	WAL	2
ESK	1	LMK	1	RTL	1	WCF	2
ETL	3	LR	1	SAL	4	WF	6
EVK	1	LSK	1	SEK	1	WGK	2
EVL	4	MAL	8	SF	20	WL	8
FFK	1	MF	16	SGF	4	WML	2
FK	1	MFK	1	SGL	2	WNF	2
FWR	1	MFR	1	SHL	2	WPF	2
GAL	2	MGF	2	SIF	2	WQK	2
GF	12	MGL	2	SIK	2	WR	1
GGL	4	MIK	2	SIL	4	WTR	1
GGR	2	MIL	2	SK	6	WWF	2
GK	6	MK	10	SL	50	YAL	1
GKF	1	MKF	2	SNL	3	YF	7
GL	29	ML	16	SPF	2	YIF	2
GLR	1	MPL	1	SPL	2	YIK	1
GNL	2	MR	4	SQL	2	YK	4
GR	3	MSK	2	SR	2	YL	8
GSL	2	MSL	10	SS	2	YPL	2
GTL	4	MTK	2	SSF	6	YS	2
GTR	2	MTL	2	SSL	2	YSK	3
GVK	1	MVK	2	STF	2	YTL	2
GWL	2	NAL	2	STK	1	YVR	1

The enzymes used in virtual hydrolysis are pepsin, trypsin, and a mixture of pepsin and trypsin.

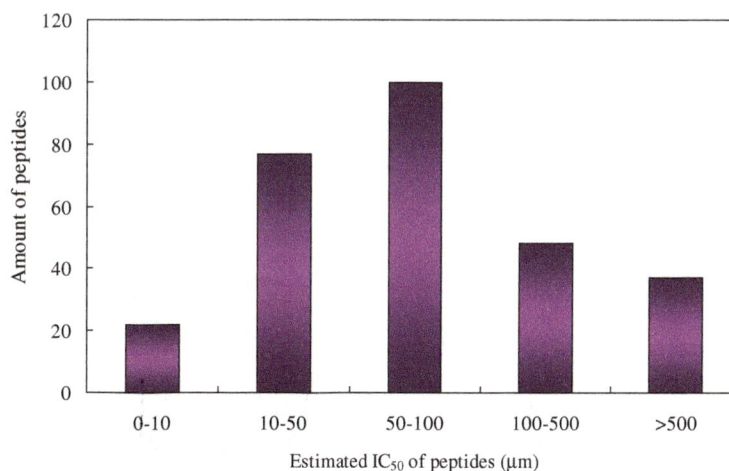

Figure 2 Distribution of estimated IC$_{50}$ of di- and tri-peptides with ACE inhibitory activity. The peptides were derived from *P. esculenta* proteins. All IC$_{50}$ values were predicted by LibDock scores according to the equation, LibDock score = 10.063 log(1/IC50) + 68.08.

Table 4 Di- and tri-peptides derived from *P. esculenta* with estimated IC$_{50}$ less than 50 μM

Peptide	Estimated IC$_{50}$	Peptide	Estimated IC$_{50}$	Peptide	Estimated IC$_{50}$
WNF	0.12	YIF	13.1	DSL	27.9
HKF	0.42	RPF	13.3	IR	28.9
WCF	0.47	APF	13.4	TGF	29.5
WPF	0.85	REL	13.7	NWL	30.7
WWF	1.18	YK	13.9	FFK	30.9
YVR	1.45	AEF	14.1	GYF	32.2
NRF	1.71	AYF	14.7	ISF	32.3
TYF	1.75	KTL	14.8	FK	33.2
IHR	2.06	WGK	14.9	SPF	33.4
WML	2.15	DWL	14.9	VML	33.7
WAF	3.27	SSF	15.0	KPL	36.1
IWL	5.22	NSF	15.4	ISL	36.5
RSF	5.44	VKL	16.1	QF	36.7
IRF	5.72	MAL	16.3	TML	36.9
CVF	7.15	NR	17.0	QEL	37.0
YIK	7.33	MTL	17.1	MKF	38.0
MGF	7.38	IMF	17.5	YAL	38.6
YF	7.48	WR	17.7	EEF	40.4
KRF	8.11	ER	18.1	IGR	40.5
SR	9.05	AEL	19.4	TEF	41.1
GNL	9.09	QDF	19.5	TTK	41.6
GKF	9.51	YTL	19.7	SGF	41.6
WTR	10.2	PIL	19.9	KVF	42.0
PRL	10.2	KDL	20.2	SIL	43.0
SYL	10.4	SVF	20.8	EVL	43.7
WF	10.6	YSK	21.5	DMF	46.6
IIL	10.8	VVF	22.5	RTL	46.9
WAL	11.3	ITK	22.6	QAL	47.0
AQF	11.5	DR	23.4	TPF	47.0
RAL	11.8	WQK	23.9	HSL	47.1
PGF	12.2	PKL	24.9	NAL	47.1
NVL	12.7	EIF	25.2	SHL	48.0
HTL	13.1	VGF	27.8	STF	49.4

Table 5 Estimated and measured log(1/IC$_{50}$) of the nine synthesized peptides derived from *P. esculenta*

Peptide	Estimated log(1/IC$_{50}$)	Measured log(1/IC$_{50}$)	Error
GYF	4.49	4.31 ± 0.02	0.18
FK	4.79	4.27 ± 0.01	0.52
WAL	4.95	4.38 ± 0.04	0.57
QF	4.44	4.21 ± 0.03	0.23
AYF	4.83	4.18 ± 0.01	0.65
EL	3.19	3.43 ± 0.02	−0.24
GLR	4.07	3.61 ± 0.02	0.46
HK	3.53	3.86 ± 0.01	−0.33
ILK	4.09	3.72 ± 0.03	0.37

aromatic amino acid with a polar functional group in C-terminus [47]; and the physicochemical attributes of amino acids such as hydrophobicity, bulkiness, and electronic properties had impacts on the bioactivity of peptides [48]. Accordingly, benzene ring in Phe can also increase the bulkiness and bring about the stability of binding between ACE and peptide and sequentially result in high ACE inhibitory activity.

There are 40 peptides among 99 peptides (40.4%) with hydrophobic amino acid at N-terminal, 38 peptides with neutral amino acid at N-terminal (38.4%), and 21.9% peptides with positively or negatively charged amino acid at N-terminal (Figure 4). N-terminal amino acid of ACE inhibitory peptides also favors the hydrophobic interactions with ACE [7,30]. The peptides with hydrophobic amino acid at N-terminal showing higher ACE inhibitory activity have some superiority in amount in the present study, which verified such view.

Conclusions

A virtual method of hydrolysis and screening of ACE inhibitory peptides with high activity such as IC$_{50}$ value < 50 μM was constructed in this work. Ninety-nine (99) peptides were obtained from 22 proteins of *P. esculenta*. Besides, the efficiency and the validity of such method were verified by comparing the predicted IC$_{50}$ and measured

and trypsin. Leu and Phe are C-terminal residues formed by pepsin hydrolysis, and C-terminal Lys and Arg are formed by trypsin reaction. Hydrophobic C-terminal (Phe and Leu) is dominant in amount and accounts for more than 80% peptides (44.4% and 36.4%, respectively). There are some accepted concepts about the structure-activity relationship of ACE inhibitory peptides, such as that peptides with hydrophobic amino acid in C-terminus showed a highly potent ACE inhibitory activity [46]. Highly active peptide in general should be composed of large, hydrophobic, and

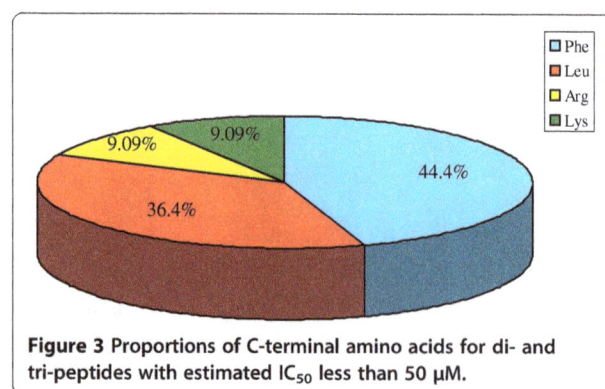

Figure 3 Proportions of C-terminal amino acids for di- and tri-peptides with estimated IC$_{50}$ less than 50 μM.

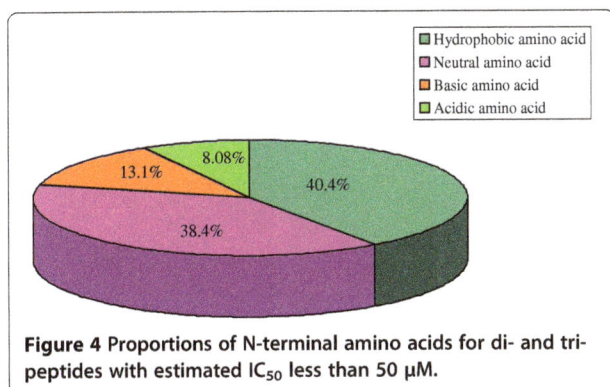

Figure 4 Proportions of N-terminal amino acids for di- and tri-peptides with estimated IC$_{50}$ less than 50 μM.

IC$_{50}$ of some synthesized peptides among the 99 peptides. The results demonstrated that the virtual hydrolysis and screening method is an efficient way that greatly cuts down the experimental labor to get highly active ACE inhibitory peptides. Moreover, *P. esculenta* proteins were proved as a good resource of ACE inhibitory peptides, which could be a beneficial ingredient for functional foods or pharmaceuticals against hypertension. Further research on larger antihypertension peptides derived from *P. esculenta* and *in vivo* activity testing will be carried out.

Competing interests
The authors declare that they have no competing interests.

Authors' contributions
YL carried out the establishment of pool of di- and tri-peptides derived from *P. esculenta* proteins and molecular docking experiments. MG carried out the measurement of ACE inhibitory activity. JX designed the study and revised the manuscript. HW performed the statistical analysis and partly revised the manuscript. LZ participated in part of the method establishment. DW conceived of the study and participated in the design and coordination. All authors read and approved the final manuscript.

Acknowledgements
This work was supported by 'National Natural Science Foundation of China (No. 31301413)', 'National Major Science and Technology Projects of China (No. 2012ZX09304009)', and the 'Fundamental Research Funds for the Central Universities', People's Republic of China.

Author details
[1]State Key Laboratory of Bioreactor Engineering, Department of Food Science and Engineering, East China University of Science and Technology, 130# Meilong Rd., P.O. Box 283, Shanghai 200237, People's Republic of China. [2]Zhejiang Key Lab of Exploitation and Preservation of Coastal Bio-resource, Wenzhou 325005, People's Republic of China.

References
1. Jung WK, Mendis E, Je JY, Park PJ, Son BW, Kim HC, Choi YK, Kim SK (2006) Angiotensin I-converting enzyme inhibitory peptide from yellowfin sole (*Limanda aspera*) frame protein and its antihypertensive effect in spontaneously hypertensive rats. Food Chem 94(1):26–32
2. Silva DG, Freitas MP, da Cunha EFF, Ramalho TC, Nunes CA (2012) Rational design of small modified peptides as ACE inhibitors. Med Chem Comm 3 (10):1290–1293
3. Kearney PM, Whelton M, Reynolds K, Muntner P, Whelton PK, He J (2005) Global burden of hypertension: analysis of worldwide data. Lancet 365 (9455):217–223
4. Li GH, Le GW, Shi YH, Shrestha S (2004) Angiotensin I-converting enzyme inhibitory peptides derived from food proteins and their physiological and pharmacological effects. Nutr Res 24(7):469–486
5. Guang C, Phillips RD (2009) Plant food-derived angiotensin I converting enzyme inhibitory peptides. J Agric Food Chem 57(12):5113–5120
6. De Leo F, Panarese S, Gallerani R, Ceci L (2009) Angiotensin converting enzyme (ACE) inhibitory peptides: production and implementation of functional food. Curr Pharm Des 15(31):3622–3643
7. Iroyukifujita H, Eiichiyokoyama K, Yoshikawa M (2000) Classification and antihypertensive activity of angiotensin I-converting enzyme inhibitory peptides derived from food proteins. J Food Sci 65(4):564–569
8. Reneland R, Lithell H (1994) Angiotensin-converting enzyme in human skeletal muscle. A simple in vitro assay of activity in needle biopsy specimens. Scand J Clin Lab Investig 54(2):105–111
9. Hartl FU (1996) Molecular chaperones in cellular protein folding. Nature 381:571–580
10. Sturrock E, Natesh R, Van Rooyen J, Acharya K (2004) Structure of angiotensin I-converting enzyme. Cell Mol Life Sci 61:2677–2686
11. Lin L, Lv S, Li B (2012) Angiotensin-I-converting enzyme (ACE)-inhibitory and antihypertensive properties of squid skin gelatin hydrolysates. Food Chem 131(1):225–230
12. Sweitzer NK (2003) What is an angiotensin converting enzyme inhibitor? Circulation 108(3):e16–e18
13. Antonios TF, MacGregor GA (1995) Angiotensin converting enzyme inhibitors in hypertension: potential problems. J Hypertens 13:S11–S16
14. Ondetti MA, Williams NJ, Sabo E, Pluscec J, Weaver ER, Kocy O (1971) Angiotensin-converting enzyme inhibitors from the venom of *Bothrops jararaca*. Isolation, elucidation of structure, and synthesis. Biochemistry 10 (22):4033–4039
15. Abubakar A, Saito T, Kitazawa H, Kawai Y, Itoh T (1998) Structural analysis of new antihypertensive peptides derived from cheese whey protein by proteinase K digestion. J Dairy Sci 81(12):3131–3138
16. Nakamura Y, Yamamoto N, Sakai K, Okubo A, Yamazaki S, Takano T (1995) Purification and characterization of angiotensin I-converting enzyme inhibitors from sour milk. J Dairy Sci 78(4):777–783
17. Andújar-Sánchez M, Cámara-Artigas A, Jara-Pérez V (2004) A calorimetric study of the binding of lisinopril, enalaprilat and captopril to angiotensin-converting enzyme. Biophys Chem 111(2):183–189
18. Wu J, Ding X (2001) Hypotensive and physiological effect of angiotensin converting enzyme inhibitory peptides derived from soy protein on spontaneously hypertensive rats. J Agric Food Chem 49(1):501–506
19. Mallikarjun Gouda K, Gowda LR, Rao AA, Prakash V (2006) Angiotensin I-converting enzyme inhibitory peptide derived from glycinin, the 11S globulin of soybean (Glycine max). J Agric Food Chem 54(13):4568–4573
20. Suh H, Whang J, Lee H (1999) A peptide from corn gluten hydrolysate that is inhibitory toward angiotensin I converting enzyme. Biotechnol Lett 21 (12):1055–1058
21. Vercruysse L, Van Camp J, Morel N, Rougé P, Herregods G, Smagghe G (2010) Ala-Val-Phe and Val-Phe: ACE inhibitory peptides derived from insect protein with antihypertensive activity in spontaneously hypertensive rats. Peptides 31(3):482–488
22. Quist EE, Phillips RD, Saalia FK (2009) Angiotensin converting enzyme inhibitory activity of proteolytic digests of peanut (*Arachis hypogaea* L.) flour. LWT-Food Sci Technol 42(3):694–699
23. Majumder K, Wu J (2009) Angiotensin I converting enzyme inhibitory peptides from simulated *in vitro* gastrointestinal digestion of cooked eggs. J Agric Food Chem 57(2):471–477
24. Wang C, Tian J, Wang Q (2011) ACE inhibitory and antihypertensive properties of apricot almond meal hydrolysate. Eur Food Res Technol 232(3):549–556
25. Vermeirssen V, Camp JV, Verstraete W (2004) Bioavailability of angiotensin I converting enzyme inhibitory peptides. Br J Nutr 92(03):357–366
26. Jimsheena V, Gowda LR (2011) Angiotensin I-converting enzyme (ACE) inhibitory peptides derived from arachin by simulated gastric digestion. Food Chem 125(2):561–569
27. Seppo L, Jauhiainen T, Poussa T, Korpela R (2003) A fermented milk high in bioactive peptides has a blood pressure-lowering effect in hypertensive subjects. Am J Clin Nutri 77(2):326–330
28. Mathews D, Adibi S (1976) Peptide absorption. Gastroenterology 71(1):151
29. Wu J, Aluko RE, Nakai S (2006) Structural requirements of angiotensin I-converting enzyme inhibitory peptides: quantitative structure-activity relationship study of di-and tripeptides. J Agric Food Chem 54(3):732–738

30. Wijesekara I, Qian ZJ, Ryu B, Ngo DH, Kim SK (2011) Purification and identification of antihypertensive peptides from seaweed pipefish (*Syngnathus schlegeli*) muscle protein hydrolysate. Food Res Int 44(3):703–707

31. Berman HM, Westbrook J, Feng Z, Gilliland G, Bhat T, Weissig H, Shindyalov IN, Bourne PE (2000) The Protein Data Bank. Nucleic Acids Res 28(1):235–242

32. Pripp AH (2007) Docking and virtual screening of ACE inhibitory dipeptides. Eur Food Res Technol 225(3–4):589–592

33. Su X, Du L, Li Y, Li T, Li D, Wang M, He J (2009) Production of recombinant protein and polyclonal mouse antiserum for ferritin from Sipuncula *Phascolosoma esculenta*. Fish Shellfish Immunol 27(3):466–468

34. Du L, Fang M, Wu H, Xie J, Wu Y, Li P, Zhang D, Huang Z, Xia Y, Zhou L (2013) A novel angiotensin I-converting enzyme inhibitory peptide from *Phascolosoma esculenta* water-soluble protein hydrolysate. J Funct Foods 5(1):475–483

35. Wu H, Liu Y, Guo M, Xie J, Jiang X (2014) A virtual screening method for inhibitory peptides of angiotensin I converting enzyme. J Food Sci 79: C1635–C1642, doi:10.1111/1750-3841.12559

36. Kerwin SM (2010) ChemBioOffice Ultra 2010 suite. J Am Chem Soc 132(7):2466–2467

37. Brooks BR, Bruccoleri RE, Olafson BD, States DJ, Swaminathan S, Karplus M (1983) CHARMM: a program for macromolecular energy, minimization, and dynamics calculations. J Comput Chem 4(2):187–217

38. Cushman D, Cheung H (1971) Spectrophotometric assay and properties of the angiotensin-converting enzyme of rabbit lung. Biochem Pharmacol 20(7):1637–1648

39. Iwaniak A, Minkiewicz P, Darewicz M (2014) Food-originating ACE inhibitors, including antihypertensive peptides, as preventive food components in blood pressure reduction. Compr Rev Food Sci Food Safety 13(2):114–134

40. Kobayashi Y, Yamauchi T, Katsuda T, Yamaji H, Katoh S (2008) Angiotensin-I converting enzyme (ACE) inhibitory mechanism of tripeptides containing aromatic residues. J Biosci Bioeng 106(3):310–312

41. Sagardia I, Roa-Ureta RH, Bald C (2013) A new QSAR model, for angiotensin I-converting enzyme inhibitory oligopeptides. Food Chem 136(3):1370–1376

42. Wu J, Aluko RE, Nakai S (2006) Structural requirements of angiotensin I-converting enzyme inhibitory peptides: quantitative structure-activity relationship modeling of peptides containing 4-10 amino acid residues. QSAR Combinat Sci 25(10):873–880

43. Pripp AH, Isaksson T, Stepaniak L, Søhaug T (2004) Quantitative structure-activity relationship modelling of ACE-inhibitory peptides derived from milk proteins. Eur Food Res Technol 219(6):579–583

44. Kim SY, Je JY, Kim SK (2007) Purification and characterization of antioxidant peptide from hoki (*Johnius belengerii*) frame protein by gastrointestinal digestion. J Nutr Biochem 18(1):31–38

45. Ruiz-Giménez P, Marcos JF, Torregrosa G, Lahoz A, Fernández-Musoles R, Valles S, Alborch E, Manzanares P, Salom JB (2011) Novel antihypertensive hexa- and heptapeptides with ACE-inhibiting properties: from the *in vitro* ACE assay to the spontaneously hypertensive rat. Peptides 32(7):1431–1438

46. Kapel R, Rahhou E, Lecouturier D, Guillochon D, Dhulster P (2006) Characterization of an antihypertensive peptide from an Alfalfa white protein hydrolysate produced by a continuous enzymatic membrane reactor. Process Biochem 41(9):1961–1966

47. Pripp AH (2005) Initial proteolysis of milk proteins and its effect on formation of ACE-inhibitory peptides during gastrointestinal proteolysis: a bioinformatic, *in silico*, approach. Eur Food Res Technol 221(5):712–716

48. Wu J, Aluko RE (2007) Quantitative structure-activity relationship study of bitter di-and tri-peptides including relationship with angiotensin I-converting enzyme inhibitory activity. J Pept Sci 13(1):63–69

8

Production of a biodiesel-like biofuel without glycerol generation, by using Novozym 435, an immobilized *Candida antarctica* lipase

Carlos Luna[1], Cristóbal Verdugo[2], Enrique D Sancho[3], Diego Luna[1,4*], Juan Calero[1], Alejandro Posadillo[4], Felipa M Bautista[1] and Antonio A Romero[1]

Abstract

Background: Novozym 435, a commercial lipase from *Candida antarctica*, recombinant, expressed in *Aspergillus niger*, immobilized on macroporous acrylic resin, has been already described in the obtention of biodiesel. It is here evaluated in the production of a new biofuel that integrates the glycerol as monoglyceride (MG) together with two fatty acid ethyl esters (FAEE) molecules by the application of 1,3-selective lipases in the ethanolysis reaction of sunflower oil.

Results: Response surface methodology (RSM) is employed to estimate the effects of main reaction. Optimum conditions for the viscosity, selectivity, and conversion were determined using a multifactorial design of experiments with three factors run by the software Stat Graphics version XV.I. The selected experimental parameters were reaction temperature, oil/ethanol ratio and alkaline environment. On the basis of RSM analysis, the optimum conditions for synthesis were 1/6 oil/EtOH molar ratio, 30°C, and 12.5 μl of NaOH 10 N aqueous solutions, higher stirring than 300 rpm, for 2 h and 0.5 g of biocatalyst.

Conclusions: These obtained results have proven a very good efficiency of the biocatalyst in the studied selective process. Furthermore, it was allowed sixteen times the successive reuse of the biocatalyst with good performance.

Keywords: Biodiesel; Enzyme biocatalysis; *Candida antarctica* lipase B (CALB); N435; Response surface methodology; Transesterification

Background

Currently, fossil fuel is globally the main primary source of energy. However, as its availability is becoming increasingly limited, it is accepted and assumed that the era of cheap and easily accessible fossil fuel is coming to its end. Thus, the production of biodiesel from renewable raw materials has become very important in recent years as a potential alternative to partially satisfy the future energy demands in the transport sector [1,2].

In this respect, transesterification is currently the most attractive and widely accepted methodology used for biodiesel production [3]. This usually involves the use of homogeneous base catalysts operating under mild conditions. In order to shift the equilibrium towards the production of fatty acid methyl esters (FAME), an excess of methanol is normally utilized in the process to produce biodiesel. However, glycerol is always produced as a contaminant in addition to alkaline impurities that need to be removed. The glycerol by-product is the main drawback of this method because it lowers the overall efficacy of the process and its removal requires several consecutive water washing steps and hence creates a demand for a lot of water [4,5].

A series of alternative methods are being investigated to avoid the problems associated with the generation of glycerol in the conventional process. They all are based on achieving various glycerol derivatives in the same transesterification process. In this way, the complex and expensive additional separation process of glycerol is eliminated and the overall yield of the process increased. These novel methodologies are able to prepare methyl

* Correspondence: qo1lumad@uco.es
[1]Department of Organic Chemistry, University of Cordoba, Campus de Rabanales, Bldg. Marie Curie, 14014, Cordoba, Spain
[4]Seneca Green Catalyst S.A., Bldg Centauro, Technological Science Park of Cordoba, Rabanales XXI, 14014, Córdoba, Spain
Full list of author information is available at the end of the article

esters of fatty acids from lipids using different acyl acceptors instead of methanol in the transesterification process that result in glycerol derivatives as co-products [6]. For example, the transesterification reaction of triglycerides with dimethyl carbonate (DMC) [7], ethyl acetate [8], or methyl acetate [9] generates a mixture of three molecules of FAME or fatty acid ethyl esters (FAEE) and one of glycerol carbonate (GC) or glycerol triacetate (triacetin). These mixtures, including the glycerol derivative molecules, have physicochemical properties similar to biodiesel-like biofuel [10]. In the present case, the atom efficiency is also improved because the total number of atoms involved in the reaction is obtained in the final mixture.

On the other hand, we have recently developed a protocol for the preparation of a new biodiesel-like biofuel, which integrates glycerol into monoglycerides via 1,3-regiospecific enzymatic transesterification of sunflower oil using free [11] and immobilized [12-14] porcine pancreatic lipase (PPL). The operating conditions of such enzymatic process were more efficient compared to the preparation method of conventional biodiesel and did not generate any acidic or alkaline impurities. Thus, the Ecodiesel biofuel [12-15] obtained through the partial ethanolysis of triglycerides with 1,3-selective lipases is constituted by a mixture of two parts of FAEE and one of monoacylglyceride (MG). These glycerol derivative MGs are soluble components in the FAEE mixture suitable for use as a biodiesel-like biofuel. In the current case, ethanol was used as a cheap reagent, instead of the more expensive substrates such as dimethyl carbonate or methyl acetate. This procedure takes advantage of the 1,3-selective nature of the most known lipases, which allows to terminate the process in the second step of the alcoholysis to yield the mixture of 2 mol of FAEE and 1 mol of MG as products (Figure 1). In this way, the glycerol is kept in the form of monoglyceride, which avoids the production of glycerol as by-product, reducing the environmental impact of the process.

In summary, the enzymatic process to obtain this biodiesel-like biofuel operates under much smoother conditions, impurities are not produced, and the biofuel produced exhibits similar physicochemical properties to those of conventional biodiesel. Last but not least, MGs enhance biodiesel lubricity, as it was demonstrated by recent studies [16-18]. Besides, the ethanol that is not spent in the enzymatic process remains in the reaction mixture, and the blend obtained after the reaction can be used directly as fuel. In this respect, very recent studies [19-21] have shown that blends of diesel fuel and biodiesel containing ethanol reduce power output slightly than regular diesel. No significant difference in the emissions of CO_2, CO, and NO_x between regular diesel and biodiesel or ethanol and diesel blends was observed. Furthermore, the use of these blends resulted in a reduction of particulate matter. Consequently, such blends can be used in a diesel engine without any modification despite the slightly reduced power output compared to pure diesel. Thus, the Ecodiesel is currently utilized as a blend of fatty acid alkyl esters with ethanol, alone or with any proportion of diesel fuel [21,22].

The high cost of traditional industrial lipases restricted their use in biofuel production, but the current availability of the recombinant purified in sufficiently high quantities has helped to achieve the economic viability as the crucial factors affecting productivity of enzymatic biodiesel synthesis are the suitable raw materials and the selected lipase. The stability and catalytic efficiency of the latter can be improved by optimizing reaction conditions such as substrate concentrations, temperature, water activity, and alkaline concentration of the enzyme's microenvironment [23]. In this respect, although Ecodiesel was initially obtained using pig pancreatic lipases (PPL), remarkable results have been also obtained with a low-cost purified microbial lipase, Lipopan 50 BG (Novozymes A/S, Bagsværd, Denmark) [15], from *Thermomyces lanuginosus* microorganism, usually used as bread emulsifier (bread improver) [24]. To our knowledge, this lipase has not been described as a biocatalyst in any chemical process. The application of an available lipase on an industrial scale is a significant

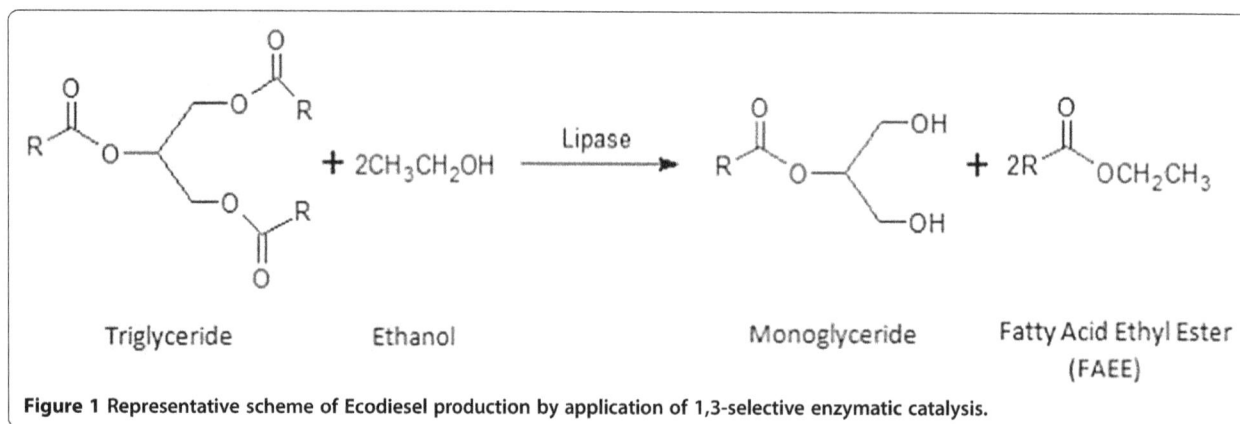

Figure 1 Representative scheme of Ecodiesel production by application of 1,3-selective enzymatic catalysis.

approximation to get an economically feasible biofuel production by enzymatic method. However, this lipase has a main drawback: it cannot be reused, since the purified lipase extract is meant to be in a soluble form.

The aim of the present study is to evaluate Novozym 435, a commercial lipase from *Candida antarctica*, recombinant (CALB), expressed in *Aspergillus niger* and immobilized onto an acrylic macroporous resin [25]. Novozym 435 has previously been used in the synthesis of conventional biodiesel as well as in other transesterification processes such as interesterification [26] and in lipase-catalyzed biodiesel production by isopropanolysis of soybean oil [27]. In this respect, Novozym 435 is also described very recently in lipase-catalyzed simultaneous biosynthesis of biodiesel and glycerol carbonate, by using dimethyl carbonate as acyl acceptor to obtain a biodiesel-like biofuel without glycerol generation [28]. Thus, in this study, it is intended to put in value the 1,3-selective behavior of these commercial lipases to make feasible the profitable production of alternative biofuels [5], using an enzymatic approach.

Methods

In this respect, in order to evaluate the influence of crucial parameters (lipase amount, temperature, oil/ethanol volumetric relationship, and alkaline environment) in the transesterification reaction, a multifactorial design of experiments and response surface methodology (RSM) using a multifactorial design of experiments with three factors run by the software Stat Graphics version XV.I were used. This study has been developed to optimize the catalytic behavior of this 1,3-selective lipase (N435) in the partial ethanolysis of sunflower oil, to obtain a biofuel that integrates glycerol as MG together with the different FAEEs in the enzymatic ethanolysis process as well as with the excess of unreacted ethanol. This biofuel mixture currently named Ecodiesel is able to directly operate diesel engines, alone or in whichever mixture with diesel fuel, without anymore separation or purification process.

Commercial sunflower oil was locally obtained. The chromatographically pure ethyl esters of palmitic acid, stearic acid, oleic acid, linoleic acid, and linolenic acid were commercially obtained from AccuStandard (New Haven, CT, USA), and hexadecane (cetane) was obtained from Sigma-Aldrich (St. Louis, MO, USA). Other chemicals like absolute ethanol and sodium hydroxide were pure analytical compounds (99.5%) obtained commercially from Panreac (Castellar Del Valles, Spain). Novozym 435, the lipase B from *C. antarctica* (CALB), was kindly provided by Novozymes A/S. This commercial lipase is produced by submerged fermentation of a genetically modified *A. niger* microorganism and adsorbed on a macroporous acrylic resin. The specific activity of this commercial lipase in the hydrolysis of tributyrin was 3,000 U/g.

In this respect, additional experiments were conducted with a pure lipase B from *C. antarctica*, recombinant from *Aspergillus oryzae*, powder, from Sigma-Aldrich. The specific activity of this commercial lipase in the hydrolysis of tributyrin was >1,000 U/g. In this way, the existence of possible differences in the catalytic behavior of CALB lipases by the influence of the support could be detected.

Results and discussion

Comparative chromatograms of standardized reaction products

To identify the most characteristic components of biofuels obtained by enzymatic alcoholysis, as well as to compare their rheological properties, several commercial standards of reference for FAME, FAEE, MG, and triacylglycerol (TG) were used, as shown in Figure 2. Here, a representative sample of monoglycerides of sunflower oil is also included, which was easily achieved by the substitution of methanol or ethanol by glycerol, in a conventional alcoholysis process with KOH as homogeneous catalyst following standard experimental conditions.

Here, we can see that the different esters of fatty acids (FAEs), which compose the lipid profile of the sunflower oil, display retention times (RTs) slightly higher than those of cetane (*n*-hexane) used as internal standard. Thus, whereas RT of cetane is around 10 min, all RTs of FAEs appear in the range of 16 to 26 min. These are composed of methyl, ethyl, and glycerol esters (the latter constitute MGs) of palmitic, stearic, linoleic, and oleic acids. Thus, palmitic acid (C16:0) derivatives are grouped in a narrow range of RT, 16 to 17 min. Derivatives of oleic (C18:1) and linoleic acid (C18:2) are grouped in RT of 19 to 21 min, with the exception of glycerol ester of oleic acid, or what is the same, the MG of the oleic acid has a different behavior, with a RT = 26 min. Glycerol RT appears at 5 min, before cetane. The absence of this compound in the obtained chromatograms clearly demonstrates the selective nature of the studied enzymatic transesterification reaction.

In Figure 2, the presence of diacylglycerol (DG) with higher retention times, 40 to 60 min, can also be seen, which do not allow its integration into the GC chromatogram, so that it is necessary to determine DG together and TG by using an internal standard such as cetane here employed. It should be noted that the differences in RT values between MG and DG are much higher than those existing between MG and FAME or FAEE, such as it is expected by the differences between their corresponding molecular weights. At the same time, it is clear that the FAMEs, FAEEs, and MGs display somewhat higher RT values than cetane, but within the molecular weight range, which allows considering similar chemical-physical properties between the FAE and the hydrocarbons that constitute diesel.

MicroVolts (uV)

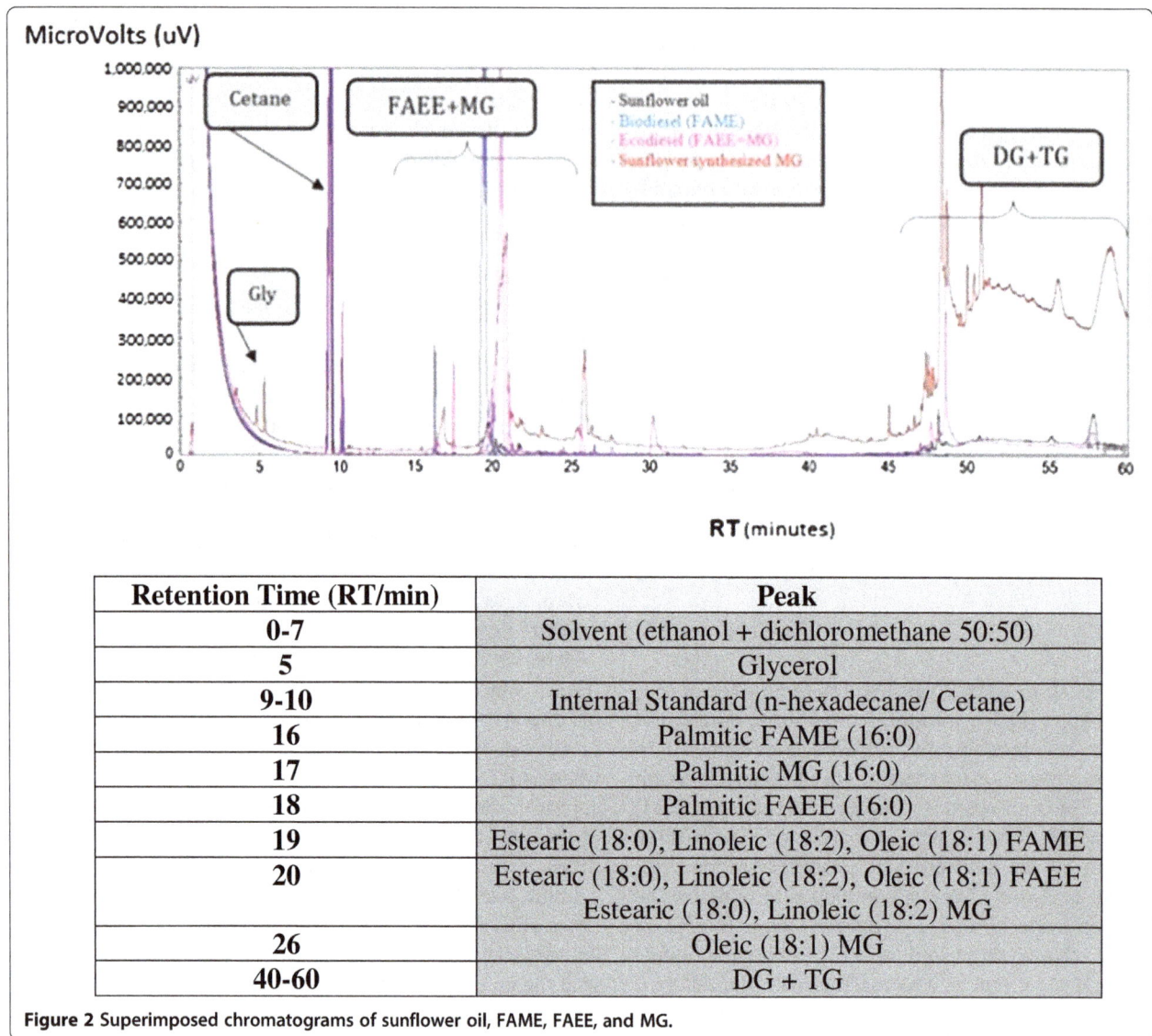

Retention Time (RT/min)	Peak
0-7	Solvent (ethanol + dichloromethane 50:50)
5	Glycerol
9-10	Internal Standard (n-hexadecane/ Cetane)
16	Palmitic FAME (16:0)
17	Palmitic MG (16:0)
18	Palmitic FAEE (16:0)
19	Estearic (18:0), Linoleic (18:2), Oleic (18:1) FAME
20	Estearic (18:0), Linoleic (18:2), Oleic (18:1) FAEE Estearic (18:0), Linoleic (18:2) MG
26	Oleic (18:1) MG
40-60	DG + TG

Figure 2 Superimposed chromatograms of sunflower oil, FAME, FAEE, and MG.

Since the retention times of different derivatives of fatty acids are considered very closely related to the chemical-physical properties of these compounds, the great similarity of RT values obtained is a clear demonstration of the similarity among the rheological properties of the different MGs with their corresponding FAMEs or FAEEs, which are crucial to allow its use as fuel able to substitute for petroleum products. Consequently, conversion is a reaction parameter where all molecules (FAEE, MG, and DG) obtained in the ethanolysis of TG are included (as %); it will be considered as a very different parameter, with respect to selectivity, where FAEEs and MGs are only included (as %), all of them having RT values lower than 26 min. These molecules exhibit RT values similar to hydrocarbons present in conventional diesel, so that they all could exhibit similar physicochemical and rheological properties. However, a high conversion, even 100%, could contain a high proportion of DG molecules, with high molecular weight and high viscosity values. Consequently, a very high selectivity, indicating a very high percentage of FAEEs and MGs, could promote a viscosity value close to the petroleum diesel, so that the highest conversion value is not enough a guaranty of lower viscosity values. Thus, both parameters will be provided as GC analysis results of reaction products.

Taking into account that retention times of a complex mixture of hydrocarbons constituting fossil diesel fuel are ranging from 1 to 25 min, it is used as a reference value for different biofuels (FAME, FAEE, MG) as selectivity value, all those FAEs that present RT values coincident with the hydrocarbons constituting diesel, or those with RT lower than 25 min, as it is expected that they also present similar physicochemical and rheological properties to the conventional diesel.

Analysis of variance and optimization of the reaction parameters by RSM

The analysis of variance methods has become very attractive in reaction parameter optimization and in the evaluation of the effects of the parameters in the TG transesterification reaction [9,15,27] due to their effectiveness in the analysis of variables. Thus, results obtained operating under the 36 runs - each one with different experimental conditions, selected by the multifactorial design of experiments with three factors, developed by the software Stat Graphics version XV.I., where two of them are developed at three levels and the other at two levels, indicated in Table 1 - are shown in Table 2.

The quantity of supported lipase (N435) in all these experiments was fixed to 0.5 g. All experiments were duplicated in order to avoid experimental errors. Data was fitted to a quadratic polynomial model using the software. The quadratic polynomial model was highly significant and sufficient to explain the relationship between conversion/selectivity/kinematic viscosity and important experimental variables. Thus, the results of factorial design suggested that the major factors affecting the transesterification, for the production of biofuels integrating glycerol as monoacylglycerols, were temperature and oil/ethanol ratio (v/v).

The values of correlation coefficients, R^2, were 0.816 for conversion, 0.914 for selectivity, and 0.937 for kinematic viscosity, respectively, which imply a good fit between models and experimental data in Pareto graphics, with respect to conversion, selectivity, and viscosity, as indicated in Figure 3a. The adjusted correlation coefficients R^2 were 0.762, 0.889, and 0.918 for conversion, selectivity, and kinematic viscosity, respectively. Obtained results pointed out that the temperature and oil/ethanol (v/v) ratios were also important parameters influencing the conversion and viscosity in the systems ($p < 0.05$).

The software also allows to obtain equations (Equations 1, 2, and 3), remarkably simpler as compared to initial ones after the elimination of non-influent parameters in the model for conversion, selectivity, and kinematic viscosity. These equations describe the model created and give solutions for the dependent variable based on the independent variable combinations, selecting the most significant in the response. Thus, taking into account that R is the oil/ethanol ratio (v/v), alkaline environment is obtained by the addition of different microliters of NaOH 10 N, and T is the reaction temperature,

$$[\text{Conversion } (\%) = 93.57 + 5.06 \times T + 7.82 \\ \times R - 6.42 \times T^2 - 6.53 \times T \\ \times R - 3.25 \times R \times \text{pH}]$$
(1)

$$[\text{Selectivity } (\%) = 35.97 + 6.26 \times T - 2.25 \times \text{pH} \\ + 4.50 \times R + 6.93 \times T \times \text{pH} \\ - 12.22 \times T \times R + 6.13 \times \text{pH}^2 \\ - 6.61 \times \text{pH} \times R]$$
(2)

$$[\text{Viscosity(cSt)} = 17.27 + 1.65 \times T - 2.32 \times R \\ + 0.97 \times T^2 + 1.95 \times T \times R \\ + 1.20 \times R \times \text{pH}].$$
(3)

The surface plots in Figure 3b, described by the regression model, were drawn to display the effects of the independent variables on conversion, selectivity, and kinematic viscosity. Here, the influence of the different variables in the conversion of the systems can be clearly seen. This model showed that the optimum values for the parameters to maximize transesterification yield (conversion, selectivity, kinematic viscosity) were intermediate temperatures (30°C), maximum amount of aqueous NaOH 10 N added (50 µl), and the maximum oil/ethanol (v/v) ratio = 6:1 studied. Conversions up to 100%, selectivities around 70%, and values of kinematic viscosity about 10 to 15 mm^2 s^{-1} could be achieved under these conditions, which in theory will render feasible the utilization of the obtained biofuel in blends with diesel. For example, by the addition of only 35% of diesel fossil, to this biofuel, a viscosity reduction at 4.8 mm^2 s^{-1}, a value within the acceptance limits of the EN 14214, is obtained.

Effect of the amount of lipase

The effect of the amount of lipase in the reaction media is very important to select the correct reaction conditions. This parameter was evaluated in order to choose the necessary amount of lipase that maximizes yield without mass transfer limitations. Previous optimized values obtained from RSM (12 ml oil/3.5 ml EtOH, 30°C, 12.5 µl NaOH 10 N, stirring speed higher than 300 rpm for 2 h) were chosen.

We can see in Figure 4 how the reaction yield and viscosity were improved, so better values could be obtained using more quantities of biocatalyst, but larger amounts of enzymes will have a detrimental effect on the economics of the process.

Table 1 Process parameters in factorial design: coded and actual values

Variables	Unit	Levels		
		−1	0	1
Temperature	°C	20	30	40
Oil/ethanol ratio (v/v)	ml/ml	12/1,75	-	12/3.5
Alkaline environment	µl (NaOH 10 N)	8 (12.5)	10 (5)	12 (50)

Table 2 Experiment matrix of factorial design and the response obtained for conversion, selectivity, and viscosity

Run	Parameters					
	Temperature	Oil/ethanol ratio	pH	Conversion (%)	Selectivity (%)	Kinematic viscosity (mm² s⁻¹)
1	1	−1	0	81.4	45.99	20.3
2	1	−1	−1	84.6	46.1	21.3
3	0	−1	−1	78.6	29.2	21.0
4	−1	−1	0	59.1	11.7	20.8
5	−1	1	1	100	40.4	13.3
6	1	1	1	100	39.6	22.3
7	−1	−1	−1	63.6	23.3	22.1
8	0	−1	1	89.5	34.1	19.0
9	0	1	0	95.7	31.4	15.2
10	0	−1	0	100	44.2	20.1
11	0	1	1	95.2	37.5	15.7
12	1	−1	1	100	72.5	18.6
13	1	1	−1	100	46.6	18.2
14	1	1	0	86.1	36.1	19.8
15	−1	1	0	94.9	46.8	13.4
16	−1	1	−1	100	70.9	10.8
17	0	1	−1	100	52.6	12.5
18	−1	−1	1	77.5	17.8	20.2
Repeated experiments						
19	0	−1	0	100	43.4	22.5
20	−1	1	1	100	39.7	13.2
21	−1	−1	−1	64.4	22.6	20.1
22	−1	1	0	95.8	46.1	13.4
23	1	1	−1	100	47.5	16.2
24	0	1	−1	100	51.9	14.5
25	−1	1	−1	100	69.6	10.9
26	0	1	0	96.5	36.5	14.6
27	1	1	1	100	38.8	22.2
28	0	−1	−1	80.1	28.6	19.0
29	1	−1	1	100	71.4	18.5
30	−1	−1	0	60.2	11.0	20.3
31	1	−1	0	82.2	45.2	20.3
32	1	1	0	86.9	33.4	19.5
33	0	−1	1	90.3	33.4	18.1
34	0	1	1	97.0	36.8	15.1
35	1	−1	−1	85.4	43.4	21.6
36	−1	−1	1	69.7	16.3	20.6

Study of enzyme activity in successive reactions

Since we have been doing these successive reactions, the reaction conditions have been changed such that the procedure enables the study according to OVAT ('variable at one time') methodology in which initial conditions are set and variables to study have been changed one by one. This method does not provide information on the combined effect of various reaction parameters, but it is useful to determine the influence of isolated variables. Because our research group had previous information about other lipase enzymes [12-15] and what were the most important reaction variables, it was decided to use this experimental

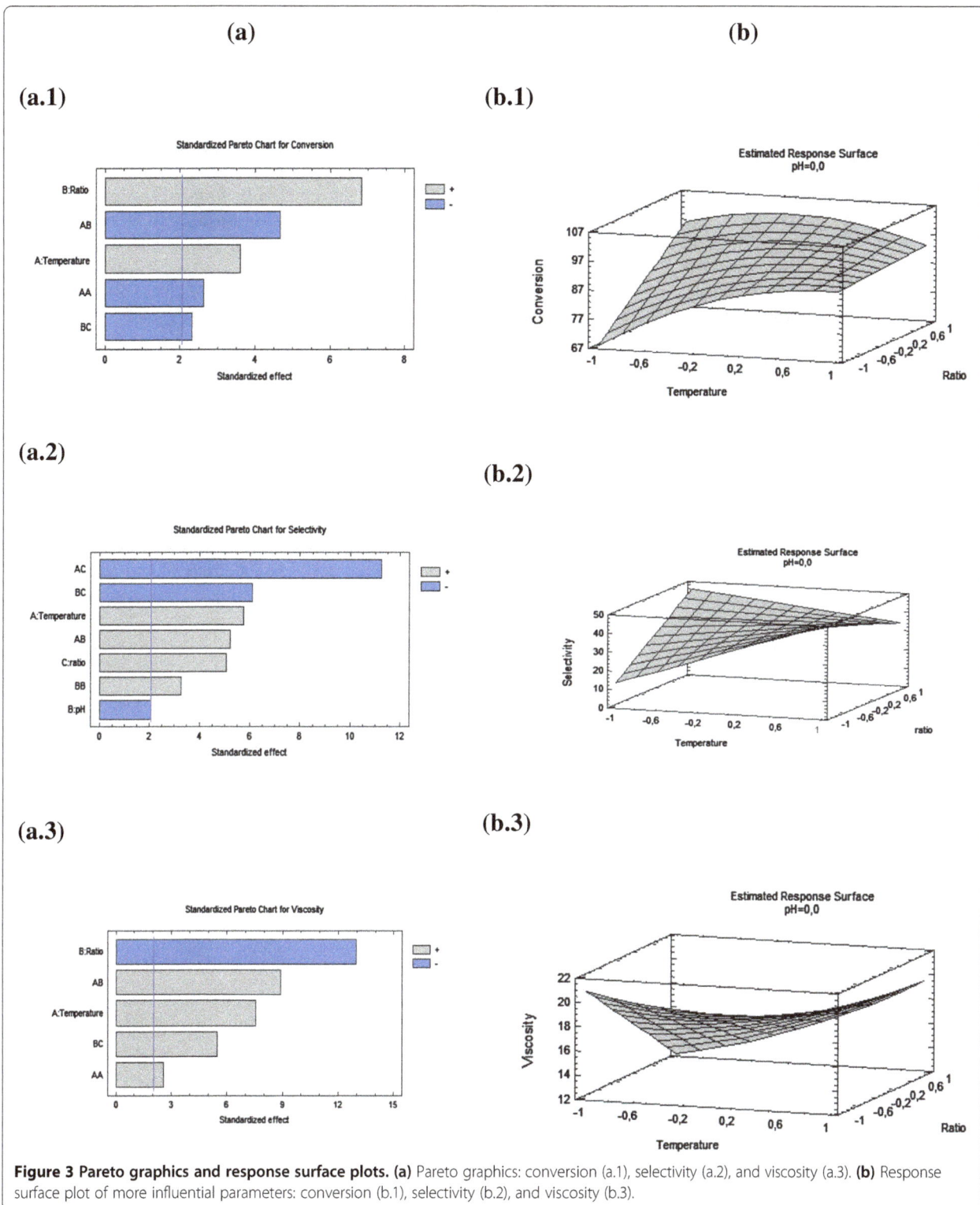

Figure 3 Pareto graphics and response surface plots. (a) Pareto graphics: conversion (a.1), selectivity (a.2), and viscosity (a.3). **(b)** Response surface plot of more influential parameters: conversion (b.1), selectivity (b.2), and viscosity (b.3).

methodology in order to evaluate in more detail the influence of those variables in the selected intervals for each study.

In this way, reaction tests of N435 lipase systems have been carried out under the optimum conditions (alkaline environments, temperatures, and relative oil/ethanol ratio) previously determined by the RSM studies. These experimental conditions were similar to those previously obtained with *T. lanuginosus*, Lipopan 50 BG Lipopan [15]. In this respect, Table 3 shows a collection of the achieved

Graphic 2

Graphic 1

Figure 4 Influence of lipase amount of supported CALB in the performance of ethanolysis reactions, developed under standard conditions. Conversion and selectivity (graphic 1) and kinematic viscosity in cSt (graphic 2).

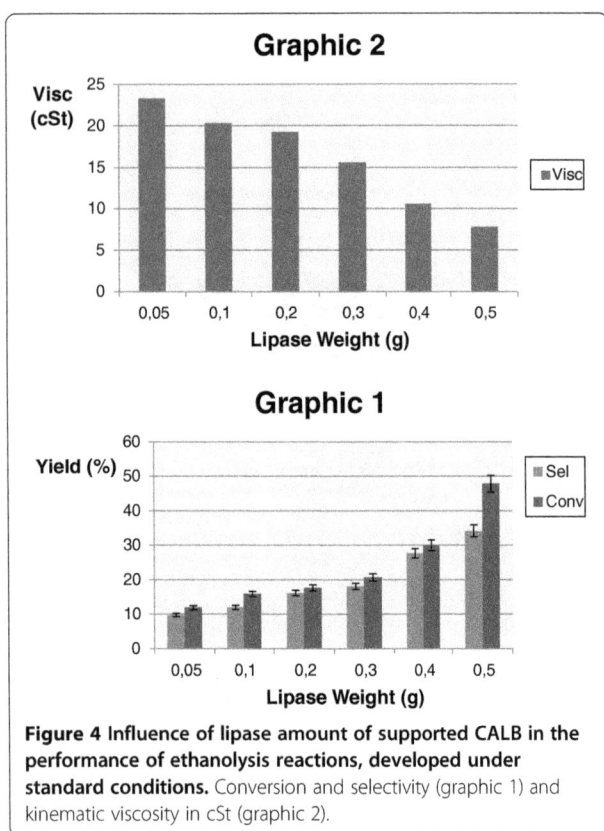

results in the transesterification reaction with this biocatalyst. Thus, in addition to obtain specific information about the influence of certain parameters (alkaline environment, temperature, etc.) on the behavior of the N435, its ability to be reused could be checked.

In this sense, from the conversion, given that we start from 0.01 mol of TG, with 2 h as reaction time and 0.5 g of catalyst system (lipase CALB immobilized on acrylic polymer), we can calculate the transformation enzymatic capacity: turnover frequency (TOF number) expressed in micromoles of TG transformed per minute and per gram of catalyst (supported lipase). In order to be able to obtain the enzyme activity of the immobilized CALB, some research experiments with a purified powdered commercial lipase B from *C. antarctica*, recombinant from *A. oryzae*, have been conducted. Thus, operating under identical experimental conditions with those used in experiments carried out with N435, but using only 0.01 g of purified CALB in free form, a conversion of 50.6% was obtained. Excluding the possible influence of support effects in the lipase activity, and taking into account that 0.5 g of CALB immobilized enzyme produced a 47.7% conversion, it can be inferred that N435 has a catalytic activity corresponding to the 1.9% of the obtained catalytic activity by the free CALB.

Therefore, to obtain a reference value, and assuming that the activity of the immobilized enzyme is similar to that of the free enzyme, it can be considered that the N435 may contain about 1.9 wt% of immobilized enzyme. In any case, it works with this equivalent enzyme

Table 3 Achieved results in the transesterification reaction with biocatalyst

Run number	NaOH amount	Temperature	EtOH/oil	Viscosity	Selectivity	Conversion	TOF	Enzyme activity
1	25	30	3.5/12 (6:1)	10.5	34.9	47.7	79.5	0.24
2	25	30	3.5/12 (6:1)	13.7	23.6	32.9	54.83	0.35
3	25	30	3.5/12 (6:1)	12.8	24.1	31.8	53.00	0.36
4	25	30	3.5/12 (6:1)	13.3	25.6	33.9	56.50	0.33
5	25	30	3.5/12 (6:1)	12.8	29.1	35.6	59.33	0.32
6	12.5	35	3.5/12 (6:1)	13.8	14.5	20.1	33.50	0.57
7	12.5	30	3.5/12 (6:1)	12.3	45.4	57.9	96.50	0.20
8	12.5	25	3.5/12 (6:1)	14.6	21.4	25.8	43.00	0.44
9	12.5	40	3.5/12 (6:1)	20.9	7.6	4.0	6.67	2.85
10	12.5	30	1.75/12 (3:1)	18,8	17,9	27,4	45.67	0.42
11	12.5	20	3.5/12 (6:1)	18.1	17.5	22.4	37.33	0.51
12	12.5	30	2.3/12 (4:1)	20.0	17.6	21.4	35.67	0.53
13	12.5	30	2.9/12 (5:1)	10.6	34.1	47.7	79.50	0.24
14	12.5	30	3.5/12 (6:1)	10.4	39.1	51.9	86.50	0.22
15	12.5	30	4.1/12 (7:1)	17.9	14.1	16.8	28.00	0.68
16	12.5	30	4.7/12 (8:1)	16.3	17.1	21.8	36.33	0.52

Viscosity (cSt), conversion (%), selectivity (%), turnover frequencies (TOF in μmol min^{-1} g cat^{-1}), and enzyme activities (U/mg lipase) of the obtained biodiesel by using the same biocatalyst (0.5 g of N435) in successive reactions under the evaluated different experimental conditions: temperature (°C), NaOH amount (μl of NaOH 10 N solution), and EtOH/oil ratio (ml/ml or mol/mol).

activity. From this parameter, a reference value of the catalytic activity of the immobilized enzymes, expressed in milligrams of enzyme required to convert 1 μmol of TG per minute (U/mg), can also be calculated.

If we subdivide the results in Table 3 in terms of the variable that has been changed each time in the study with OVAT methodology and if we organize these results in their corresponding graphics (Figure 5), the influence of each variable on the transesterification

reaction yield is clearly displayed and we reached similar conclusions to those obtained in the analysis of variance (ANOVA) study by RSM; the optimum reaction parameters for the enzymatic ethanolysis reaction with lipase N435, CALB (*C. antarctica* lipase B) immobilized on acrylic resin, are 30°C and 12.5 μl of 10 N NaOH, with 1:6 oil/ethanol molar ratio.

Also, we can obtain from the data in Table 3 that N435 may be reused repeatedly. Thus, 16 reactions were

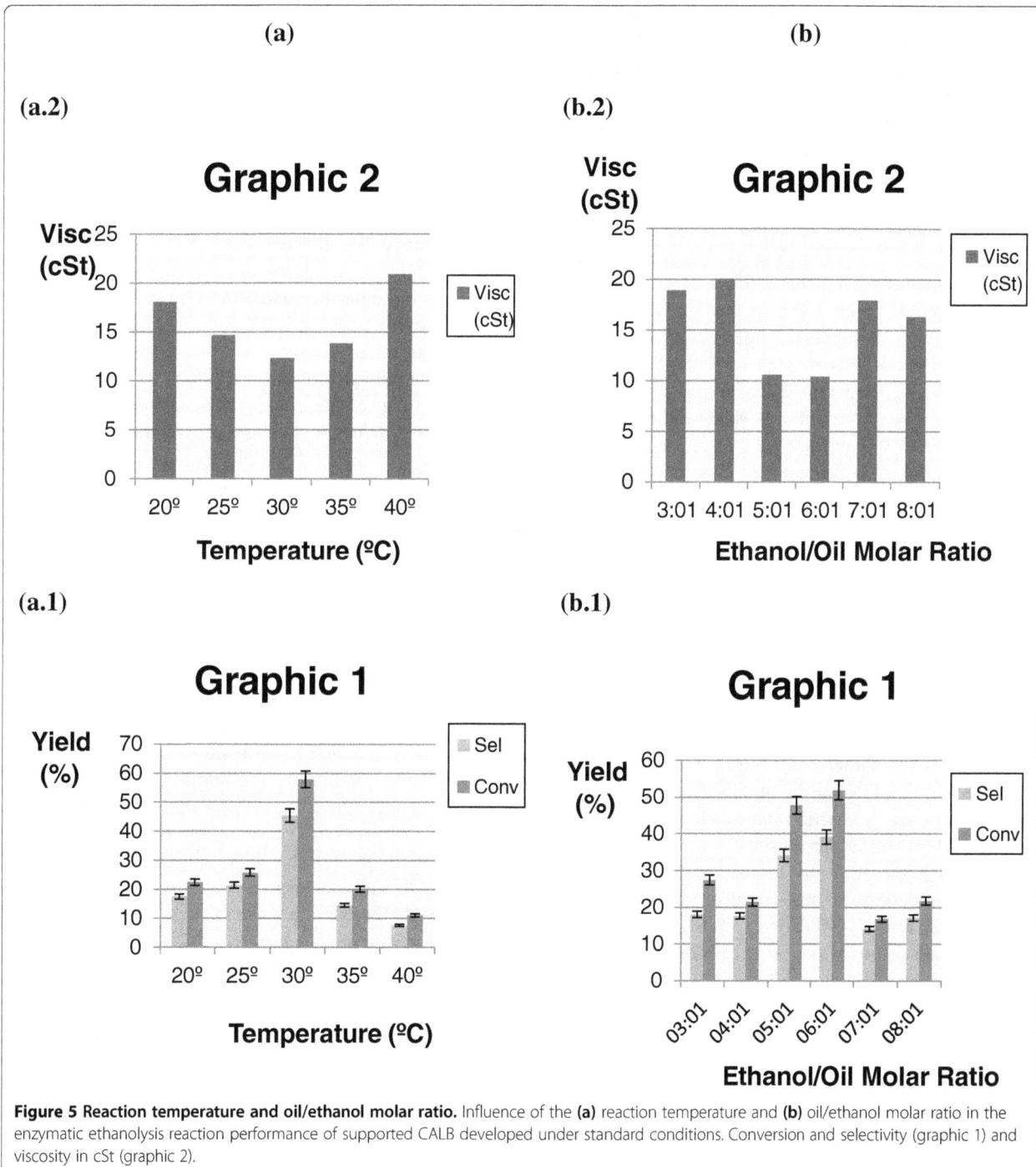

(a) **(b)**

(a.2) **(b.2)**

(a.1) **(b.1)**

Figure 5 Reaction temperature and oil/ethanol molar ratio. Influence of the **(a)** reaction temperature and **(b)** oil/ethanol molar ratio in the enzymatic ethanolysis reaction performance of supported CALB developed under standard conditions. Conversion and selectivity (graphic 1) and viscosity in cSt (graphic 2).

developed successively without an appreciable loss in catalytic activity in the ethanolysis reaction of sunflower oil, operating under different experimental conditions. However, an important drawback is associated to its acrylic polymeric character. Thus, to recover the immobilized lipase on macroporous resin, after each successive reaction, it is necessary to carry out the centrifugation of reaction products.

Finally, this research, developed in order to improve a new methodology to integrate the glycerol as different monoacylglycerol molecules, is in connection with precedent researches performed to determine the optimal experimental conditions of the selective ethanolysis reaction. In this respect, the results here obtained resemble those described with the commercial PPL lipases [12-14] and with a low-cost purified lipase, [15] from *T. lanuginosus*, Lipopan 50 BG (Novozymes A/S), widely used in the bakery industry as an emulsifier [24]. In this respect, the possibility of application of commercial immobilized lipases in this process is now explored to exploit the opportunity of its reuse, thus lowering its operational cost and consequently increasing the economic potential of their industrial application. Thus, in this study, the behavior of a commercial supported lipase, Novozym 435, is determined because this may involve some technological advantages in its development, so that this study has the aim to prove its viability, its yield, and its possibility of reuse, what will mean an important advance in its biotechnological applications in the biofuel enzymatic production [5].

Currently, results obtained in this study indicate that this commercial supported lipase is specially efficient in obtaining 1,3-selective ethanolysis processes, where glycerol is maintained as MG in the biofuel mixture, with the different obtained FAEEs and together to the excess of unreacted ethanol. In this way, a new kind of biodiesel currently named Ecodiesel is achieved, constituted by a mixture of monoacylglycerols and FAEEs mainly (1/2 nominally), which can be used in different blends with diesel fuel, without anymore separation or purification process. This new biofuel can be obtained with the help of N435 at very short reaction times (1 to 2 h) and under soft reaction conditions. Besides, not only a higher atomic yield is achieved, with respect to the conventional biodiesel reaction (because no glycerol by-product is generated), but also a purification step of residual glycerol is not necessary, so it can be used directly after its production.

Despite the good results obtained in the transesterification reaction of sunflower oil with ethanol to produce FAEE and MG, unfortunately, the main drawback of this commercial biocatalyst is the low stability of the organic polymeric little spheres which constitute the catalytic powder that easily disintegrates in the first ethanolysis reaction and becomes a gelatinous material which is very uncomfortable to work in the successive reuses, so that centrifugation of reaction products is necessary for catalyst recovery to avoid important losses of catalyst. This additional centrifugation step may constitute a major drawback for the application of the process on an industrial scale.

Experimental
Ethanolysis reactions
These reactions were performed according to the experimental procedure previously described [12-15] to determine the optimal conditions for obtaining the selective ethanolysis reaction, such as alkaline environment, amount of lipase, the oil/ethanol molar ratio (v/v), and temperature. Thus, enzymatic assays are carried out with 9.4 g (12 ml, 0.01 mol) of commercial sunflower oil at controlled temperatures (20°C to 40°C) in a 25-ml round bottom flask. Reaction mixtures were stirred with a conventional magnetic stirrer at a stirring speed higher than 300 rpm to avoid mass transfer limitations, along a reaction time of 2 h. Variable oil/alcohol volume ratios at different alkaline environments and different quantities of lipase are studied. The different oil/ethanol ratios (v/v) are obtained by introducing absolute ethanol volumes in the range of 1.75 to 3.5 ml, and the influence of different amounts of lipase is studied in the range of 0.05 to 0.5 g. The influence of alkaline environment values was achieved by adding different volumes (12.5 to 50 µl) of 10 N NaOH aqueous solution. In this regard, a blank reaction in the presence of the highest quantity of solution of NaOH was performed to rule out a potential contribution from the homogeneous NaOH catalyzed reaction. Less than 10% conversion of the starting material was obtained, so that a homogenous base catalysis contribution can be considered as negligible under the investigated conditions.

In this respect, the NaOH solution, to operate as a transesterification catalyst, needs usually to be used under higher temperatures and under higher catalyst concentrations (1 to 2 wt%), enough to generate sodium methoxide or ethoxide, the true catalyst of the homogeneous transesterification reaction. Thus, the use of a very small concentration of NaOH works as a promoter of the process, through its effect on the lipase biocatalytic activity. Accordingly, NaOH ions are responsible for the pH environment that affects the enzyme yield, because depending on the pH, the enzymatic protein offers different structures, more or less effective in the biocatalytic action.

All variables were studied and optimized according to a factorial experimental design and a response surface methodology.

Analytical method
Reaction products were monitored by capillary column gas chromatography, using a Varian 430-GC gas chromatograph

(Varian Inc., Palo Alto, CA, USA), connected to a HT5 capillary column (25 m × 0.32 mm ID × 0.1 μm, SGE, Supelco) with a flame ionization detector (FID) at 450°C and splitless injection at 350°C. Helium is used as carrier gas with a flow of 1.5 ml/min. A heating ramp from 90°C to 200°C at a rate of 7°C/min has been applied, followed by another ramp from 200°C to 360°C at a rate of 15°C/min, maintaining the temperature of the oven at 360°C for 10 min using as internal standard *n*-hexadecane (cetane) to quantify the content of ethyl esters and the different glycerides (mono-, di-, and triglycerides) with the help of several commercial standard fatty acid esters. This method allows us to make a complete analysis of the sample in a single injection and in a time not longer than 60 min, which simplifies the process and increases the speed of analysis [12-15].

Considering that sunflower oil is constituted by a mixture of fatty acids in variable proportion (mainly linoleic, oleic, palmitic, and stearic acids), the results obtained are expressed as the relative amounts of the corresponding ethyl esters (FAEE, fatty acid ethyl esters), monoglycerides (MG), and diglycerides (DG) that are integrated in the chromatogram. The amount of triglycerides (TG) which has not reacted is calculated from the difference to the internal standard (cetane). Thus, the conversion includes the total amount of triglyceride transformed (FAEE + MG + DG) in the ethanolysis process, and selectivity makes reference to the relative amount of FAEE + MG obtained. The latter are the ones having retention times close to the cetane standard, which is the reference hydrocarbon for diesel fuel.

Viscosity measurements

The transesterification reactions of oils and fats are basically carried out to obtain an important reduction in the viscosity of these materials, as they share similar values in all of other chemical-physical significant parameters with the fossil diesel except the viscosity. In this respect, most oils exhibit viscosities in the range of 30 to 45 mm²/s cSt values, while the fossil diesel is in the range of 2.5 to 6 cSt values. Thus, due to the importance of viscosity for the correct running of diesel engines, this parameter becomes a critical factor to change the chemical-physical properties of vegetable oils before their use as biofuel. The transesterification process of oils and fats is actually developed in order to obtain a noticeable lowering of viscosity in oils to employ the resulting product as biofuel in current existing diesel engines. Thus, accurate viscosity measurements are critical to assess the quality of biofuels produced, since unsuitable viscosity values can decisively affect the correct working conditions of the diesel engine. Therefore, the characterization of this parameter is essential to evaluate the result obtained in the process of ethanolysis.

Viscosities were determined in a capillary viscometer Oswald Proton Cannon-Fenske Routine Viscometer 33200, size 150. This is based on determining the time needed for a given volume of fluid passing between two points marked on the instrument. The kinematic viscosity is given by the ratio between the dynamic viscosity (h, in Poise, g/cm s) and the density (r, in g/cm³) $v = h/r$ (in cm²/s or centistokes (cSt), mm²/s). Samples, previously centrifuged at 3,500 rpm for 10 min and filtered at 50°C, are immersed in a thermostatic bath at 40°C for 15 min, making sure that the temperature is stable. Then, samples are introduced into the viscometer and this, in turn, in the water bath, making sure that it is rigorously positioned vertically, with the bottom end at a minimum distance of 2 cm from the floor of the bath [12-15].

Experimental design

The effect of process parameters in the enzymatic transesterification reaction to obtain the optimum conditions for the viscosity, selectivity, and conversion was studied using a multifactorial design of experiments with three factors run by the software Stat Graphics version XV.I. Two of them are developed at three levels and the last one at two levels, so that it gives 36 runs. The experiments were performed in random order. The experimental parameters selected for this study were reaction temperature, oil/ethanol ratio (v/v), and different alkaline environments obtained by the addition of variable volumes (in μl) of NaOH 10 N. Table 1 shows the coded and actual values of the process parameters used in the design matrix.

Statistical analysis

The experimental data obtained from experimental design were analyzed by response surface methodology (RSM) [9,15,27]. A mathematical model, following a second-order polynomial equation, was developed to describe the relationships between the predicted response variable (viscosity, conversion, and selectivity) and the independent variables of reaction conditions, as shown in the Equation 4, where Y is the predicted response variable; β_0, β_i, β_{ii}, and β_{ij} are the intercept, linear, quadratic, and interaction constant coefficients of the model, respectively; and X_i and X_j ($i = 1, 3$; $j = 1, 3$; $i \neq j$) represent the coded independent variables.

$$Y = \beta_0 + \sum_{i=1}^{3}\beta_0 x_i + \sum_{i=1}^{3}\beta_{ii} x_i^2 + \sum \sum_{i<j=1}^{3} \beta_{ij} x_i x_j \quad (4)$$

Response surface plots were developed using the fitted quadratic polynomial equation obtained from regression analysis, holding one of the independent variables at constant values corresponding to the stationary point and changing the order of two variables. The quality of the fit of the polynomial model equation was evaluated by the coefficient of determination R^2, and its regression coefficient significance was checked with F-test. Confirmatory experiments were carried out in order to validate the model, using combinations of independent variables which were not part of the original experimental design but within the experimental region.

Conclusions

Novozym 435 lipase was evaluated in the 1,3-selective ethanolysis of sunflower oil to integrate the glycerol as monoacylglycerol molecules. On the basis of RSM analysis, the optimum conditions for synthesis were 1:6 molar oil/EtOH ratio, 30°C, and 12.5 μl NaOH 10 N aqueous solutions, with stirring speed higher than 300 rpm for 2 h and 0.5 g of biocatalyst. Accordingly, its ability to be repeatedly reused could open a new way for the production of alternative biodiesel using an enzymatic approach, which is technically feasible and economically viable. The main drawback is the permanent need for centrifugation to recover the biocatalyst for the next reuse.

Abbreviations
ANOVA: analysis of variance; CALB: *Candida antarctica* lipase B; DG: diacylglycerol; FAE: fatty acid esters; FAEE: fatty acid ethyl esters; FAME: fatty acid methyl esters; MG: monoacylglycerol; OVAT: variable at one time; RSM: response surface methodology; TG: triacylglycerol; TOF: turnover frequency.

Competing interests
The authors declare that they have no competing interests.

Authors' contributions
CL, CV, EDS, DL, JC, AP, FMB, and AAR have made substantive intellectual contributions to this study, making substantial contributions to the conception and design of it as well as to the acquisition, analysis, and interpretation of data. All of them have been also involved in the drafting and revision of the manuscript. All authors read and approved the final manuscript.

Acknowledgements
Grants from the Spanish Ministry of Economy and Competitiveness (Project ENE 2011-27017), Spanish Ministry of Education and Science (Projects CTQ2010-18126 and CTQ2011-28954-C02-02), FEDER funds and Junta de Andalucía FQM 0191, PO8-RMN-03515 and P11-TEP-7723 are gratefully acknowledged by the authors. We are also grateful to Novozymes A/S, Denmark, for the kind supply of the macroporous resin immobilized lipase from *Candida antarctica* (Novozym 435).

Author details
[1]Department of Organic Chemistry, University of Cordoba, Campus de Rabanales, Bldg. Marie Curie, 14014, Cordoba, Spain. [2]Crystallographic Studies Laboratory, Andalusian Institute of Earth Sciences, CSIC, Avda. Las Palmeras, n°4, 18100, Armilla, Granada, Spain. [3]Department of Microbiology, University of Cordoba, Campus de Rabanales, Ed. Severo Ochoa, 14014, Cordoba, Spain. [4]Seneca Green Catalyst S.A., Bldg Centauro, Technological Science Park of Cordoba, Rabanales XXI, 14014, Córdoba, Spain.

References
1. Demirbas A (2009) Political, economic and environmental impacts of biofuels: a review. Appl Energy 86:108–117
2. Luque R, Herrero-Davila L, Campelo JM, Clark JH, Hidalgo JM, Luna D, Marinas JM, Romero AA (2008) Biofuels: a technological perspective. Energy Environ Sci 1:542–564
3. Oh PP, Lau HLN, Chen JH, Chong MF, Choo YM (2012) A review on conventional technologies and emerging process intensification (PI) methods for biodiesel production. Renew Sust Energy Rev 16:5131–5145
4. Saleh J, Dube MA, Tremblay AY (2011) Separation of glycerol from FAME using ceramic membranes. Fuel Process Technol 92:1305–1310
5. Calero J, Luna D, Sancho ED, Luna C, Posadillo A, Bautista FM, Romero AA, Berbel J, Verdugo C (2014) Technological challenges for the production of biodiesel in arid lands. J Arid Environ 102:127–138
6. Ganesan D, Rajendran A, Thangavelu V (2009) An overview on the recent advances in the transesterification of vegetable oils for biodiesel production using chemical and biocatalysts. Rev Environ Sci Biotech 8:367–394
7. Ilham Z, Saka S (2010) Two-step supercritical dimethyl carbonate method for biodiesel production from *Jatropha curcas* oil. Bioresour Technol 101:2735–2740
8. Kim SJ, Jung SM, Park YC, Park K (2007) Lipase catalyzed transesterification of soybean oil using ethyl acetate, an alternative acyl acceptor. Biotechnol Bioprocess Eng 12:441–445
9. Tan KT, Lee KT, Mohamed AR (2011) A glycerol-free process to produce biodiesel by supercritical methyl acetate technology: an optimization study via response surface methodology. Bioresour Technol 102:3990–3991
10. Casas A, Ruiz JR, Ramos MJ, Perez A (2010) Effects of triacetin on biodiesel quality. Energy Fuels 24:4481–4489
11. Verdugo C, Luque R, Luna D, Hidalgo JM, Posadillo A, Sancho ED, Rodriguez S, Ferreira-Dias S, Bautista F, Romero AA (2010) A comprehensive study of reaction parameters in the enzymatic production of novel biofuels integrating glycerol into their composition. Bioresour Technol 101:6657–6662
12. Caballero V, Bautista FM, Campelo JM, Luna D, Marinas JM, Romero AA, Hidalgo JM, Luque R, Macario A, Giordano G (2009) Sustainable preparation of a novel glycerol-free biofuel by using pig pancreatic lipase: partial 1,3-regiospecific alcoholysis of sunflower oil. Process Biochem 44:334–342
13. Luna D, Posadillo A, Caballero V, Verdugo C, Bautista FM, Romero AA, Sancho ED, Luna C, Calero J (2012) New biofuel integrating glycerol into its composition through the use of covalent immobilized pig pancreatic lipase. Int J Mol Sci 13:10091–10112
14. Luna C, Sancho E, Luna D, Caballero V, Calero J, Posadillo A, Verdugo C, Bautista FM, Romero AA (2013) Biofuel that keeps glycerol as monoglyceride by 1,3-selective ethanolysis with pig pancreatic lipase covalently immobilized on AlPO₄ support. Energies 6:3879–3900
15. Verdugo C, Luna D, Posadillo A, Sancho ED, Rodriguez S, Bautista F, Luque R, Marinas JM, Romero AA (2011) Production of a new second generation biodiesel with a low cost lipase derived from *Thermomyces lanuginosus*: optimization by response surface methodology. Catal Today 167:107–112
16. Xu YF, Wang QJ, Hu XG, Li C, Zhu XF (2010) Characterization of the lubricity of bio-oil/diesel fuel blends by high frequency reciprocating test rig. Energy 35:283–287
17. Haseeb A, Sia SY, Fazal MA, Masjuki HH (2010) Effect of temperature on tribological properties of palm biodiesel. Energy 35:1460–1464
18. Wadumesthrige K, Ara M, Salley SO, Ng KYS (2009) Investigation of lubricity characteristics of biodiesel in petroleum and synthetic fuel. Energy Fuels 23:2229–2234
19. Çelikten I (2011) The effect of biodiesel, ethanol and diesel fuel blends on the performance and exhaust emissions in a diesel engine. GU J Sci 24:341–346
20. Cheenkachorn K, Fungtammasan B (2009) Biodiesel as an additive for diesohol. Int J Green Energy 6:57–72
21. Jaganjac M, Prah IO, Cipak A, Cindric M, Mrakovcic L, Tatzber F, Ilincic P, Rukavina V, Spehar B, Vukovic JP, Telen S, Uchida K, Lulic Z, Zarkovic N (2012) Effects of bioreactive acrolein from automotive exhaust gases on human cells in vitro. Environ Toxicol 27:644–652
22. Pang XB, Mu YJ, Yuan J, He H (2008) Carbonyls emission from ethanol-blended gasoline and biodiesel-ethanol-diesel used in engines. Atmos Environ 42:1349–1358
23. Szczesna-Antczak M, Kubiak A, Antczak T, Bielecki S (2009) Enzymatic biodiesel synthesis - key factors affecting efficiency of the process. Renew Energy 34:1185–1194
24. Moayedallaie S, Mirzaei M, Paterson J (2010) Bread improvers: comparison of a range of lipases with a traditional emulsifier. Food Chem 122:495–499
25. Xu Y, Nordblad M, Woodley JM (2012) A two-stage enzymatic ethanol-based biodiesel production in a packed bed reactor. J Biotechnol 162:407–414

26.	Yara-Varon E, Joli JE, Torres M, Sala N, Villorbina G, Mendez JJ, Canela-Garayoa R (2012) Solvent-free biocatalytic interesterification of acrylate derivatives. Catal Today 196:86–90
27.	Chang C, Chen JH, Chang CMJ, Wu TT, Shieh CJ (2009) Optimization of lipase-catalyzed biodiesel by isopropanolysis in a continuous packed-bed reactor using response surface methodology. N Biotechnol 26:187–192
28.	Min JY, Lee EY (2011) Lipase-catalyzed simultaneous biosynthesis of biodiesel and glycerol carbonate from corn oil in dimethyl carbonate. Biotechnol Lett 33:1789–1796

9

Analysis of alkali ultrasonication pretreatment in bioethanol production from cotton gin trash using FT-IR spectroscopy and principal component analysis

Jersson Plácido* and Sergio Capareda

Abstract

Background: Cotton gin trash (CGT) is a lignocellulosic residue that can be used in the production of cellulosic ethanol. In a previous research, the sequential use of ultrasonication, liquid hot water, and ligninolytic enzymes was selected as pretreatment for the production of ethanol from CGT. However, an increment in the ethanol production is necessary. To accomplish that, this research evaluated the effect of pretreating CGT using alkaline ultrasonication before a liquid hot water and ligninolytic enzymes pretreatments for ethanol production. Three NaOH concentrations (5%, 10%, and 15%) were employed for the alkaline ultrasonication. Additionally, this work is one of the first applications of Fourier transform infrared (FT-IR) spectrum and principal component analysis (PCA) as fast methodology to identify the differences in the biomass after different types of pretreatments.

Results: The three concentrations employed for the alkaline ultrasonication pretreatment produced ethanol yields and cellulose conversions higher than the experiment without NaOH. Furthermore, 15% NaOH concentration achieved twofold increment yield versus the treatment without NaOH. The FT-IR spectrum confirmed modifications in the CGT structure in the different pretreatments. PCA was helpful to determine differences between the pretreated and un-pretreated biomass and to evaluate how the CGT structure changed after each treatment.

Conclusions: The combination of alkali ultrasonication hydrolysis, liquid hot water, and ligninolytic enzymes using 15% of NaOH improved 35% the ethanol yield compared with the original treatment. Additionally, we demonstrated the use of PCA to identify the modifications in the biomass structure after different types of pretreatments and conditions.

Keywords: Alkali-hydrolysis; Cotton gin trash; Ethanol production; FT-IR; Ligninolytic enzymes; Principal components analysis; Ultrasonication

Background

Cotton is one of the major crops grown in the world. In 2006 to 2007, the worldwide production was 24 million tons and it continues to increase by 2% each year [1]. The residues from cotton production are of two types: cotton plant trash (CPT) and cotton gin trash (CGT) [2]. CPT is the residue that stays in the field after the harvest of cotton; while CGT is the residue coming from the ginning process. CGT is composed of pieces of sticks, leaves, bolls,

* Correspondence: plac324@tamu.edu
Department of Biological and Agricultural Engineering, Texas A&M University, Room 201 Scoates Hall, TAMU 2117, College Station, Texas 77841, USA

and soil cleaned from lint during ginning. In fact, 218 kg of cotton generates 68 to 91 kg of CGT [1]. Annually, the production of this waste in the USA is around 2.26 million tons [3].

The United States regulation requires by 2020 the production of 36 billion gallons of biofuels and from these, 21 billion should be produced from lignocellulosic materials or other new advanced fuels [4]. Agro-industrial waste (i.e., CGT) is one of the most significant sources of lignocellulosic materials, and bio-ethanol is one of the most essential bio-fuels produced from this kind of wastes. In general, bio-ethanol production from lignocellulosic material includes three principal steps: 1) pretreatment, 2) saccharification, and 3) fermentation. To produce bioethanol from agro-industrial feedstocks, different kinds of pretreatments have been investigated. These are generally divided into physical, physicochemical, chemical, and biological [5]. In CGT, the pretreatment principally used is physicochemical (steam explosion) followed by chemical (acid or basic hydrolysis) and biological (fungal or enzymatic) pretreatment [6,7]. Current biomass pretreatment process utilizes energy intense methodologies (high pressures and temperatures) and harsh chemical compounds.

To overcome the issues related with the traditional pretreatment process, new pretreatment strategies have been evaluated and developed. One of these strategies is the combination of ultrasonication, liquid hot water, and ligninolytic enzymes [8]. The ultrasonication and liquid hot water modified the lignin and cellulose structure; meanwhile, the ligninolytic enzymes treatment realized a detoxification and delignification process. This combination generated an ethanol yield of 30% and cellulose conversion of 23% [8]; however, both results need to be increased. Cellulose conversion and ethanol yield can be improved by modifying the pretreatment's conditions. Several works proved the efficiency of basic hydrolysis to decrease the lignin content in biomass [9]. Additionally, in sweet sorghum bagasse, the combination of alkaline hydrolysis simultaneously with ultrasonication augmented the final ethanol yield and cellulose conversion [10]. The use of ultrasonication and alkali-hydrolysis has not been tested in CGT; thus, the synergic effect of these technologies may raise delignification, cellulose conversion, and ethanol yield from CGT. The effect of these pretreatments over the CGT structure can be determined using compositional analysis [11] and/or the biomass' Fourier transform infrared (FT-IR) spectrum. In this moment, FT-IR is principally applied to study qualitatively the modifications in the structure and is not utilized for quantitative analysis. However, FT-IR is a fastest technique compared with the traditional compositional examination and can be used as a tool to identify qualitatively modification in biomass structure after different pretreatments. As a complementary tool to FT-IR spectroscopy, multivariate statistical techniques have been employed to identify the modifications in the FT-IR spectra and evaluate the difference between the different treatments. One of these multivariate methods is the principal component analysis (PCA); this technique reduces the dimensionality of the data by explaining the variance-covariance structure of a set of variables using few linear combinations of these variables. The use of PCA facilitates the visualization of the spectra changes and the identification of the most important features of the FT-IR spectra as the peak shifts and nonsymmetries [12]. The use of FT-IR coupled with PCA in pretreated and un-pretreated CGT has not been evaluated

in any other research. In the future, this type of methodology can be applied in quality or process control, and if it is coupled with regression techniques, it can be useful to make quantitative evaluations of the biomass composition.

This research evaluated the application of alkali ultrasonication pretreatment as a methodology to increase the ethanol yield produced from CGT using the combination of ultrasonication, liquid hot water, and ligninolytic enzyme. Additionally, FT-IR and PCA were utilized as a tool to analyze the modifications in the biomass structure after pretreatment methodologies.

Methods
Substrate
The samples of CGT were obtained from the Varisco-Court Gin CO near College Station, in Brazos Valley County, Texas. The CGT samples were ground in a Wiley mill (Philadelphia, PA, USA) to achieve an average particle size of approximately 1 mm in diameter.

Pretreatments
The experiment followed the selected sequence of pretreatments such as ultrasonication, liquid hot water, and ligninolytic enzymes [8]. However, the ultrasonication step was modified to simultaneously perform a basic hydrolysis using different concentrations of NaOH. Table 1 lists the experiments utilized in this paper. The experimental design was completely randomized with the NaOH concentration as factor with four levels (15%, 10%, 5%, and 0% w/v) and a control of un-pretreated CGT. All the experiments were developed in three replicates using cellulose conversion and ethanol yield as response variables. The ultrasonication process employed a solution of 10% solids of CGT biomass and the corresponding NaOH concentration for 1 h. The ultrasonicator (Hielscher Ultrasonic Processors, Ringwood, NJ, USA) was set at the highest value of amplitude (100%) and cycle (1). After the alkali ultrasonication, the solution was neutralized and centrifuged at 5,000 rpm for 5 min. The pelletized biomass was not washed before the hot water treatment, thus small quantities of NaOH are still present in the hot water treatment. The hot water pretreatment used Erlenmeyer flasks with 10% solution solids at 121°C, 15 psi for 1 h

Table 1 Pretreatments evaluated in the experiment

Pretreatments sequence	Abbreviation
Ultrasonication + liquid hot water + ligninolytic enzymes	(U+HW+E)
Ultrasonication/NaOH5% + liquid hot water + ligninolytic enzymes	(U-NaOH5%+HW+E)
Ultrasonication/NaOH10% + liquid hot water + ligninolytic enzymes	(U-NaOH10%+HW+E)
Ultrasonication/NaOH15% + liquid hot water + ligninolytic enzymes	(U-NaOH15%+HW+E)

in an autoclave. The ligninolytic enzymes pretreatment consists of the commercial laccase mediator system Prima-Green® EcoFade LT100 from GENENCOR International Inc. (Palo Alto, CA, USA). The enzymatic reactions were performed in 250-ml Erlenmeyer flasks with 50 ml of phosphate buffer 25 mM pH 6, an initial enzyme load of 3 g, with 10% solids of CGT at 30°C, 150 rpm for 96 h (Innova, New Brunswick Scientific, NJ, USA).

Saccharification process

The saccharification process employed the combination of two types of commercial cellulases: Accellerase 1500 and Accellerase XY (GENENCOR, Palo Alto, CA, USA). The experiment had an initial enzyme loading of 0.3 ml/g of Accellerase 1,500 + 0.1 ml/g of Accellerase XY. The process utilized three replicates in 250-ml Erlenmeyer flasks with 50 ml of a solution of 50-mM sodium acetate buffer at pH 4.8 for 96 h at 50°C and 125 rpm in an incubator/shaker (Innova, New Brunswick Scientific, NJ, USA). Cellulose conversion was calculated by using the next equation: % glucose conversion = $[(c \times V)/m] \times 100$ %, where c is the concentration (g/L) of sugars in the sample hydrolyzed, as determined by high performance liquid chromatography (HPLC), V is the total volume (L) hydrolyzed, and m is the initial weight (g) of glucose or xylose determined through the National Renewable Energy Laboratory (NREL) protocols. The statistical tests were performed in the software SAS system 9.3.

Fermentation process

Ethanol Red (*Saccharomyces cerevisiae*) provided by Fermentis (Lesaffre Yeast Corp., Milwaukee, WI, USA) was employed for the fermentation process. The activation of the strain was in 0.5 g of dry yeast in 10 mL of the inoculum broth. The composition of the inoculum broth had 0.2 g glucose, 0.05 g peptone, 0.03 g yeast extracts, 0.01 g KH_2PO_4, and 0.005 g $MgSO_4 \cdot 7H_2O$. The inoculums were shaken at 200 rpm in an incubator shaker at 38°C for 25 to 30 min. The fermentation process was performed in 125-ml Erlenmeyer flasks with 50 ml of the slurry supplemented with 0.3 g of yeast extract. The slurry was then incubated with 1 ml of freshly activated dry yeast (Ethanol Red) and run for a period of 72 h at 32°C, pH 4, and 100 rpm. The ethanol yield was calculated from the ratio between the average produced ethanol and the theoretical ethanol production of 51.1 g of ethanol generated per 100 g of glucose in the biochemical conversion of the sugar. The response variable was the ethanol yield, and it was analyzed using the software SAS system 9.3 employing one-way ANOVA and the LSD test.

High-performance liquid chromatography

After each of the processes (saccharification and fermentation), the samples were centrifuged at 10,000 rpm for 10 min, and the supernatants were filtered through 0.45-μm hydrophilic PTFE syringe filters (Millipore, Billerica, MA, USA). These samples were then analyzed for glucose, mannose, xylose, arabinose, galactose, and cellobiose concentration using HPLC (Waters 2690, Separations Module, Waters Corporation, Milford, MA, USA) equipped with an auto-sampler, Shodex SP 810 (Shodex, New York, NY, USA) packed column, and a refractive index (RI) detector. Each sample ran for 25 min at a flow rate of 1 ml/min, 60°C using HPLC water as mobile phase.

Compositional analysis

To determine the composition of the CGT biomass before and after the pretreatments, the analytical protocols developed at the NREL of the US Department of Energy were followed. This entailed the determination of (a) total solids in biomass and total dissolved solids in liquid process samples, (b) extractives in biomass, and (c) structural carbohydrates and lignin in biomass [11]; the past protocols were developed using dried biomass. FT-IR spectroscopy (Shimadzu, IR Affinity-1 with a MIRacle universal sampling accessory; Kyoto Prefecture, Japan) was used to evaluate the properties of the CGT with and without pretreatments. The infrared spectra collected range was 4,000 to 700 cm^{-1} with a resolution of 4 cm^{-1}. The compounds were analyzed, and their wavenumbers are given in Table 2. The spectra examination was developed using PCA in the range between 800 and 1,800 cm^{-1}. The PROC PRINCOMP statement of the SAS system 9.3 was employed for the PCA calculations using the correlation matrix of the data.

Results and discussion
Cellulose conversion

The effect of alkali ultrasonication employed two response variables, cellulose conversion and ethanol yield. Figure 1 shows the bar plot of the cellulose conversion for each pretreatment. The experiments with pretreatments evidence

Table 2 Wavenumbers of IR vibration frequencies used for CGT characterization

Compound	Functional group	Wave number (cm^{-1})
Cellulose	Beta-D-cellulose	898
Cellulose	Intense polysaccharide	1,030; 1,050
Cellulose	C-O-C anti-symmetric stretch (b-1,4 glycosyl)	1,170 to 1,150
Cellulose	Cellulose II and amorphous cellulose strong broad band	1,090
Lignin	Phenolic OH region and aliphatic CH stretch	1,370
Lignin	Aromatic skeletal vibration and CH deformation	1,514; 1,595
Hemicellulose	Ester carbonyls, C = O	1,240; 1,732

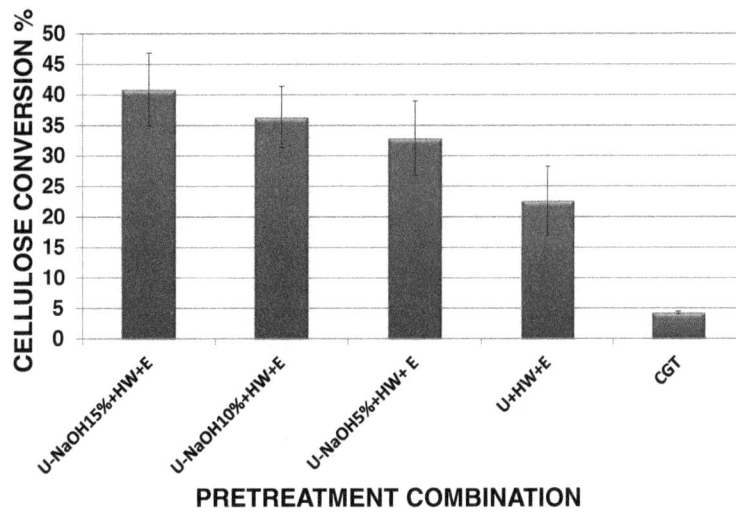

Figure 1 Bar plot of pretreatment combination vs. cellulose conversion.

an increment between 16% and 35% in the cellulose conversion over the un-pretreated biomass. Additionally, the treatments with alkali ultrasonication revealed a cellulose conversion greater than the treatment with only ultrasonication; this increment fluctuated between 11% and 18%. The statistical analysis (Table 3) indicates that statistically the alkali hydrolysis pretreatments are not different from each other. However, the treatments U-NaOH 15% + HW + E and U-NaOH 10% + HW + E were statistically different compared with the U + HW + E. The 5% treatment did not get significant differences against the U + HW + E. It indicates that 5% treatment increased the cellulose conversion similar to U-NaOH 15% + HW + E, but the increment is not enough to be different than the pretreatment without alkali hydrolysis. The U-NaOH 15% + HW + E's cellulose conversion (40%) of CGT was larger than the microbial pretreatment (18%) [13] and the sulfuric acid pretreatment [6] both over cotton stalks. Additionally, the cellulose conversion in this research is comparable with the achieved by steam explosion with a severity factor of 2 (42%) in CGT [3]. Nevertheless, the U-NaOH 15% + HW + E conversion is lower than the results accomplished in CGT with a severity factor of 4.68 (66.88%) [3] and the steam explosion of CGT and

recycled paper sludge (73.8%) [7]. The NaOH hydrolysis has been evaluated in other cotton wastes as textile wastes [14] and cotton stalks [6]. In both wastes, the use of NaOH improved the cellulose conversion, similar to the results displayed in this study.

The cellulose conversion using alkali ultrasonication on CGT (40%) was larger than the conversion found by Silverstein et al. [6] in cotton stalks (21%) but it was lower than Kaur et al. [15] over cotton stalks (63%) and Jeihanipour and Taherzadeh [14] in textiles wastes (99%). The difference between CGT and textile wastes is associated with their composition. Textile wastes comprise small lignin content which allows an easy access to the cellulose. Alkali hydrolysis has been utilized in several types of biomass (sugarcane bagasse, sweet sorghum bagasse, corn stover, etc.) to reduce the lignin content [16]. This pretreatment is normally related with the removal of lignin because NaOH breaks the ester bonds cross-linkage in lignin and xylan [9]. This breaking augments the biomass porosity allowing an easier access of the enzymes to the cellulose. However, alkali hydrolysis normally does not produce considerable modification in the cellulose structure, and this type of modification is necessary to increase the cellulose conversion [6]. The

Table 3 LSD's test for the cellulose conversion and ethanol yield

Pretreatment	LSD's statistic[a]	Cellulose conversion	Pretreatment	LSD's statistic[1]	Ethanol yield
U-NaOH 15% + HW + E	A	0.409 ± 0.059	U-NaOH 15% + HW + E	A	0.639 ± 0.139
U-NaOH 10% + HW + E	A	0.363 ± 0.050	U-NaOH 10% + HW + E	AB	0.585 ± 0.0563
U-NaOH 5% + HW + E	AB	0.329 ± 0.061	U-NaOH 5% + HW + E	B	0.514 ± 0.077
U + HW + E	B	0.226 ± 0.056	U + HW + E	C	0.28867 ± 0.086
CGT	C	0.042 ± 0.002	CGT	D	0.08000 ± 0.104

[a]Means with the same letter are not significantly different from each other.

combination of alkaline pretreatment and ultrasonication have been evaluated in sugarcane bagasse and rice straw. In sugarcane, alkali ultrasonication boosted the cellulose conversion approximately 50% against the un-pretreated biomass and 40% versus the pretreatment without ultrasonication [17]. The difference between the pretreated and un-pretreated biomass coincides with the results of this research (Figure 1). In rice straw, the alkali ultrasonication exhibited cellulose conversion greater than the un-pretreated biomass and the alkali pretreatment; nevertheless, the difference between the alkali pretreatment and the alkali ultrasonication was small [18]. The cellulose conversion can be enhanced by different strategies such optimizing the severity of the liquid hot water pretreatment or improving the conditions of the ligninolytic enzymes and ultrasonication pretreatment (reaction time, temperatures, etc.).

Ethanol yield

The second variable analyzed for the pretreatments was the ethanol yield (Figure 2). This variable evidenced a clear difference (55% to 20%) between the pretreated biomass and the un-pretreated biomass. The ethanol yield produced by the alkali ultrasonication treatments increased compared with the U + HW + E treatment. It indicates a beneficial effect in the use of alkali hydrolysis with the ultrasonication pretreatment for the ethanol production. The greatest ethanol yield was obtained in the U-NaOH 15% + HW + E and U-NaOH 10% + HW + E pretreatments with 63% and 58%, respectively. The statistical analysis of the experiment (Table 3) describes that U-NaOH 15% + HW + E and U-NaOH 10% + HW + E were not statistically different. Meanwhile, all the alkaline hydrolysis treatments were different other than U + HW + E.

The U-NaOH 15% + HW + E's ethanol yield was higher than the yield reported by Shen and Agblevor [7] on a mix of CGT and recycled paper sludge (40%). Additionally, our results are comparable with the ones described for CGT using steam explosion and a severity of 3.47 (58.1%) [3]. However, the ethanol yield of this research exhibited low values compared with the yields described by Jeoh and Agblevor [3] and Agblevor et al. [19] in CGT with conversions around 78% to 95% [19]. The increment in the ethanol yield using NaOH and ultrasonication has been found in other feedstock such as hazelnut husks [10], cotton stalk [15], and sugar cane bagasse [10]. The ethanol yields produced in sugar cane bagasse (81%) and hazelnut husks (76.7%) were higher than the U-NaOH 15% + HW + E pretreatment (63%); meanwhile, the alkali-ultrasonicated cotton stalk had an ethanol yield (41%) [15] that is lower than the pretreatment combination selected in this research.

Compositional analysis

Table 4 illustrates the composition of the biomass at the end of the pretreatments. The principal differences were in the glucose and acid insoluble material. The use of alkali ultrasonication produced a decrease in the lignin content, and this was related with the concentration of NaOH. The highest reduction occurred in the treatment with 15% of NaOH; meanwhile, the 10% and 5% treatments exhibited similar values. In studies with sweet sorghum, the alkali ultrasonication reduced the lignin content 6% to 10% [10], which has a similar level as that found in CGT. The glucose percentage in alkali ultrasonication treatments increased in comparison with the untreated CGT and the U + HW + E treatment. Similar to the lignin percentage, the greatest change was in the U-NaOH 15% + HW + E treatment. In this case, the

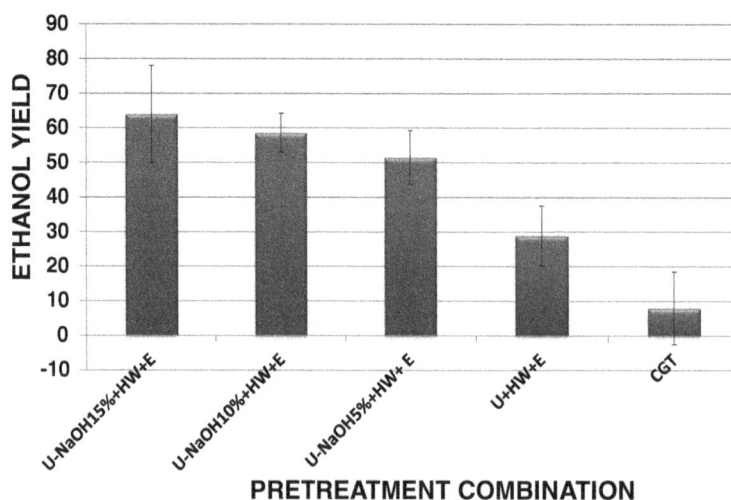

Figure 2 Bar plot of pretreatment combination vs. ethanol yield.

Table 4 Structural composition of untreated and pretreated cotton gin trash (CGT)

Compound	CGT	U + HW + E	U-NaOH 5% + HW + E	U-NaOH 10% + HW + E	U-NaOH 15% + HW + E
Water and ethanol extractives	19.6	15.3	14.9	14.3	14.5
Acid insoluble material	25.5	22.9	20.5	19.5	18.7
Arabinose	1.5	1.2	1.3	1.4	1.9
Xylose	5.7	5.4	5.3	5.1	5.2
Mannose	1.1	1.3	1.1	1.3	1.3
Galactose	1.7	1.6	1.3	1.4	1.2
Glucose	24.9	29.8	31.3	33.1	35.5
Ash	10.7	10.3	11.7	12.1	12.5

increment was around 10% compared with the original biomass and 6% versus the U + HW + E treatment. In this variable, the treatment with 10% showed a glucan content greater than the 5% treatment. The other sugars exhibited slightly diminutions compared with the untreated CGT and U + HW + E treatment. This type of small reductions in the hemicellulose components have been reported in barley straw [20] and sorghum [10]. The use of large concentrations of NaOH has been reported as a methodology to solubilize the hemicellulose components; however, this methodology has reaction times (24 h) greater than the alkali ultrasonication pretreatment (1 h). The short reaction time in the alkali ultrasonication avoided a larger hemicellulose solubilization.

FT-IR was performed in this study to monitor the composition changes in the CGT biomass according with the pretreatments employed. The analyzed peaks (Table 2) and new strong signals were followed in all the samples. Figure 3 shows the FT-IR spectra of the pretreatment sequence of U-NaOH 15% + HW + E. The addition of NaOH produced an increment in the signals at 800 to 900 cm^{-1} and 1,300 to 1,500 cm^{-1}; the

outstanding peak was the signal at 1,431 cm^{-1} followed by the 1,402 and 1,327 cm^{-1}. These peaks are related with the presence of NaOH because they augmented depending of the NaOH concentration and were only observed in the pretreatments with alkali hydrolysis. These signals reduced through the pretreatments; in fact, after the laccase pretreatment and the enzymatic hydrolysis, these peaks were not seen. The reduction in the NaOH peaks indicates that this compound did not have considerable concentrations that could affect the saccharification and fermentation processes.

The absorbance in the cellulose's peaks (898, 1,030, 1,050, 1,090, and 1,170 to 1,150 cm^{-1}) increased at the end of the four experiments. The highest increment in all the pretreatments was the 1,090 cm^{-1} peak followed by the signals at 898, 1,030, and 1,050 cm^{-1}. U-NaOH 15% + HW + E (Figure 3) obtained the highest increment among the treatments in all the cellulose picks followed by U-NaOH 10% + HW + E pretreatment. The absorbance in the cellulose's signal was increasing according to the pretreatments that were added, showing a synergic effect of the pretreatments over the cellulose

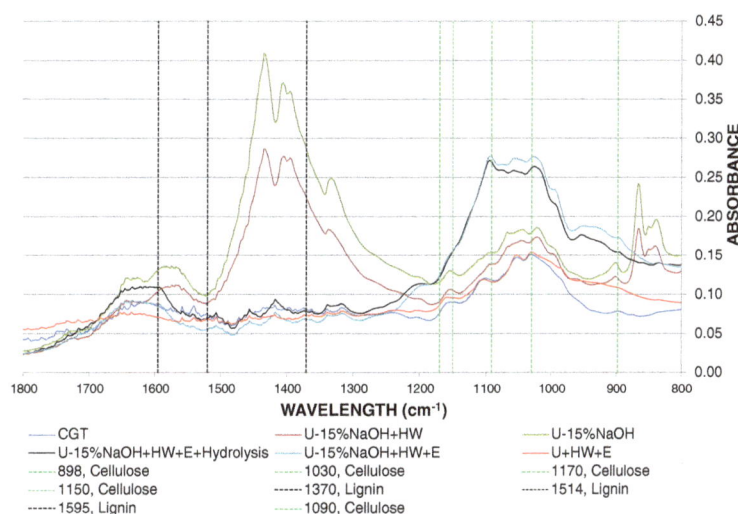

Figure 3 FT-IR spectra for the different steps in the U-NaOH15%+HW+E pretreatment.

structure. In addition to the cellulose peak analysis, the cellulose total crystallinity index (TCI) was utilized to evaluate deeper the cellulose structure. The TCI has been used to express the relative amount of crystalline material in cellulose, and it can be defined using the FT-IR using the absorbance at A1430/A898 [10]. At the end of the four trials, the TCI decreased in 46%, 63%, 66%, and 67% for the pretreatments U + HW + E, U-NaOH 5% + HW + E, U-NaOH 10% + HW + E, and U-NaOH 15% + HW + E, respectively. The reduction in the TCI coincides with the final results observed in the ethanol yield (Figure 2). The decrease in the TCI after alkaline ultrasonication pretreatment was also noticed in the work of Goshadrou et al. [10] over sweet sorghum bagasse, but in sorghum the shrinkage was lower (13%) than CGT 67%.

In the fully pretreated biomass, the absorbance in the lignin signals (1,370, 1,514, and 1,595 cm^{-1}) reduced compared with the un-pretreated biomass. The hemicellulose signals (1,240 and 1,732 cm^{-1}) obtained the greatest variation among the four experiments evaluated. In these peaks, U + HW + E and U-NaOH 15% + HW incremented the values; meanwhile, U-NaOH 10% + HW + E and U-NaOH 5% + HW reduced the absorbances. These differences can be attributed to the CGT composition and the pretreatment interactions. The increment in the cellulose peak, the diminution in the crystallinity index, and lignin content are some of the reasons why this pretreatment combination produced the greatest cellulose yield and ethanol conversion. The peaks of 898 and 1,030 cm^{-1} displayed a considerable drop compared with the pretreated biomass; meanwhile, the other peaks did not show any significant change. These differences can be used to follow the hydrolysis reaction of CGT biomass using the FT-IR.

Principal component analysis of the FT-IR spectrum

This is the first study that use FT-IR spectrum and principal components to analyze the changes produced in the CGT biomass after different pretreatments for ethanol production. The PCA correlated the changes in the absorbance in the FT-IR spectrum and discriminated or grouped the variations in the biomass structure after the different pretreatments. The PCA variables were the spectrum wave numbers in the range 800 to 1,800 cm^{-1}. The pretreatments (U, HW, and E) and the saccharification process were the variables applied for the grouping. The PCA used 626 observations, 525 variables, and the covariance matrix. The variance explained by the PCA using the two initial principal components (PRIN1, PRIN2) was 91%, 76% from PRIN1 and 15% from PRIN2. PRIN1 and PRIN2 loadings plots (Figure 4) identify which wave numbers were the most important in the variability explained for each principal component [12,21]. PRIN1 did not display negative loadings for any of

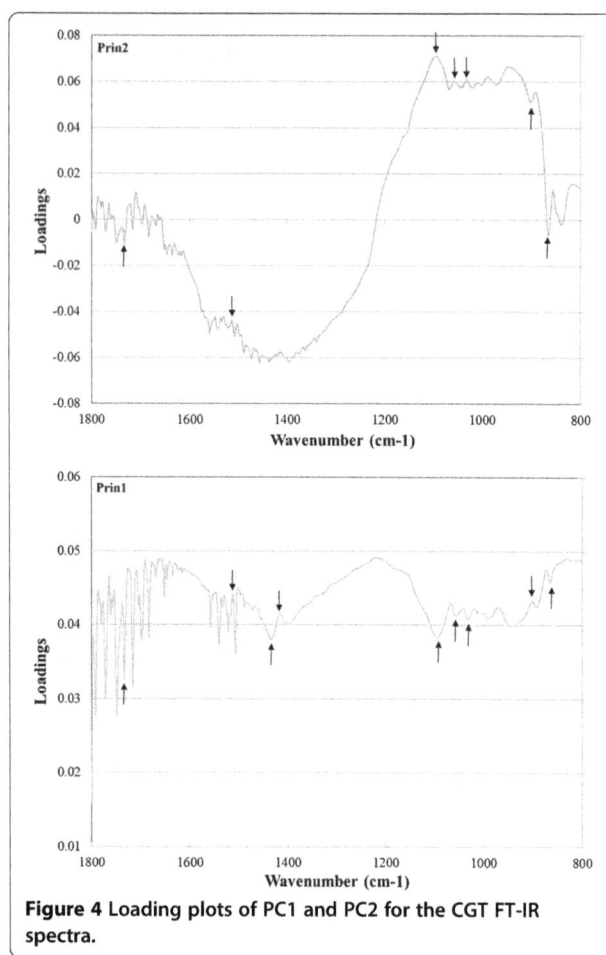

Figure 4 Loading plots of PC1 and PC2 for the CGT FT-IR spectra.

the wave numbers; meanwhile, PRIN2 has negative and positive loadings. In both cases, different minimum and maximum points were possible to identify; the most notable point was at 1,090 cm^{-1} signal. This signal exhibited high peaks in both principal components, in PRIN1 the signal was the lowest point between 800 and 1,200 cm^{-1}; meanwhile, in PRIN2, it was the highest point in the complete plot. This wave number has been connected with different types of cellulose; this is an indication of the effects of the treatments over the CGT cellulose. Other significant signal was the 864 cm^{-1}; this signal was a minimum in both principal components and corresponds with one of the NaOH signals found in the FT-IR spectrum (Figure 3). Other signals linked with the NaOH were detected in the PRIN1 loading plot; these signals correspond to 1,431 and 1,416 cm^{-1}, the first one is a minimum peak, while the second one is a maximum point. These peaks are clearly observed in the FT-IR spectra of the pretreated biomass with alkali ultrasonication and liquid hot water (Figure 3).

The cellulose signals found in the loading plots were the signals at 898 (B-D-cellulose), 1,030, and 1,050 (intense polysaccharide), and 1,090 cm^{-1} (cellulose II and amorphous cellulose) [10,22]. These points were perceived in PRIN1 and PRIN2 loading plots; in PRIN1, 898 and 1,090 cm^{-1} were minimums, while 1,030 and 1,050 cm^{-1} were maximums, which is an opposite behavior compared with PRIN2. These signals showed large values in the PRIN2 than PRIN1, which represents the importance of PRIN2 for the cellulose signals explanation. The most significant lignin signal in the loading plots was the signal at 1,514 cm^{-1} which is related with the aromatic skeletal vibration and CH deformation. In the same way, the most influential hemicellulose signal in the loading plot was the 1,732 cm^{-1}, this signal was a minimum in both cases and relates the modifications in the esters found in the hemicellulose.

Using the scores plot of the pretreatments (Figure 5), the PCA could group the pretreatments in four clusters, each one associated with the three pretreatments and the saccharified biomass. The groups that displayed the highest separation were the complete pretreated biomass and the saccharified biomass. The clusters were clearer in the treatments U-NaOH 15% + HW + E and U-NaOH 10% + HW + E. In these cases, the pretreatments were located sequentially through the PRIN2 axis, this sequence was the same as the experimental order (ultrasonication, hot water, and enzyme). The saccharified biomass spectrum was placed between the HW pretreatment and the enzyme pretreatment. This behavior was also observed in the full FT-IR spectrum where the line of the saccharified biomass descends compared with the U-NaOH 15% + HW + E. This reduction is detected in the wave numbers that correspond to the cellulose signals (Figure 3). The scores plot for the 10% concentration has the saccharification and the full pretreatment clusters closer than the 15% concentration plot. In the pretreatment U-NaOH 5% + HW + E, the ultrasonication and liquid hot water pretreatment groups had similar scores, and the separation between them was not clear (Figure 5). Meanwhile, the separation between the other two clusters improved versus the 10% plot. The scores plot of the U + HW + E discriminates clearly the sonication pretreatment; nevertheless, the other three groups were not evidently distinguished. In the future, the use of PCA in the CGT's FT-IR spectra can be implemented in the quality control of the biomass for bioethanol production and to predict the behavior of the CGT after different kinds of pretreatments [23].

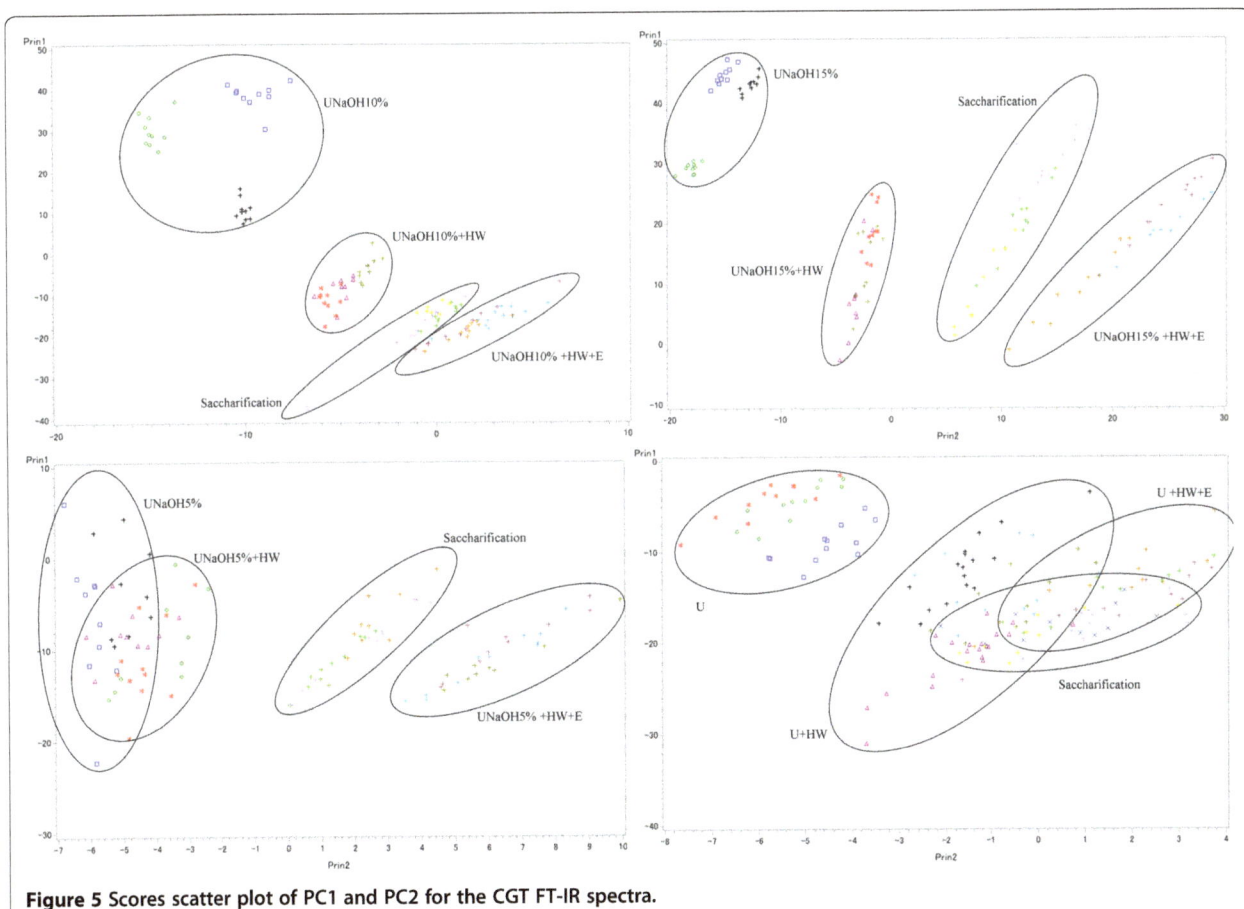

Figure 5 Scores scatter plot of PC1 and PC2 for the CGT FT-IR spectra.

Conclusions

In this research, the addition of alkali ultrasonication pretreatment to liquid hot water and ligninolytic enzyme pretreatments increased the cellulose conversion in 11% to 18% and the ethanol yield in 23% to 35% versus the treatment without alkali ultrasonication. From these pretreatments, the U-NaOH 15% + HW + E pretreatment exhibited the highest cellulose conversion (41%) and ethanol yield (64%). The use of FT-IR and principal components was effective as a tool to identify the variations in the signal of the cellulose, hemicellulose, and lignin from CGT after the different pretreatments. Additionally, the PCA could separate and identify the CGT biomass from different types of pretreatments and identify the signals with the most significant variation inside the spectra. In the future, this type of discrimination technique can be used in the bioethanol industry for quality control and prediction analysis.

Competing interests

The authors declare that they have no competing interest.

Authors' contributions

JP was responsible of experiment's conception and design, acquisition of data, analysis and interpretation of data, and manuscript drafting. SC was responsible of the experiment's conception and design and final manuscript approval. Both authors read and approved the final manuscript.

Acknowledgements

The Colombian government and the Fulbright Association are acknowledged for the financial support via the granting of FULBRIGHT-COLCIENCIAS for the PHD studies of Jersson Plácido in the USA. Dr. Raghupathy Karthikeyan and the Water Quality Engineering laboratory are acknowledged for their technical and methodological support as well as the BioEnergy Testing and Analysis Laboratory at the Biological and Agricultural Engineering Department and the Texas AgriLife Research of Texas A&M University for funding this research. GENENCOR International is acknowledged for providing the enzymes employed in the study.

References

1. Sharma-Shivappa RR, Chen Y (2008) Conversion of cotton wastes to bioenergy and value-added products. T ASABE 51:2239–2246
2. Rogers GM, Poore MH, Paschal JC (2002) Feeding cotton products to cattle. Vet Clin North Am Food Anim Pract 18:267–294
3. Jeoh T, Agblevor FA (2001) Characterization and fermentation of steam exploded cotton gin waste. Biomass Bioenergy 21:109–120
4. Sissine F (2007) Energy Independence and Security Act of 2007: a summary of major provisions 2011 https://wiki.umn.edu/pub/ESPM3241W/S09TopicSummaryTeamFourteen/CRS_Report_for_Congress.pdf.
5. Sarkar N, Ghosh SK, Bannerjee S, Aikat K (2012) Bioethanol production from agricultural wastes: an overview. Renew Energy 37:19–27
6. Silverstein RA, Chen Y, Sharma-Shivappa RR, Boyette MD, Osborne J (2007) A comparison of chemical pretreatment methods for improving saccharification of cotton stalks. Bioresour Technol 98:3000–3011
7. Shen J, Agblevor FA (2008) Optimization of enzyme loading and hydrolytic time in the hydrolysis of mixtures of cotton gin waste and recycled paper sludge for the maximum profit rate. Biochem Eng J 41:241–250
8. Plácido J, Imam T, Capareda S (2013) Evaluation of ligninolytic enzymes, ultrasonication and liquid hot water as pretreatments for bioethanol production from cotton gin trash. Bioresour Technol 139:203–208
9. Chaudhary G, Singh LK, Ghosh S (2012) Alkaline pretreatment methods followed by acid hydrolysis of *Saccharum spontaneum* for bioethanol production. Bioresour Technol 124:111–118
10. Goshadrou A, Karimi K, Taherzadeh MJ (2011) Bioethanol production from sweet sorghum bagasse by *Mucor hiemalis*. Ind Crop Prod 34:1219–1225
11. Sluiter A, Hames B, Ruiz R, Scarlata C, Sluiter J, Templeton D, Crocker D (2011) Determination of structural carbohydrates and lignin in biomass. Technical report National Renewable Energy Laboratory/TP-510-42618:2011:15 http://www.nrel.gov/biomass/pdfs/42618.pdf.
12. Popescu M-C, Simionescu BC (2012) Multivariate statistical analysis of mid-infrared spectra for a G1 allyl-terminated carbosilane dendrimer. Spectrochim Acta A Mol Biomol Spectrosc 92:398–405
13. Shi J, Chinn MS, Sharma-Shivappa RR (2008) Microbial pretreatment of cotton stalks by solid state cultivation of *Phanerochaete chrysosporium*. Bioresour Technol 99:6556–6564
14. Jeihanipour A, Taherzadeh MJ (2009) Ethanol production from cotton-based waste textiles. Bioresour Technol 100:1007–1010
15. Kaur U, Oberoi HS, Bhargav VK, Sharma-Shivappa R, Dhaliwal SS (2012) Ethanol production from alkali- and ozone-treated cotton stalks using thermotolerant *Pichia kudriavzevii* HOP-1. Ind Crop Prod 37:219–226
16. Chen M, Zhao J, Xia L (2009) Comparison of four different chemical pretreatments of corn stover for enhancing enzymatic digestibility. Biomass Bioenergy 33:1381–1385
17. Velmurugan R, Muthukumar K (2012) Sono-assisted enzymatic saccharification of sugarcane bagasse for bioethanol production. Biochem Eng J 63:1–9
18. Kim I, Han J-I (2012) Optimization of alkaline pretreatment conditions for enhancing glucose yield of rice straw by response surface methodology. Biomass Bioenergy 46:210–217
19. Agblevor F, Batz S, Trumbo J (2003) Composition and ethanol production potential of cotton gin residues. Appl Biochem Biotechnol 105:219–230
20. Duque A, Manzanares P, Ballesteros I, Negro MJ, Oliva JM, Saez F, Ballesteros M (2014) Study of process configuration and catalyst concentration in integrated alkaline extrusion of barley straw for bioethanol production. Fuel 134:448–454
21. Monti F, Dell'Anna R, Sanson A, Fasoli M, Pezzotti M, Zenoni S (2013) A multivariate statistical analysis approach to highlight molecular processes in plant cell walls through ATR FT-IR microspectroscopy: the role of the α-expansion PhEXPA1 in *Petunia hybrida*. Vib Spectrosc 65:36–43
22. Krasznai DJ, Champagne P, Cunningham MF (2012) Quantitative characterization of lignocellulosic biomass using surrogate mixtures and multivariate techniques. Bioresour Technol 110:652–661
23. Ferreira D, Barros A, Coimbra MA, Delgadillo I (2001) Use of FT-IR spectroscopy to follow the effect of processing in cell wall polysaccharide extracts of a sun-dried pear. Carbohydr Polym 45:175–182

10

Purification of DP 6 to 8 chitooligosaccharides by nanofiltration from the prepared chitooligosaccharides syrup

Huizhong Dong[1,2], Yaosong Wang[1,2], Liming Zhao[1,2*], Jiachun Zhou[1,2], Quanming Xia[1,2], Lihua Jiang[1,2] and Liqiang Fan[1,2]

Abstract

Background: Chitooligosaccharides (COS) with degrees of polymerization (DP) 6 to 8 are degraded from chitosan, which possess excellent bioactivities. However, technologies that could purify them from hydrolysis mixtures in the narrow DP range (984 to 1,306 Da) are absent. The objective of this research is to purify DP 6 to 8 COS by nanofiltration on the basis of appropriate adjustments of the feed condition.

Methods: Syrup containing DP 6 to 8 COS at different concentrations (19.0 to 46.7 g/L) was prepared. A commercial membrane (QY-5-NF-1812) negatively charged was applied. Experiments were carried out in full recycle mode, so that the observed COS retentions were investigated at various transmembrane pressures (6.0 to 20.0 bar), temperatures (10°C to 50°C), and pHs (5.0 to 9.0). Then, the feasibility of separation of DP 6 to 8 COS was further studied by concentration ratio under optimum conditions.

Results: The results indicate that the purification of DP 6 to 8 COS by nanofiltration NF is feasible. It was found that the permeate flux was 95.0 L/(m^2 h) at 10.0 bar, while it reached to 140.0 L/(m^2 h) at 20.0 bar, and it increased with feed temperature, but the membrane pores were also swelled by heating and led to an irreversible wastage of target oligomers. Additionally, the retention behaviors of chitooligosaccharides are significantly influenced by pH.

Conclusions: Although glucosamine and dimer were permeatable at low pH, their retention ratios were remarkably varied from 0.458 to 0.864 when pH was 9.0. With the interaction of hydrogen bonds, structural curling and overlapping of chitooligosaccharides were formed. Consequently, the rejection of chitooligosaccharides at various pHs is variable. Spray-dried products were finally characterized by the matrix-assisted laser desorption/ionization time-of-flight mass spectrum. The spectrum identified the distributions of hexamer, heptamer, and octamer. Combined with high-performance liquid chromatography profiles, the purity and yield of DP 6 to 8 chitooligosaccharides were up to 82.2% and 73.9%, respectively.

Keywords: Nanofiltration; Chitooligosaccharides; Degree of polymerization; Separation; MALDI-TOF-MS

Background

Chitooligosaccharides (COS) are defined as the partially degraded products of chitosan or low-molecular-weight chitosan (LMWC) with a degree of polymerization (DP) ranges from 2 to 20 [1]. As shown in Figure 1, their molecular structures are linear oligosaccharides composed of 2-amino-2-deoxy-D-glucopyranose and 2-acetamino-2-deoxy-D-glucopyranose (Figure 1(A)) units which are linked by β(1 → 4) glycosidic bonds [2].

According to many previous reports, COS possess a series of attractive bioactive properties, including antibacterial [3], anticoagulant [4], antimicrobial [5], antioxidant [6], anticancer [7], hypolipidemic [8], and immune-stimulating [9] effects. Based on these excellent advantages mentioned before, COS are responsible for practical applications in beverage processing [10], functional ingredients [11], and biomedicines [12], which are different from chitosan and

* Correspondence: zhaoliming@ecust.edu.cn
[1]State Key Laboratory of Bioreactor Engineering, R&D Center of Separation and Extraction Technology in Fermentation Industry, East China University of Science and Technology, Shanghai 200237, China
[2]Shanghai Collaborative Innovation Center for Biomanufacturing Technology, Shanghai 200237, China

A

B

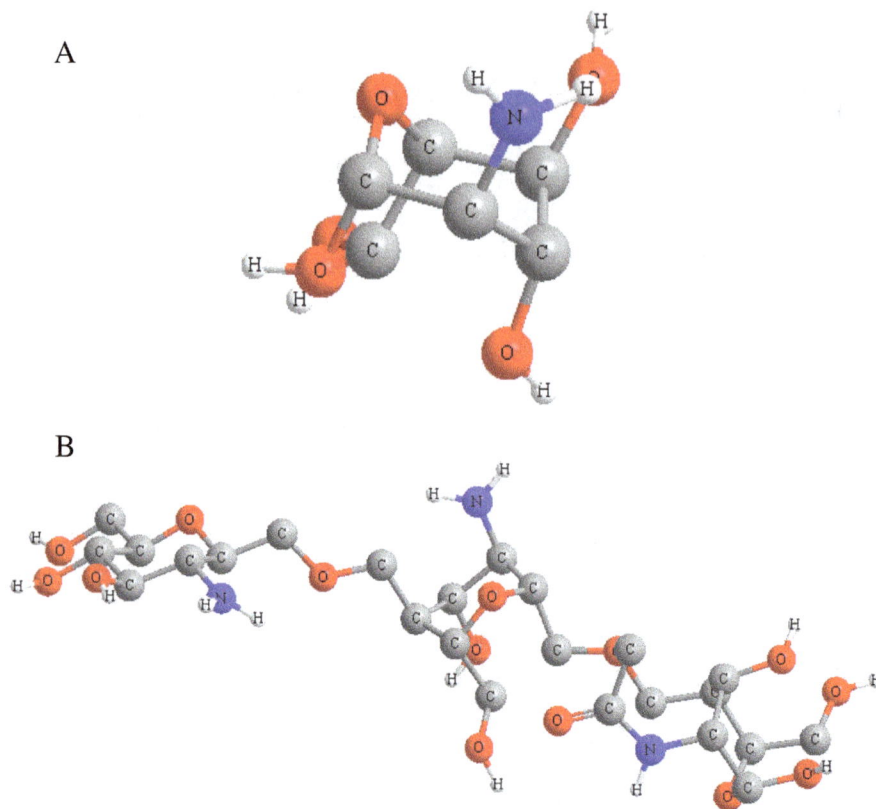

Figure 1 Molecular structures of glucosamine and DP 3 chitooligosaccharide. Structurally, chitooligosaccharides, including glucosamine and trimer, are linear oligosaccharides composed of 2-amino-2-deoxy-D-glucopyranose and 2-acetamino-2-deoxy-D-glucopyranose units which are linked by β(1 → 4) glycosidic bonds. **(A)** The molecular structure of glucosamine. **(B)** The molecular structure of DP 3 chitooligosaccharide.

chitin except for their contributions on food packaging [13]. Nowadays, hybrid enzymatic hydrolysis has become the ideal technology for COS preparation due to its high efficiency and little structural modification [14]. Unfortunately, the products after enzymatic degradation were just intermediate ones [15]. Suitable methods should be used to separate target COS from mixtures coexisting in the solution, such as the high molecular weight of chitosan, proteins (enzymes), and inorganic salts.

Conventionally, COS is purified by various chromatographies. Fan et al. succeeded in obtaining COS by macroporous resins from fermentation broth, and the productivity of target products could go up to 90% (w/w) under optimum conditions [16]. Meanwhile, Cabrera and Custem reported that the concentrations of COS could be analyzed by matrix-assisted laser desorption/ionization time-of-flight mass spectrum (MALDI-TOF-MS) [17]. Although the chromatography technology is feasible to remove irrelevant components and refine COS in terms of DP, it is only accepted in laboratory analysis but far from scale-ups. In practice, several negative problems inevitably appear during continuous using of chromatography, including the demands on cleaning and regeneration of column packing,

adsorption capacities of COS, and relatively high production cost [18,19].

Recently, there is a growing attention on nanofiltration (NF) applications, especially for COS preparation. Kim et al. reported that several instruments were used to prepare COS, the details of which could be summarized in immobilized enzymatic columns and ultrafiltration (UF) membrane reactors [2]. Han et al. demonstrated the desalination feasibility of NF-40 membrane for chitobiose solution. Under acidic conditions, the interception of solutes ranks as chitobiose > glucosamine > Na$^+$ > H$^+$ [20]. Furthermore, the steric and electrostatic effects are inferred according to the sequence. Han et al. also studied the influence of three membranes (DL, DK, NTR-7540) on COS separation and drew identical conclusions [21]. In a word, NF has been proved to be an effective technology for the separation and purification of COS.

Chitosan was originally hydrolyzed by enzymes, which resulted in the coexistence of various DP COS, glucosamines, and salts. It is evident that there is a great difference in molecular weight among the solutes. Also, the wider ranges of hydrolysis products make it difficult to improve the purity and yield of COS. For instance, DP 6 to 8 COS (984 to 1,306 Da) plays an important role in cancer curing [22]. Therefore, this study is to investigate the separation performance of DP 6 to 8 COS at different solution properties and present a promising purification technology by NF.

Methods

Materials

The raw syrup containing DP 6 to 8 COS was made from chitosan by enzymatic hydrolysis. Chitosan was supplied by Yunzhou Biochemistry Co., Ltd. (Shandong, China). The enzyme mixture (chitosanase from *Streptomyces griseus* - EC 3.2.1.132; cellulase from *Trichoderma* - EC 3.2.1.4) was obtained from Golden-Shell Biochemical Co. (Zhejiang, China). DP 2 to 8 COS and glucosamine standards were purchased from Huicheng Biochemical Co., Ltd. (Shanghai, China). All chemicals used in the NF operation and high-performance liquid chromatography (HPLC) analysis were analytical grade or chromatographic grade, respectively. Deionized water (conductivity <3.0 µs/cm) for membrane cleaning was produced by ion exchange.

Preparation of chitooligosaccharide syrup

Chitosan (91.5% degree of deacetylation) was dissolved in acetic acid (1%, *w/v*) with stirring, and pH was adjusted to 5.3, and then kept at 45°C. The chitosan concentration was 5% (*w/v*). Combined enzymes (75 U/g) based on chitosan were added into the chitosan solution and hydrolyzed for 6 h, and then, the hydrolysis was terminated by immersion in a boiling water bath. Finally, a UF membrane module

(QY-3-UF-1812, AMFOR Inc., Newport Beach, US) was applied to remove enzymes at 50°C. After cooling to ambient temperature, the syrup was diluted to various concentrations for NF separation.

Membrane

Purification experiments were operated in a pilot setup (Figure 2). The setup composes of a feed tank, a high-flux pump, a membrane vessel, and a flowmeter. Moreover, the inlet and outlet pressures were metered by two pressure gauges, and the feed temperature could be regulated by a circulating cooling water system surrounding the feed tank.

The membrane module employed was an organic polymer composite with spiral-wound structures (QY-5-NF-1812, AMFOR Inc., Newport Beach, US). The membrane module is measured as 1.8 in. (4.6 cm) in section diameter and 12 in. (30.5 cm) in length, which is so-called 1812 type module. Also, it is measured by an approximate molecular weight cut-off (MWCO) of 500 Da, which is close to the molecular weight of DP 4 COS. As described in Table 1, the membrane is negatively charged that can tolerate a maximum pressure up to 25.0 bar. The effective surface area is 0.2 m^2. The temperature and pH tolerance range cover from 0°C to 60°C and 4 to 12, respectively.

Membrane permeate flux

In this study, the permeate flux is represented as the average one in process. The average permeate flux (J_v) is calculated by Equation 1, as follows:

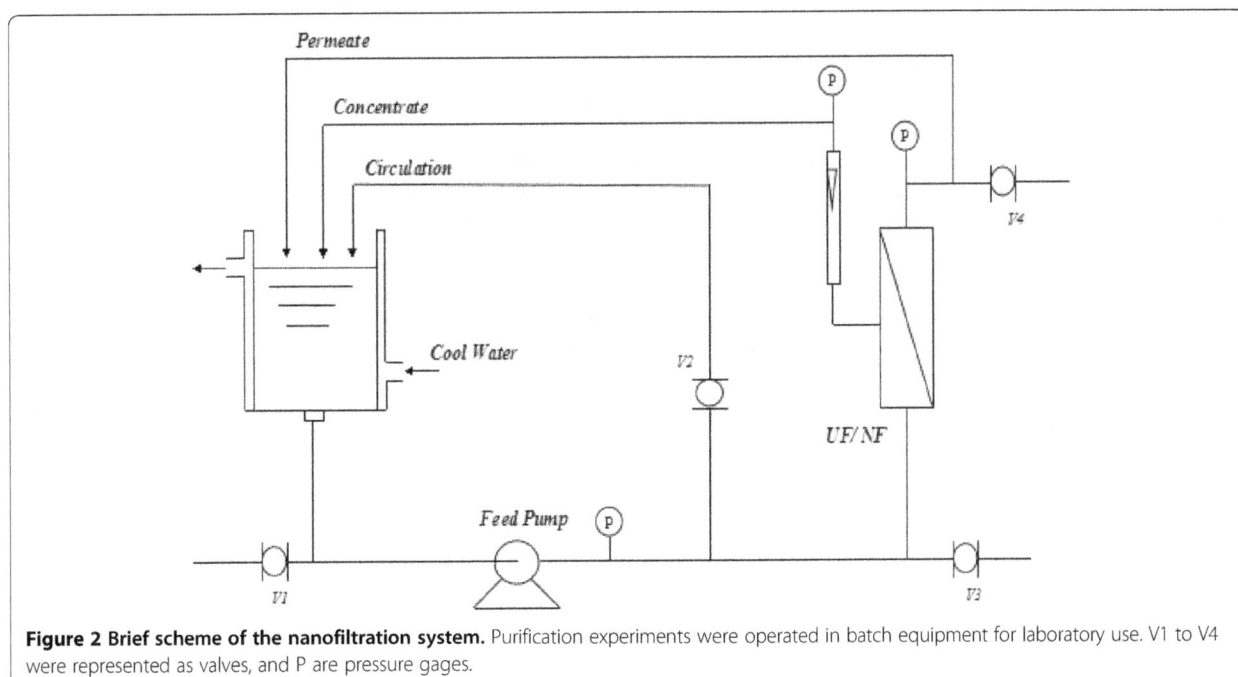

Figure 2 Brief scheme of the nanofiltration system. Purification experiments were operated in batch equipment for laboratory use. V1 to V4 were represented as valves, and P are pressure gages.

Table 1 Information of NF membrane module

Membrane index	Parameter
Type	1812
Texture	Organic polymers
Filtration area (A_m)	0.2 m²
MWCO	500 Da
Temperature	0°C to 60°C
pH	4 to 12
Pressure	<25.0 bar

$$J_v = \frac{V_p}{A_m \times t} \quad (1)$$

where J_v is the average permeate flux of the membrane [L/(m² h)]; V_p is the volume of permeate accumulated in testing time (L); A_m is the effective areas of membrane (m²); t is the testing time (h).

Observed retention ratio

Retention ratio is also one of the most important factors in membrane separation. The observed retention ratio is described by Equation 2:

$$R_{obs} = 1 - \frac{C_p}{C_b} \quad (2)$$

where R_{obs} is the observed retention ratio of solute (%); C_p is the concentration of solute in the permeate (mol/L); C_b is the concentration of solute in the feed (mol/L).

Full recycle mode of NF

Initially, 5.0 L raw syrup was added, and the recycling flow rate was adjusted to 360 L/h, which represented the optimum rotation condition for the working pump. The effects of transmembrane pressure (TMP), operation temperature, and pH on NF performance were successively carried out. All the permeate were flowed back to the feed tank, while the J_v and retention behaviors of DP 6 to 8 COS at different concentrations were measured after the renewed conditions were stable for 15 min. It is worth being noted that two of the parameters mentioned must be constant when the third is variable. During the

process, the temperature was controlled at 50°C ± 2°C by a heat exchanger surrounding the feed tank, except for the experiments to investigate the effects of temperature on the NF retention on DP 6 to 8 COS.

Purification of DP 6 to 8 chitooligosaccharides

Purification experiments for DP 6 to 8 COS were executed in batch mode. Under the optimized conditions obtained from the preliminary experiments, the concentrate stream was circulated back to the feed tank, whereas the permeate was collected individually. Considering the practical capacity of the feed tank, 7.0 L of diluted syrup (C_b = 19.0 g/L) was added firstly. Every 1.0 L of extra diluted syrup should be supplied as soon as the permeate was equally removed. Certainly, the systematic temperature during NF was maintained by cooling water. After adding all the syrup (16.0 L), the process was terminated until the volume of permeate reached 14.0 L (2.0 L syrup left in the tank). The effect of the concentration ratio on the purity of DP 6 to 8 COS was confirmed by flux and rejections.

HPLC analysis

The concentrations of glucosamine and DP 2 to 8 COS were analyzed by an HPLC system (Shimadzu 10A, Shimadzu, Kyoto, Japan) equipped with a high-performance sugar column (Shodex Asahipak NH2P-50 4E, Shodex, Kyoto, Japan) and an RI detector. The mobile phase consists of methyl cyanide and pure water with the ratio of 70:30 (v/v). The column temperature was maintained at 30°C, and the flow velocity was kept at 1.0 mL/min. All the solutes were measured in the form of single-arranged peaks. In general, the glucosamine was firstly eluted, and then, the dimer and trimer were sequentially characterized due to the adsorption strength difference to the stationary phase. The distributions of COS were quantified by integrating peak areas.

MALDI-TOF-MS analysis

MALDI-TOF-MS analysis of COS was carried out using Shimadzu AXIMA Performance matrix-assisted laser desorption/ionization time-of-flight mass spectrometry (Shimadzu, Kyoto, Japan). All spectra were measured in the reflector mode by external calibration. The laser was scanned at a scale from 500 to 1,500 Da. An aqueous

Table 2 The component analysis of raw materials for the NF process

Chitosan (%)	Total Proteins[a] (mg/L)	Glucosamine[a] (g/L)	DP 6 to 8 chitooligosaccharides[a] (g/L)
1.0	1.4 ± 0.3	0.4 ± 0.1	9.6 ± 1.2
2.0	2.7 ± 0.2	0.9 ± 0.1	19.0 ± 1.5
3.0	3.3 ± 0.4	1.4 ± 0.2	28.3 ± 1.7
4.0	2.9 ± 0.3	1.8 ± 0.2	37.6 ± 2.4
5.0	1.4 ± 0.3	2.3 ± 0.4	46.7 ± 2.1

[a]The data were determined by three parallel measurements.

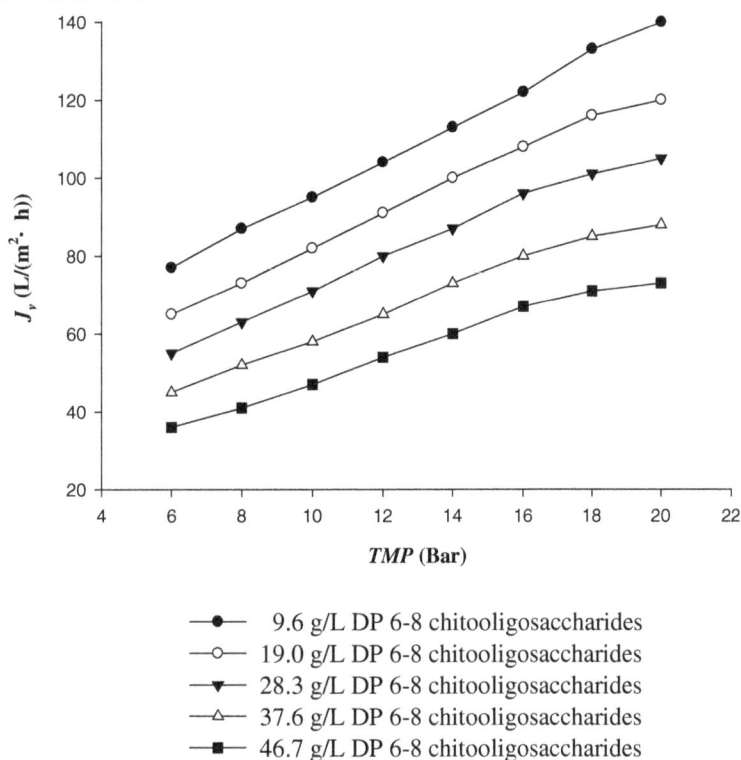

Figure 3 TMP vs. J_v of chitooligosaccharide syrup at 50°C during NF separation. J_v of chitooligosaccharide syrup with different $C_{chitosan}$ ranged by transmembrane pressure (6.0 to 20.0 bar). The permeate fluxes increased with the pressure, whereas they decreased with the increasing concentration of DP 6 to 8 chitooligosaccharides.

solution of 2,5-dihydroxybenzoic acid (DHB, 100 mg/mL) was used as the matrix.

Statistical methods

All parameters and experimental results were obtained with means ± SD of parallel tests. The data was analyzed by Statistix 9.0 software (Analytical Software, Tallahassee, FL, USA).

Results and discussion

Characterization of chitooligosaccharide syrup

As the raw material for NF process, the COS syrup was defined as the permeate of UF that was pretreated after enzymatic hydrolysis. Summarized by the previous tests, 10.0 L of chitosan hydrolysates were prepared and purified by UF, while a 2.0-L concentrate and 8.0-L permeate were separately bulked at each concentration level due to the cross-flow filtration mode.

Table 2 shows that the principle components of COS syrup varied with the changeable concentrations of chitosan (1.0% to 5.0%, w/w) after UF. It demonstrated that most of the target products permeated through the UF membrane, while the blocking layer adhered to by macromolecular solutes was slightly affected. However, there was still a little amount of proteins ($C_{protein}$ = 1.4 to 2.9 mg/L) that

remained, which was partially caused by the temperature effect and concentration polarization. Besides the protein content in the purified syrup, its calculation data was in good agreement with the R_{obs} during the UF process, which was scaled from 0.996 to 0.998 (the data was not shown). The phenomenon suggested that the tested UF membrane was sensitive to the protein rejections.

In addition, with the increment of crude concentration ($C_{chitosan}$), the contents of glucosamine (C_s) and DP 6 to 8 COS ($C_{DP\ 6\ to\ 8}$) were also dramatically increased. For example, C_g and $C_{DP\ 6\ to\ 8}$ at $C_{chitosan}$ = 2.0% were 0.9 and 19.0 g/L, respectively, while those at $C_{chitosan}$ = 5.0% changed to 2.3 and 46.7 g/L, respectively. The tendency indicates that the preparation and separation process of DP 6 to 8 COS was rarely affected by the syrup concentration. According to the data listed in Table 2, the conclusion could be drawn that the purification of hydrolysates by the UF membrane was highly performed. What is more important, the excellent elimination of unexpected impurities may greatly benefit for the next NF treatments.

Effects of transmembrane pressure on permeate flux

Figure 3 illustrates that the permeate fluxes (J_v) of COS syrup at different $C_{chitosan}$ increase with the TMP from 6.0 to 20.0 bar. Moreover, it shows a linear relationship

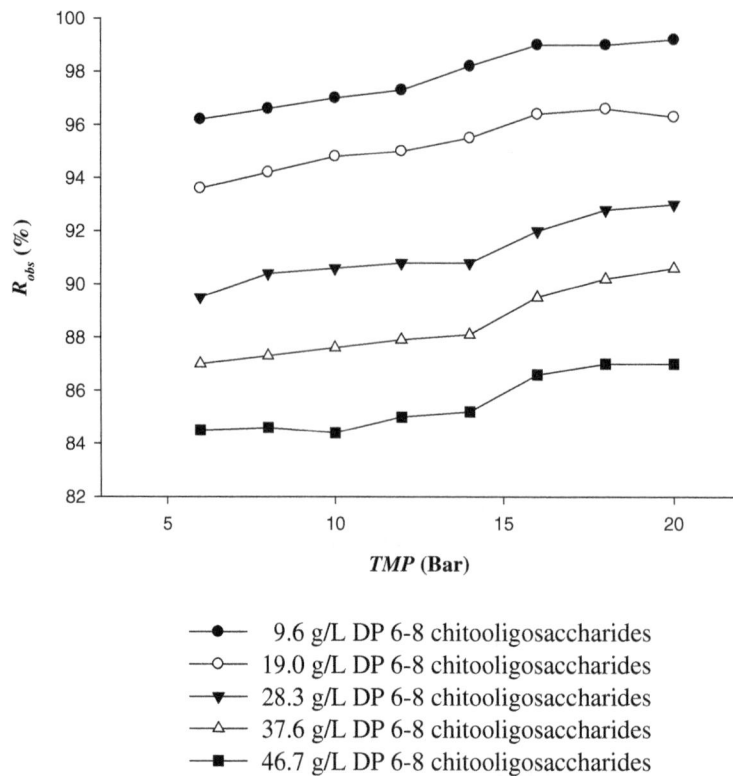

Figure 4 TMP vs. R_{obs} of DP 6 to 8 chitooligosaccharides at 50°C during NF separation. It pictured the retentions of DP 6 to 8 chitooligosaccharides when the TMP was set between 6.0 and 20.0 bar. The R_{obs} of DP 6 to 8 chitooligosaccharides were significantly judged by viscosity and pressure applied.

between TMP and J_v. For instance, the J_v was 95.0 L/(m² h) at 10.0 bar for the 9.6-g/L COS solution, whereas it elevated to 122.0 and 140.0 L/(m² h) at 16.0 and 20.0 bar at the same syrup concentration, respectively. Membrane compaction reduces the mass transfer resistance and enhances the velocity on membrane surface, and thus, leads to an increasing permeate volume of feed solution in regular intervals.

Conversely, J_v decreased with the increment of concentration of DP 6 to 8 COS. It could be well explained by the interaction of TMP and the accumulation effects. On the one hand, the solutes would be greatly accumulated by the driving pressure, which is beneficial for the formation of cake layer around the boundary of the membrane. As a result, the membrane pore would be partially blocked and result in a decrement of J_v. On the other hand, the concentration polarization was also exacerbated with the increase of the sugar concentration. When the osmotic pressure was 16.0 bar, J_v declined from 108.0 to 67 L/(m² h), although the concentration of filtrated syrup ascended from 19.0 to 46.7 g/L.

Effects of TMP on R_{obs} of DP 6 to 8 COS
As shown in Figure 4, the retentions of DP 6 to 8 COS varied with the TMP, and the R_{obs} of target oligosaccharides

were significantly affected by the density and pressure applied. The influence of feed concentration on separation behaviors was emphasized at 28.3 g/L. Due to the existence of DP 6 to 8 COS, the increased concentrations were inevitably brought to a higher permeation of solutes. However, it was also found that the retention coefficient of DP 6 to 8 COS was immobile at higher-pressure areas (18.0 to 20.0 bar). The details could be explained by the steric hindrance interaction between solutes and membrane pores.

Additionally, the R_{obs} of DP 6 to 8 COS increased with TMP, which indicate that the concentration polarization was a subordinate factor for the separation conditions in this case. As an alkaline molecule, the rejection properties of COS are significantly affected by the Donnan effects [23]. When the concentration was kept at 28.3 g/L, the R_{obs} of solute decreased with the decrease of the pressure (6.0 to 16.0 bar). According to the Donnan theory, dielectric exclusion impels the increasing sugar accumulation, which is responsible for the considerable R_{obs} of DP 6 to 8 COS. Therefore, it could be seen that the dropping of observed retention at a high-pressure range (18.0 to 20.0 bar) was similar. The corresponding results were of the same order with reported phenomena by Zhang [24]. Impressive retention proportions of DP 6 to 8 COS at all selected concentrations and applied pressures were performed.

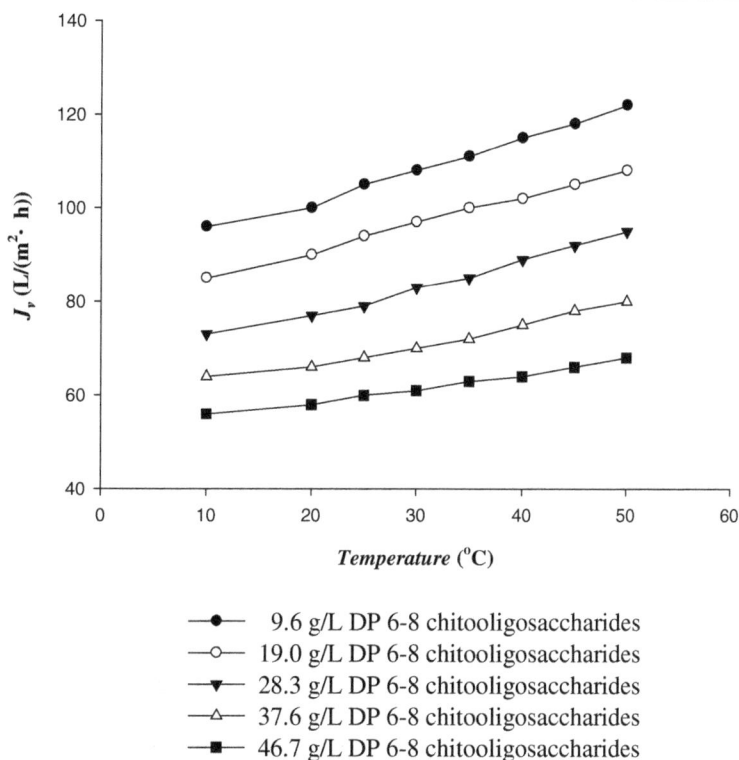

Figure 5 Temperatures vs. J_v of chitooligosaccharide syrup at 16.0 bar in NF process. It depicted the effects of J_v on operation temperature, as well as the concentration variations of DP 6 to 8 chitooligosaccharides. Membrane pores were stretched by thermal expansion at high temperature, which reduced the rate of concentration polarization and improved the membrane flux.

Effects of operation temperature on permeate flux

Figure 5 illustrates the experimental data of membrane fluxes over operation temperature, as well as the concentration variations of DP 6 to 8 COS. It hypothesized that the syrup concentration was a constant and that the permeate flux during full recycle mode in the NF separation process was proportional to the temperature, which resulted from membrane swelling and the decrease of syrup viscosity under elevated temperature. As shown in Figure 5, the permeate flux was 79.0 L/(m^2 h) at 25°C when the concentration was set at 28.3 g/L. After the temperature rose to 40°C, the permeate flux was correspondingly reached to 89.0 L/(m^2 h).

To be detailed, the membrane pores were stretched by structural modification when thermal expansion was interposed by external temperature. At the same moment, the increment of temperature meant for a decline of solution viscosity but a rise both in Reynolds number (*Re*) and mass transfer coefficient (*k*). Followed by the rules analyzed above, all solutes in the process preferred to move to the bulk part of the syrups, which reduced the rate of concentration polarization and thus improved the membrane flux.

However, excessive temperature was negative to preserve the stability of DP 6 to 8 COS. Because of the special

structures that the amino group locates on C-2 sites in every monomer unit linked by β (1 → 4) glycosidic bonds, the velocity of the Maillard reaction was motivated by high temperature. Certainly, further studies are needed to understand the mechanism of the Maillard reaction and its derivative products in permeate during the filtration process.

Effects of operation temperature on R_{obs} of DP 6 to 8 COS

The retention manners of DP 6 to 8 chitooligosaccharides in different concentrations of syrup are shown in Figure 5, as the temperature varied from 10°C to 50°C. As expected, the retentions of DP 6 to 8 chitooligosaccharides decreased with the increment of temperature. All curves coincide with the conclusion drawn from Figure 5. That is, the thermo swelling of membrane structures promoted more DP 6 to 8 chitooligosaccharides permeated through the NF membrane from the concentrate side. Specifically, the R_{obs} of DP 6 to 8 chitooligosaccharides was 0.912 in 20°C, whereas it diminished to 0.895 at 50°C with the concentration of 37.6 g/L. The transformation was apparently introduced that the R_{obs} of solutes decreased with a boosting temperature that forced the fluid properties of

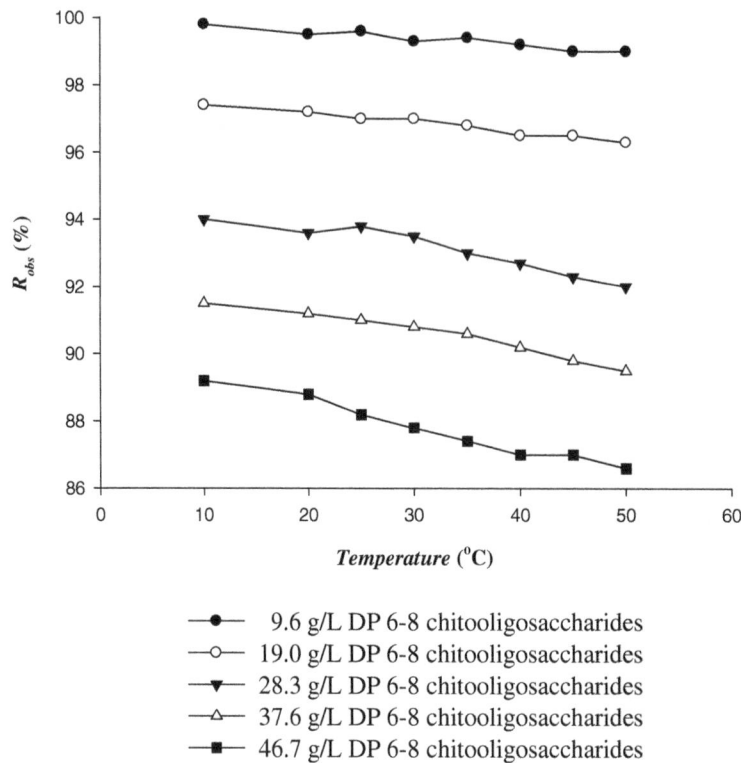

Figure 6 Temperatures vs. R_{obs} of DP 6 to 8 chitooligosaccharides at 16.0 bar in NF process. It reflected the retention behaviors of DP 6 to 8 chitooligosaccharides in different concentrations when the temperature varied from 10°C to 50°C. The retentions of DP 6 to 8 chitooligosaccharides decreased with the temperature increased.

chitooligosaccharide syrup to be a Newtonian liquid and hence resulted in a decrease in viscosity.

Besides, it is also worth noting that the charge effect shifted to become a major factor in the retention behaviors of DP 6 to 8 COS at variable concentrations (Figure 6). Under the same condition (i.e., TMP and temperature), the R_{obs} of hexamer, heptamer, and octamer summed to 0.994 at a concentration of 9.6 g/L, but it declined to 0.874 at a concentration of 46.7 g/L. That could be explained by the Donnan steric pore model (DSPM) theory assumed by Bowen et al. [25]. Given that the negative charges were evenly distributed inside the selected membrane (QY-5-NF-1812), all of the opposite-charged ions as H^+ and NH_4^+ in the solution would be attracted thus the transport of COS was facilitated. Overall, the J_v and retention trends were both restricted by the concentration of syrup and its operation temperature.

Effects of pH on R_{obs} of COS
The effects of pH ranged from 5.0 to 9.0 on rejections of glucosamine and DP 2 to 8 COS were investigated. Singles situated in the area of 13.0 to 19.0 min represent COS within DP 6 to 8. As shown in Figure 7, the R_{obs} of COS increased with pH, especially for the monomer, dimer, trimer, and tetramer. In the chromatogram of NF

concentrate at pH 5.0, it was found that nearly all of COS (DP ≤4) had permeated through the membrane, whereas higher DP were effectively rejected. On the contrary, when the syrup was adjusted to a basic condition (pH = 9.0), the R_{obs} of the monosaccharide and disaccharides arrived at 0.458 and 0.864, respectively, even that the trimer and tetramer completely escaped from the permeate. The reason for this phenomenon might be due to the inter- and intramolecular hydrogen bonds, which further resulted in the unequal curling and overlap in structures. Similar results on R_{obs} and inference for COS have been taken by Han et al. [20]. The stability of the primary structure was broken by hydrogen bonds and restabilized spontaneously. During the recombination process, the curling extent would extremely impact the viscosity and stereospecific blockage of molecules. The structural explosion guided the poor affinity of solutes with the membrane and led to a high resistance of permeation. Furthermore, the rejections of DP 6 to 8 COS, formulated from the concentrate and permeate profiles, were scarcely impacted via pH options (Figure 7). The results indicated that the applied driving force was centered on oligomers of DP ≤4 under acidic conditions and sequentially spread to higher polymers as the pH increased. DP 6 to 8 COS is expected to be completely rejected by the NF membrane at a wider

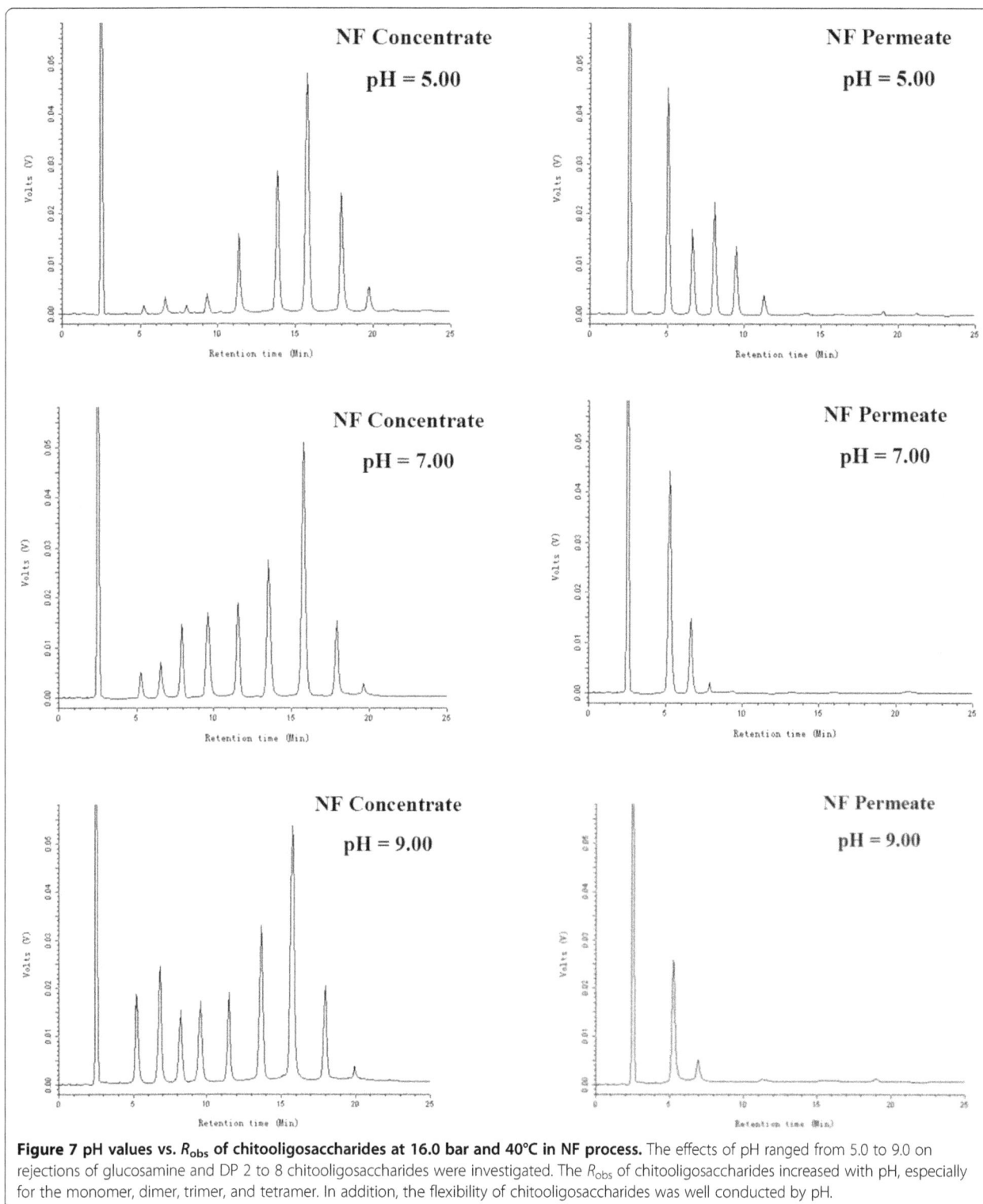

Figure 7 pH values vs. R_{obs} of chitooligosaccharides at 16.0 bar and 40°C in NF process. The effects of pH ranged from 5.0 to 9.0 on rejections of glucosamine and DP 2 to 8 chitooligosaccharides were investigated. The R_{obs} of chitooligosaccharides increased with pH, especially for the monomer, dimer, trimer, and tetramer. In addition, the flexibility of chitooligosaccharides was well conducted by pH.

pH range (5.0 to 9.0). More important, the flexibility of COS was well conducted by the adjustment of pH. Nevertheless, in view of the objective of purification, a lower pH should be adopted for the impurity elimination. Interestingly, several small-sized peaks that were followed by octasaccharides were also characterized at 19.5 min (Figure 7). Due to the randomness of enzymatic hydrolysis on chitosan, this was probably attributed to the nonamers (DP 9 COS) existing in the syrup.

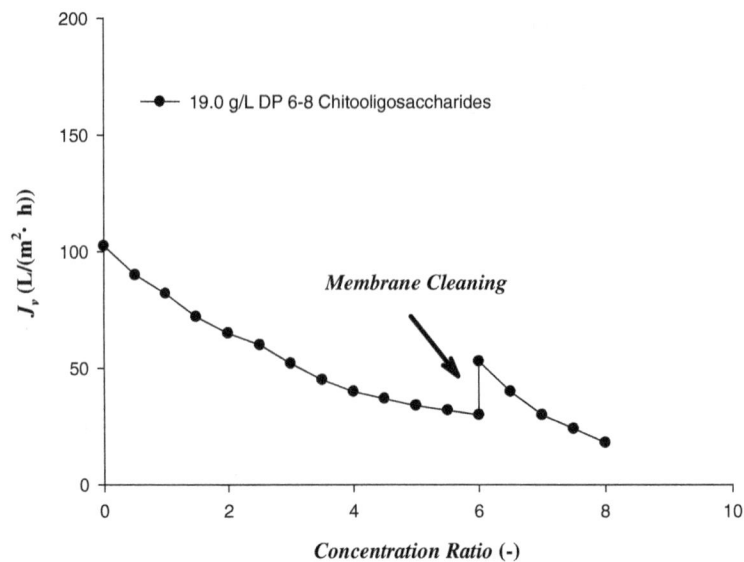

Figure 8 Concentration ratios vs. J_v of 19.0 g/L chitooligosaccharide syrup. Long-term experiments were carried out at 16.0 bar for enrichment of DP 6 to 8 chitooligosaccharides. The flux was about 102.4 L/(m² h) at the beginning, whereas it decreased heavily when the concentration rate was 6.0.

Purification of DP 6 to 8 COS

Long-term experiments were carried out at 16.0 bar to enrich the DP 6 to 8 COS with a virgin concentration of 19.0 g/L (pH = 5.0, V_0 = 16.0 L and T = 40°C). The performance of the NF membrane was evaluated by measuring the permeate flux and R_{obs} of the target products as illustrated in Figures 8 and 9. In addition, the concentration ratio of DP 6 to 8 COS was monitored as a formula of V_0/V_c. The flux value was about 102.4 L/(m² h) at the beginning of the experiment, but it decreased to 30.2 L/(m² h) when

the concentration ratio reached to 6.0 (Figure 8). Then, the membrane was cleaned for 1 h. After that, the permeate flux of syrup was recovered to 53.5 L/(m² h), which indicated that about 50% of virtual flux decline due to the increasing solid content and viscosity. However, the second drop of permeate flux was remarkable till the end of the run, as the value changed from 53.5 to 25.0 L/(m² h) just 1.5 times of the later concentration. There was a direct reflection between the effect of solid contents and experimental records. A moderate concentration ratio and timely

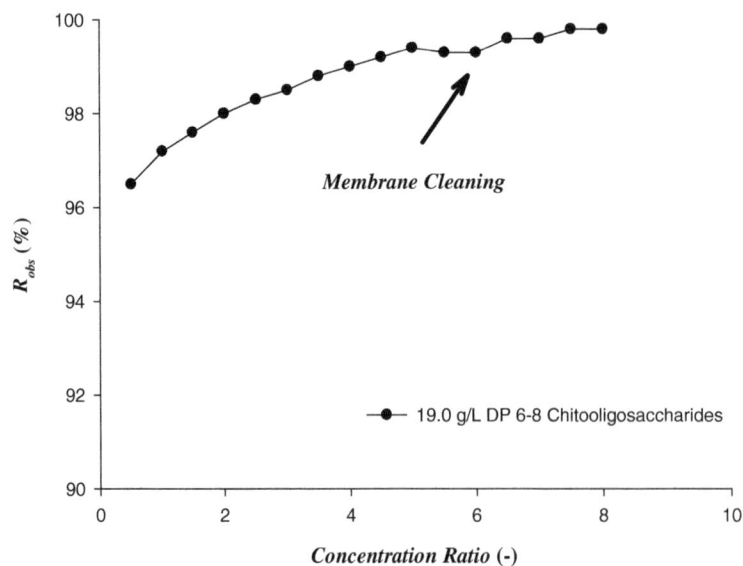

Figure 9 Concentration ratios vs. R_{obs} of DP 6 to 8 chitooligosaccharides in the syrup. The R_{obs} of DP 6 to 8 chitooligosaccharides went up when the circulating flux decreased. The phenomenon was accredited to the DSPM and steric hindrance pore (SHP) effects.

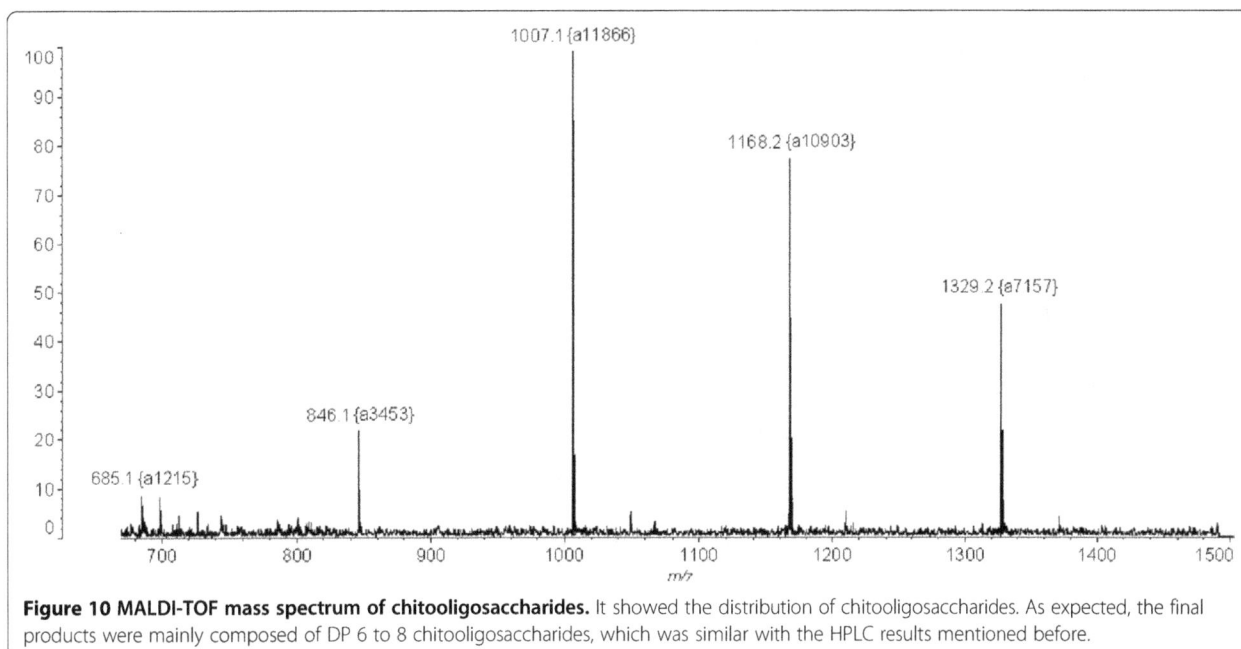

Figure 10 MALDI-TOF mass spectrum of chitooligosaccharides. It showed the distribution of chitooligosaccharides. As expected, the final products were mainly composed of DP 6 to 8 chitooligosaccharides, which was similar with the HPLC results mentioned before.

regenerations of the membrane should be arranged for acceptable efficiency in practice.

Moreover, the R_{obs} of DP 6 to 8 COS went up when the circulating flux decreased (Figure 9). The result was principally accredited to the synergy with the DSPM and steric hindrance pore (SHP) effects. On the one hand, the anions preferred to be repelled while the cations were attracted to the NF membrane. The unique property transported the amino groups, which are commonly sensed in COS, to the permeate. Ultimately, the molecules containing NH_4^+ were separated from the syrup because of the electric attraction. On the other hand, the R_{obs} of DP 6 to 8 COS was up to 1.0 after the concentration ratio increased to 8.0. The MWCO of the applied membrane is 500 Da, while the molecular weights of hexamer, heptamer, and octamer are 984, 1,145 and 1,306, respectively. There is a conspicuous difference in the sizes between membrane pores and DP 6 to 8 COS. Therefore, the observed retention behaviors illustrated that DP 6 to 8 COS were rejected sterically and a promising purity (82.2%) was eventually achieved via the NF purification designed.

MALDI-TOF-MS analysis of chitooligosaccharides

Matrix-assisted laser desorption/ionization time-of-flight (MALDI-TOF) is an outstanding tool for the comprehensive investigation of COS because the mass spectrum can exhibit the relative quantities of a mixture to be determined. Figure 9 shows the distribution of COS using MALDI-TOF-MS after NF enrichment. As expected, the final products were mainly composed of DP 6 to 8 COS, which was similar to the HPLC results mentioned before. In the spectrum, COS often contains intensive

quasi-molecular ions, which is called $[M + Na]^+$, because of its weak protonation degree. For instance, the peak at 1,007.1 m/z is attributed to the sodium form ($[M + Na]^+$) of a hexa-oligomer (Figure 10).

Conclusions

In this study, the separation behavior of COS syrups, which were enriched by different concentrations of DP 6 to 8 COS, was investigated via a bench-scale NF process.

One negatively charged membrane with MWCO of 500 Da was selected to purify DP 6 to 8 COS. During the full recycle mode, the experimental results indicated that the retentions of these components increased with pressure before 16.0 bar. Also, the operation temperature was optimized. Although the circulating flux could be improved by elevating temperature, a greater wastage of DP 6 to 8 COS was irreversibly formed during the process as well. In addition, the effects of pH on R_{obs} of COS were compared. The HPLC profiles illustrated that the R_{obs} of COS within DP ≤4 in alkali conditions was significantly higher than that in acidic environment. This phenomenon could be explained by structural curling and sterical overlap due to hydrogen bonds. However, the mechanisms should be discussed in further researches.

Under the optimum conditions (TMP = 16.0 bar, T = 40°C, and pH = 5.0), the purification of DP 6 to 8 COS was carried out. It was found that the membrane could support a reluctant flux after the concentration ratio was over 6.0 in the syrup with the concentration of 19.0 g/L DP 6 to 8 COS. MALDI-TOF mass spectrum confirmed that DP 6 to 8 COS were dominant in the final products, and the purity was up to 82.2% (*w/w*) according to HPLC profiles. As a conclusion, the NF system equipped with a selected membrane module is a promising approach in the purification of DP 6 to 8 COS from specific syrups.

Competing interests
The authors declare that they have no competing interests.

Authors' contributions
This paper is the result of joint efforts. Prof. LZ designed the whole experimental plan and confirmed the main objective of this paper. Dr. YW developed the statistical methods for experimental data. HD was responsible for optimization of the nanofiltration technology and partial investigation of the transmechanism in process. QX was responsible for the quantification of proteins and total sugars. HPLC and MALDI-TOF-MS analysis were done by Prof. JZ and Prof. LJ. LF helped us complete the paper writing and correcting some grammatical errors in details. All authors read and approved the final manuscript.

Acknowledgements
This work is financially supported by the National Natural Science Foundation of China (No. 31371725 and No. 31101381). Also, the authors are grateful to the Fundamental Research Funds for the Central Universities.

References
1. Xia W, Liu P, Zhang J, Chen J (2011) Biological activities of chitosan and chitooligosaccharides. Food Hydrocolloids 25:170–179
2. Kim S, Rajapakse N (2005) Enzymatic production and biological activities of chitosan oligosaccharides (COS): a review. Carbohydr Polym 62:357–368
3. Benhabiles MS, Salah R, Lounici H, Drouiche N, Goosen MFA, Mameri N (2012) Antibacterial activity of chitin, chitosan and its oligomers prepared from shrimp shell waste. Food Hydrocolloids 29:48–56
4. Huang RH, Du YM, Yang JH, Fan L (2003) Influence of functional groups on the in vitro anticoagulant activity of chitosan sulfate. Carbohy Res 338:483–489
5. Lin S, Lin Y, Chen H (2009) Low molecular weight chitosan prepared with the aid of cellulase, lysozyme and chitinase: characterisation and antibacterial activity. Food Chem 116:47–53
6. Kim KW, Thomas RL (2007) Antioxidative activity of chitosans with varying molecular weights. Food Chem 101:308–313
7. Zhang Y, Huo M, Zhou J, Yu D, Wu Y (2009) Potential of amphiphilically modified low molecular weight chitosan as a novel carrier for hydrophobic anticancer drug: synthesis, characterization, micellization and cytotoxicity evaluation. Carbohydr Polym 77:231–238
8. Zhang J, Zhang W, Mamadouba B, Xia W (2012) A comparative study on hypolipidemic activities of high and low molecular weight chitosan in rats. Int J Biol Macromol 51:504–508
9. Chang Y, Chang C, Huang T, Chen S, Lee J, Chung Y (2011) Effects of low molecular weight chitosans on aristolochic acid-induced renal lesions in mice. Food Chem 129:1751–1758
10. Berth G, Dautzenberg H (2002) The degree of acetylation of chitosans and its effect on the chain conformation in aqueous solution. Carbohydr Polym 47:39–51
11. Lee EH, Lee JJ, Jon SY (2010) A novel approach to oral delivery of insulin by conjugating with low molecular weight chitosan. Bioconjugate Chem 21:1720–1723
12. Li N, Wang CY, Wang M, Sun X, Nie S, Pan W (2009) Liposome coated with low molecular weight chitosan and its potential use in ocular drug delivery. Int J Pharm 379:131–138
13. Chien PJ, Sheu F, Lin HR (2007) Coating citrus (Murcott tangor) fruit with low molecular weight chitosan increases postharvest quality and shelf life. Food Chem 100:1160–1164
14. Abd–Elmohdy FA, Sayed ZE, Essam S, Hebeish A (2010) Controlling chitosan molecular weight via bio-chitosanolysis. Carbohydr Polym 82:539–542
15. Jeon Y, Kim S (2000) Production of chitooligosaccharides using an ultrafiltration membrane reactor and their antibacterial activity. Carbohydr Polym 41:133–141
16. Fan W, Yan W, Xu Z, Ni H (2012) Formation mechanism of monodisperse, low molecular weight chitosan nanoparticles by ionic gelation technique. Colloids Surf B 90:21–27
17. Cabrera JC, Custem PV (2005) Preparation of chitooligosaccharides with degree of polymerization higher than 6 by acid or enzymatic degradation of chitosan. Biochem Eng J 25:165–172
18. Simonnot MO, Castel C, Nicolai M, Rosin C, Sardin M, Jauffret H (2000) Boron removal from drinking water with a boron selective resin: is the treatment really selective. Water Res 34:109–116
19. Nadav N (1999) Boron removal from seawater reverse osmosis permeate utilizing selective ion exchange resin. Desalination 124:131–135
20. Han YP, Lin Q, He XW (2009) Research on desalination and purification characteristics of chitobiose solution with nanofiltration membrane. Membrane Science and Technology 29:105–109
21. Han YP, Lin Q, Wang XL (2012) Feasibility study on chitooligosaccharides purification by nanofiltration membranes. Ion Exchange and Adsorption 28:86–96
22. Xu W, Jiang C, Kong X, Liang Y, Rong M, Liu W (2012) Chitooligosaccharides and N-acetyl-D-glucosamine stimulate peripheral blood mononuclear cell-mediated antitumor immune response. Mol Med Rep 6:385–390
23. Zhao HF, Hua X, Yang RJ, Zhao LM, Zhao W, Zhang Z (2012) Diafiltration process on xylo-oligosaccharides syrup using nanofiltration and its modeling. Int J Food Sci Tech 47:32–39
24. Zhang Z, Yang RJ, Zhang S, Zhao H, Hua X (2011) Purification of lactose syrup by using nanofiltration in a diafiltration mode. J Food Eng 105:112–118
25. Bowen WR, Mohammad AW (1998) Diafiltration by nanofiltration: prediction and optimization. AIChE J 44:1799–1812

11

Purification and partial characterization of serine alkaline metalloprotease from *Bacillus brevis* MWB-01

Folasade M Olajuyigbe[*] and Ayodele M Falade

Abstract

Background: Proteases from bacteria are among the most important hydrolytic enzymes that have been studied due to their extracellular nature and high yield of production. Of these, alkaline proteases have potential for application in detergent, leather, food, and pharmaceutical industries. However, their usefulness in industry is limited by low activity and stability at high temperatures, extreme pH, presence of organic solvents and detergent ingredients. It is therefore very crucial to search for new alkaline proteases with novel properties from a variety of microbial sources.

Results: In the present study, 21 *Bacillus* species isolated from organic waste sites were screened for proteolytic activity on casein agar. *Bacillus brevis* MWB-01 exhibited highest proteolytic activity with a clear zone diameter of 35 mm. Production of protease from *B. brevis* MWB-01 was investigated in optimized media after 48 h of cultivation with shaking (180 rpm) at 37°C. The protease was partially purified in a two-step procedure using ammonium sulphate precipitation and gel filtration chromatography on Sephadex G-200 column. The enzyme was purified 2.1-fold with yield of 4.6%. The purified protease had optimum temperature of 40°C with relative activity of about 50% at 50°C and was uniquely stable up to 60°C after 30 min of incubation exhibiting 63% residual activity. The enzyme had optimum pH of 8.0 and remarkably showed relative activity above 70% at pH 9.0 to 11.0 and 53% at pH 12.0, respectively and was very stable over a wide pH range (6.0 to 12.0). Ca^{2+} and Mn^{2+} increased protease activity with 9.8% and 3.5%, respectively; Hg^{2+} and Zn^{2+} strongly inhibited protease activity by 89% and 86%. The almost complete inhibition of the enzyme by phenylmethylsulphonyl fluoride (PMSF) and ethylene diamine tetra acetic acid (EDTA) confirmed the enzyme as a serine metalloprotease. The enzyme had highest compatibility with Sunlight, a commercial laundry detergent.

Conclusion: The characteristics of purified protease from *B. brevis* MWB-01 reveal the enzyme as a thermotolerant serine alkaline metalloprotease compatible with detergent formulation aids. Results suggest that protease from *B. brevis* MWB-01 is a good bioresource for industrial applications.

Keywords: Serine alkaline metalloprotease; *Bacillus brevis*; Purification; Characterization

Background

Proteases have a variety of applications in detergent, pharmaceutical, leather, and food industries [1-3]. Globally, proteases constitute the largest product segment of industrial enzymes, accounting for about 60% of the total worldwide enzyme sales [4,5]. Bacterial proteases are the most widely exploited when compared with fungal, plants, animal, and fungal proteases because of their extracellular nature and high yield of production; however, their usefulness is limited by various physicochemical factors, such as enzyme instability at high temperatures, extreme pH, presence of organic solvents, anionic surfactant and oxidizing agents, and need for co-factors [6].

Ideally, alkaline proteases used in detergent formulation should demonstrate high level of activity and stability over a broad range of pH and temperature, possess broad substrate specificity and be active in the presence of detergent ingredients, such as surfactants, bleaching

* Correspondence: folajuyi@futa.edu.ng
Department of Biochemistry, Federal University of Technology, Akure 340001, Nigeria

agents, fabric softeners, and other formulation aids [7]. Hence, it is imperative to search for new alkaline proteases with desirable properties for commercial viability from various bacterial sources.

Recently, we reported purified thermostable alkaline proteases from *Bacillus licheniformis* LHSB-05 isolated from hot spring [8] and *Bacillus coagulans* PSB-07 isolated from poultry litter site [9]. In this present study, 21 proteolytic *Bacillus* species isolated from organic waste sites were identified and *Bacillus brevis* MWB-01 was selected based on zone of clearance on casein agar. *B. brevis* produced protease under submerged conditions and the protease was purified and characterized.

This paper presents characteristics of alkaline protease from *B. brevis* MWB-01 which demonstrate the viability of the enzyme as good bioresource for industrial applications.

Methods
Materials
Sephadex G-200, bovine serum albumin (BSA), ethylene diamine tetra acetic acid (EDTA), casein, β-mercaptoethanol, phenyl methyl sulphonyl fluoride (PMSF) and media components were products of Sigma-Aldrich, St Louis, MO, USA. All other chemicals used were of analytical grade and obtained from Fisher Scientific (Waltham, MA, USA). Commercial laundry detergents used were Omo and Sunlight from Unilever (Rotterdam, Netherlands); Ariel was a product of Procter and Gamble (Cincinnati, OH, USA).

Isolation and identification
Sub-soil samples were collected aseptically from organic waste sites which included beds of effluent treatment plants of selected brewery, dairy and food industries, drainage from abattoir, poultry litter site and locust bean processing farm in the south western part of Nigeria for the preparation of initial culture. These were sub-cultured to obtain pure isolates of *Bacillus* species using the method of Aslim et al. [10]. The *Bacillus* species were identified based on methods described in Bergey's Manual of Systematic Bacteriology [11,12].

Screening for proteolytic activity of *Bacillus* species
The identified bacterial isolates were plated onto casein milk agar plates and were incubated at 37°C for 24 h. 15% $HgCl_2$ in 20% HCl was added to the plates and examined for clearing zone around the bacterial growth. The diameter of the zone was measured in millimeters. A clear zone of casein hydrolysis gave an indication of protease-producing organisms. Depending on the zone of clearance, *B. brevis* MWB-01 was selected for further studies.

Production of protease
Production of protease from *Bacillus brevis* MWB-01 was carried out in a culture medium containing the following: 0.5% glucose, 0.75% peptone, and 5% salt solution made up of 0.5% $MgSO_4.7H_2O$ and 0.1% NaCl, maintained at 37°C for 72 h in a shaking incubator (180 rpm). At the end of each cultivation period, the broth was centrifuged at 10,000 rpm at 4°C for 15 min. The cell-free supernatant was collected as crude enzyme preparation and subjected to purification procedures.

Assay of protease activity
Extracellular protease activity was determined using a modified procedure of Fujiwara et al. [13] with 1.0% casein in 50 mM Tris-HCl buffer pH 8.0 as substrate. The assay mixture consisted of 0.4 ml of substrate and 0.1 ml of enzyme solution in 50 mM Tris-HCl buffer pH 8.0. The assay mixture was incubated at 40°C for 30 min and reaction was terminated by the addition of 2.5 ml of 10% (*w/v*) trichloroacetic acid (TCA). The mixture was allowed to stand for 15 min and then centrifuged at 10,000 rpm for 10 min at 4°C to remove the resulting precipitate. Protease activity was determined by estimating the amount of tyrosine in the supernatant which was done by measuring the absorbance at 280 nm. One unit of protease activity was defined as the amount of enzyme required to release 1 μg of tyrosine per milliliter per minute under the specified assay conditions.

Purification of protease from *B. brevis* MWB-01
The cell-free supernatant was fractionated by precipitation with ammonium sulphate of 80% saturation. The precipitated protein collected by centrifugation was dissolved in 50 mM Tris-HCl buffer pH 7.5, and dialyzed against the same buffer at 4°C with three buffer changes, each for 12 h using Spectra/Por dialysis membrane (MWCO 3,500; Serva, Heidelberg, Germany). The resulting dialysate was centrifuged at 10,000 rpm, 4°C for 15 min, and the supernatant was applied on Sephadex G-200 (1.5 × 24 cm) column (Sigma-Aldrich, St Louis, MO, USA) equilibrated with 50 mM Tris-HCl buffer, pH 7.5. The column was eluted at a flow rate of 0.5 ml/min. Protease activity was assayed in all eluted fractions. The fractions (25 to 27) with high protease activity corresponding to the highest peak on the chromatogram were pooled, and subsequently used for characterization studies. The concentration of protein during purification studies was determined by Bradford method [14].

Partial characterization of purified protease
Effect of temperature on the activity and stability of protease
The optimum temperature of purified protease was determined by measuring enzyme activity at varied temperatures (30°C to 70°C). The reaction mixture was incubated

Table 1 Identification of isolated *Bacillus* species

Isolate code	Colour/ pigment	Gram reaction	Cellular morphology	Catalase test	Oxidase test	Indole production	Motility test	Methyl red test	Voges-Proskaver test	Urease activity	Citrate utilization	Starch hydrolysis	Gelatin hydrolysis	Casein hydrolysis
BWB-01	Cream	+	Rods	+	+	–	+	–	+	–	+	+	+	+
BWB-02	Cream	+	Rods	+	+	+	+	–	+	+	–	+	+	+
BWB-03	Cream	+	Rods	+	+	–	+	–	+	–	+	+	+	+
BWB-04	Cream	+	Rods	+	+	–	+	–	+	–	+	+	+	+
MWB-01	Cream	+	Rods	+	+	–	+	+	–	–	–	+	+	+
MWB-02	Cream	+	Rods	+	+	–	+	–	–	–	+	+	+	+
MWB-03	Cream	+	Rods	+	+	–	+	–	–	–	–	–	+	+
PFB-01	Cream	+	Rods	+	+	–	+	–	+	–	+	+	+	+
PFB-02	Cream	+	Rods	+	+	–	+	+	–	–	–	+	+	+
PFB-03	Cream	+	Rods	+	–	–	+	–	+	–	+	+	+	+
ADB-01	Cream	+	Rods	+	+	–	+	–	+	–	–	+	–	+
ADB-02	Cream	+	Rods	+	+	–	+	–	–	–	–	+	–	+
ADB-03	Cream	+	Rods	+	+	–	+	–	+	–	+	+	+	+
ADB-04	Cream	+	Rods	+	+	–	+	+	–	–	+	+	+	+
FWB-01	Cream	+	Rods	+	+	–	+	–	–	–	–	–	+	+
FWB-02	Cream	+	Rods	+	–	–	+	–	+	–	+	+	+	+
FWB-03	Cream	+	Rod	+	+	–	+	–	+	–	–	+	–	+
FWB-04	Cream	+	Rods	+	+	–	+	–	+	–	+	+	+	+
LFB-01	Cream	+	Rods	+	–	–	+	–	+	–	+	+	+	+
LFB-02	Cream	+	Rods	+	+	–	+	–	–	–	+	+	+	+
LFB-03	Cream	+	Rods	+	+	–	+	–	–	–	–	+	+	+

Table 1 Identification of isolated *Bacillus* species (*Continued*)

Isolate code	NO₃ Reduction	Spore test	Glucose	Xylose	Lactose	Sucrose	Maltose	Mannitol	Raffinose	Fructose	Galactose	Sorbitol	Arabinose	Organism identity
BWB-01	+	+	+	+	+	+	+	+	+	+	+	+	+	*Bacillus polymyxa*
BWB-02	–	+	+	–	–	–	–	–	–	–	–	–	–	*Bacillus alvei*
BWB-03	+	+	+	–	+	+	–	+	–	+	+	–	+	*Bacillus subtilis*
BWB-04	+	+	+	–	+	+	+	–	+	–	–	–	–	*Bacillus cereus*
MWB-01	–	+	+	+	–	–	–	+	–	+	+	–	–	*Bacillus brevis*
MWB-02	–	+	+	+	–	+	–	+	+	+	–	–	+	*Bacillus megaterium*
MWB-03	+	+	+	–	–	–	–	–	–	–	–	–	–	*Bacillus pasterii*
PFB-01	+	+	+	–	+	+	–	–	+	+	+	–	–	*Bacillus amyloliquefasciens*
PFB-02	–	+	+	–	–	–	–	+	–	–	–	+	–	*Bacillus firmus*
PFB-03	+	+	+	+	–	+	+	+	–	–	+	–	+	*Bacillus licheniformis*
ADB-01	+	+	+	+	+	+	+	–	+	–	–	–	+	*Bacillus coagulans*
ADB-02	+	+	+	–	–	–	–	–	–	–	–	–	–	*Bacillus flavothermus*
ADB-03	+	+	+	–	+	+	–	+	–	+	+	–	+	*Bacillus subtilis*
ADB-04	+	+	+	+	+	–	–	+	–	–	–	–	+	*Bacillus circulans*
FWB-01	+	+	+	–	+	+	–	+	+	–	–	–	–	*Bacillus laterosporous*
FWB-02	+	+	+	+	–	+	+	+	–	–	+	–	+	*Bacillus licheniformis*
FWB-03	+	+	+	+	+	+	+	–	+	–	–	–	+	*Bacillus coagulans*
FWB-04	+	+	+	+	+	+	+	+	+	+	+	–	+	*Bacillus polymyxa*
LFB-01	+	+	+	+	–	+	+	+	–	–	+	–	+	*Bacillus licheniformis*
LFB-02	–	+	+	+	–	+	–	+	–	+	–	–	+	*Bacillus megaterium*
LFB-03	–	+	+	+	–	+	–	+	–	+	–	–	+	*Bacillus alcalophilus*

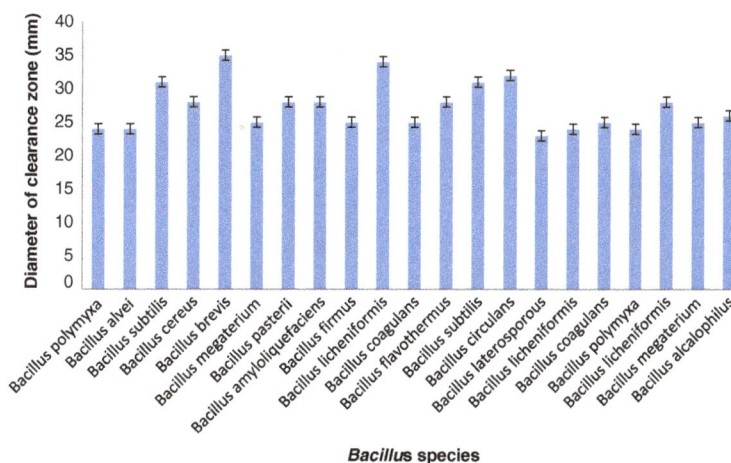

Figure 1 Zone of clearance of *Bacillus* species on casein agar. Proteolytic activity of isolated Bacillus species was detected by the presence of a clear zone which indicated casein hydrolysis. Symbols and bars represent mean values and standard deviations of triplicate determinations.

at respective temperatures for 30 min before determining protease activity according to the standard assay procedure earlier described. Thermal stability was determined by measuring the residual protease activity after 30 min and 60 min of pre-incubation of purified enzyme at temperatures ranging from 30°C to 70°C in 50 mM Tris-HCl buffer pH 8.0.

Effect of pH on activity and stability of protease
Effect of pH on activity of protease was determined by assaying for enzyme activity at different pH values ranging from 4.0 to 12.0. The pH was adjusted using 50 mM of the following buffer solutions: sodium acetate (pH 4.0 to 5.0), sodium citrate (pH 6.0), Tris-HCl (pH 7.0 to 8.0) and glycine-NaOH (pH 9.0 to 12.0). Reaction mixtures

were incubated at 40°C for 30 min and the activity of the protease was measured. To determine the effect of pH on stability of protease, the purified protease was incubated in relevant buffers of varying pH (4.0 to 12.0) without substrate for 60 min at 40°C. The residual protease activity was determined as previously described.

Effect of inhibitors on protease activity
Effect of inhibitors (phenylmethylsulphonyl fluoride [PMSF], β-mercaptoethanol [β-ME] and ethylene diamine tetra acetic acid [EDTA]) at 5 mM on protease activity was determined by pre-incubating the purified enzyme solution with inhibitor for 30 min at 40°C before the addition of substrate following the standard assay procedure.

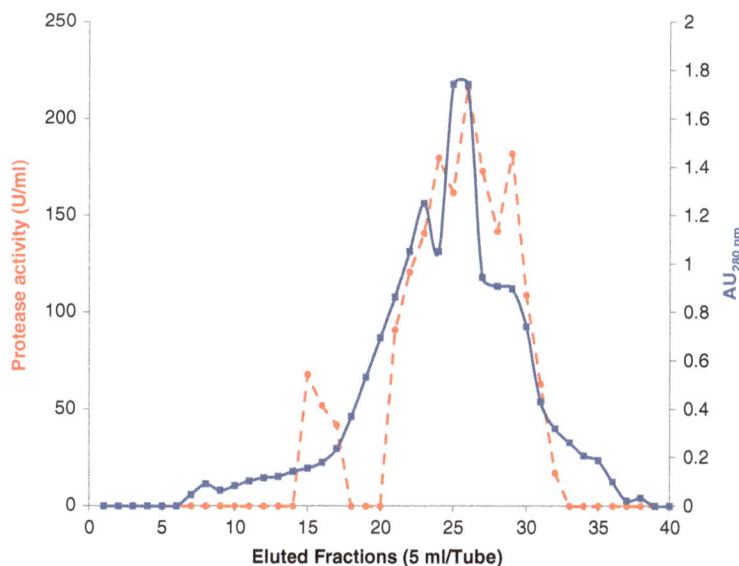

Figure 2 Protease activity in eluted fractions vs chromatogram of alkaline protease from *Bacillus brevis* MWB-01.

Table 2 Purification table of protease from *Bacillus brevis* MWB-01

Purification steps	Volume (ml)	Activity (U/ml)	Total activity (U)	Protein concentration (mg/ml)	Total protein (mg)	Specific activity (U/ml)	Fold	Yield (%)
Crude enzyme preparation	800	202	161,600	8.15	6412	25.2	1.0	100
Ammonium sulphate precipitation dialysed	30	310	9,300	7.63	228.9	40.6	1.6	5.8
Gel filtration chromatography on Sephadex G-200	50	150	7,500	2.77	138.5	54.2	2.1	4.6

Effect of metal ions on protease activity

The effects of metal ions on enzyme activity (Ca^{2+}, Mg^{2+}, Al^{3+}, Mn^{2+}, Zn^{2+}, and Hg^{2+}) at 5 mM was investigated by pre-incubating the purified protease with each of the metallic chlorides without substrate for 30 min at 40°C. The residual protease activity was measured as previously described.

Substrate specificity of purified protease

Substrate specificity of purified protease from *B. brevis* MWB-01 was studied by examining proteolytic activity on protein substrates. The substrates studied were casein, bovine serum albumin (BSA), egg albumin, and gelatin. Purified protease (0.5 ml) was added to 2.0 ml of 50 mM Tris-HCl buffer pH 8.0 containing 1% substrate. After incubation at 40°C for 30 min, the reaction was stopped by adding 2.5 ml of 10% TCA. Protease activity was determined by following the standard assay procedure.

Detergent compatibility of purified protease

The compatibility of *B. brevis* protease with commercial laundry detergents, Sunlight (Unilever), Ariel (Procter and Gamble), and Omo (Unilever), was studied. The effect of 5 mg/ml of each detergent on the stability of purified protease was determined. The diluted detergent solution was heated to 60°C for 1 h to denature the enzymes present in the detergent and left to cool at room temperature for 1 h. The purified protease was incubated with the diluted detergent for 1 h at 40°C and the residual activity was determined. The enzyme activity of the control was taken as 100%.

Results and discussion

Identification and screening for proteolytic activity of *Bacillus* species

Twenty one *Bacillus* species were isolated and identified using the culture, motility, morphological, and biochemical parameters (Table 1). All the species exhibited varying degrees of zones of clearance on 10% casein agar. The proteolytic activity was detected by the presence of a clear zone which indicated casein hydrolysis. *B. brevis* MWB-01exhibited the highest clearance zone measuring average diameter of 35.0 mm (Figure 1).

Purification of extracellular protease from *B. brevis* MWB-01

The protease was purified from soluble dialysate by gel filtration chromatography as described under methods section. The chromatogram shows that the protease was eluted as a single major peak (Figure 2). Fractions

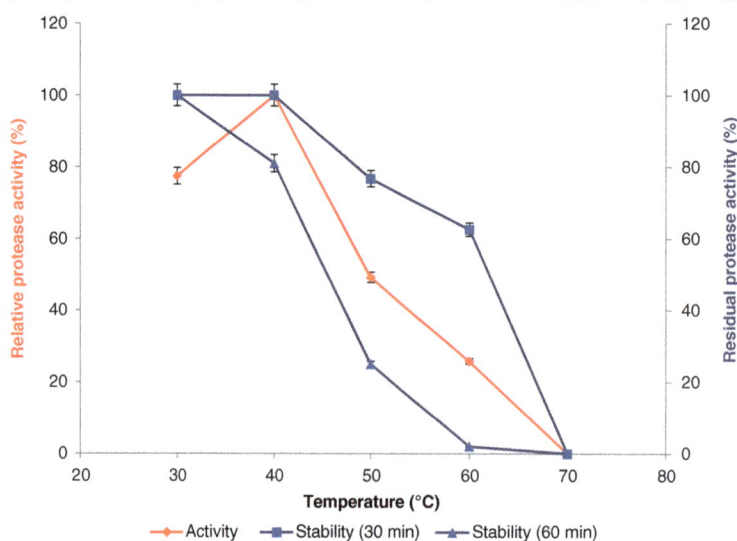

Figure 3 Effect of temperature on activity and stability of purified protease.

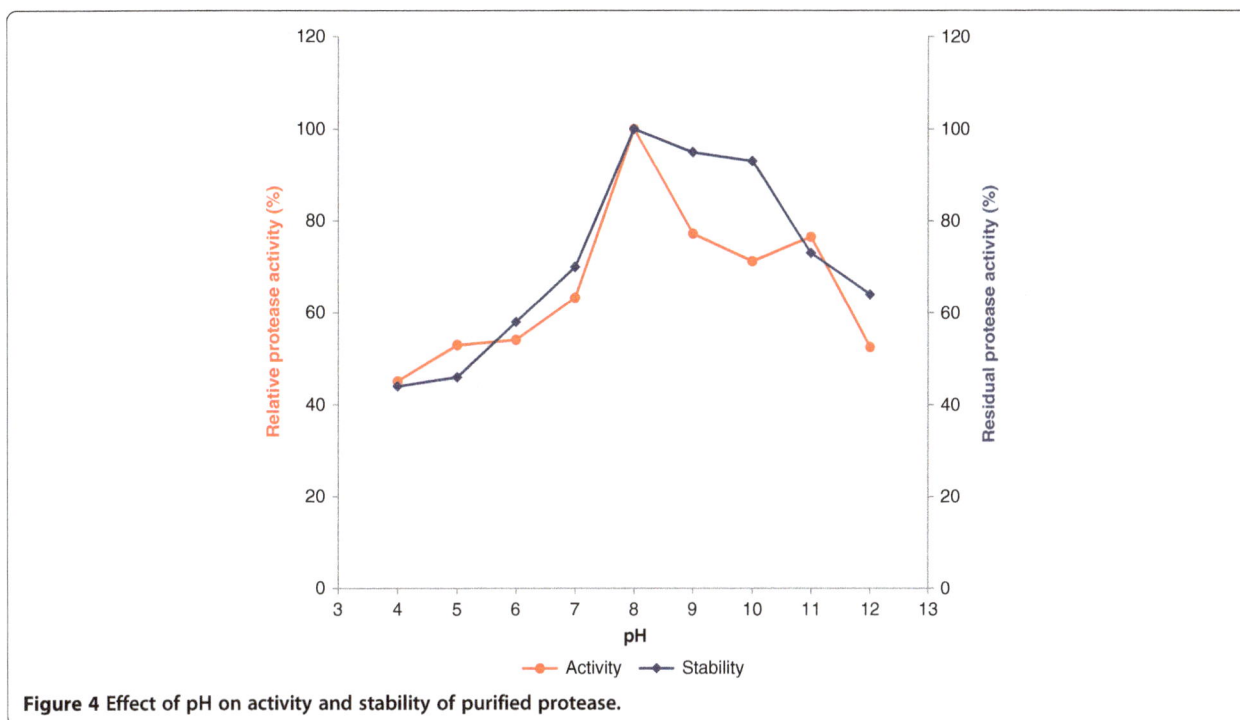

Figure 4 Effect of pH on activity and stability of purified protease.

(25 to 27) with protease activity (Figure 2) were pooled and concentrated for characterization studies. The purification procedure is summarized in Table 2, showing that the enzyme was purified 2.1-fold with a specific activity of 54.2 U/mg protein after Sephadex G-200 gel filtration chromatography. The yield of the enzyme after purification was 4.6%.

Partial characterization of purified protease
Optimum temperature and thermal stability of protease
The activity of the purified protease from *B. brevis* was determined at different temperatures ranging from 30°C to 70°C. The optimum temperature for the activity of the purified protease was 40°C with 50% relative activity at 50°C (Figure 3). This is really unusual and surprising since some previous reports on proteases from *Bacillus* species have temperature optima of 50°C to 60°C [5,15-21]. The result obtained from the present study is however similar to those of reports by Feng et al. [22] and Park and Cho [23] on proteases from *Bacillus pumilus* strain and *Bacillus sp. JSP1* which had optimum temperature of 40°C. Banerjee et al. [24] reported a protease from a strain *B. brevis* isolated from hot springs which had optimum activity at 60°C. The *B. brevis* under study was isolated from a mesophilic environment (beds of effluent treatment plant of a dairy industry). The observed difference in characteristics between same species of *Bacillus* might be strain related [25] and can also be attributed to the critical role which the source of isolation plays in determining the function of microbial

species [26,27]. A decrease in relative activity of purified protease under study above 40°C might be due to the autolysis or denaturation of the enzyme at higher temperature [28]. The purified protease from *B. brevis* MWB-01 uniquely demonstrated 63% relative stability at 60°C after 30 min of preincubation at this high temperature in the absence of Ca^{2+} (Figure 3). It is very interesting that this purified protease exhibited good thermal stability despite a lower optimum temperature.

Optimum pH and stability of protease
The purified protease had optimum pH of 8.0 and was active over a broad pH range of 5.0 to 12.0 exhibiting above 70% relative activity at pH 9.0 to 11.0 and about 54% relative activity at pH 5.0, 6.0, and 12.0, respectively (Figure 4). This is a very remarkable characteristic of purified protease from *B.brevis* MWB-01 considering reports on some proteases that have higher optimum pH but relatively low activity at pH values below and above the optimum pH Asker et al. [29] reported an alkaline protease from *Bacillus megaterium* with optimum pH of

Table 3 Effect of inhibitors on activity of purified protease from *Bacillus brevis* MWB-01

Inhibitor (5 mM)	Relative activity (%)
Control	100
EDTA	7.5 ± 0.5
β-Mercaptoethanol	9.6 ± 0.7
Phenylmethylsulphonyl fluoride (PMSF)	8.3 ± 0.5

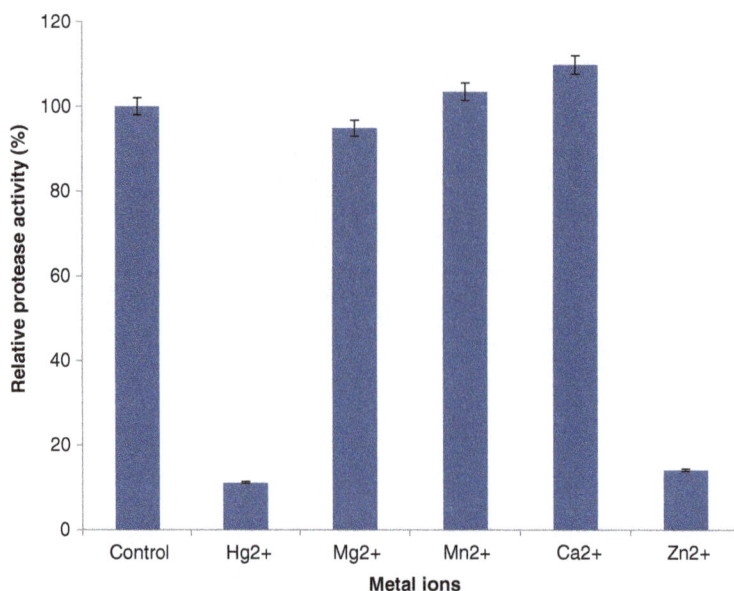

Figure 5 Effect of metal ions on activity of purified protease.

7.5 which showed relative activity of 8.0% and 10.0% at pH 5.5 and 9.0. Genkal and Tari [30] also reported an alkaline protease from a *Bacillus* sp. with optimum activity at pH 11.0 and low relative activity of 20% at pH 13.0. *Bacillus pseudofirmus* AL-89 had optimum pH of 11.0 and showed low activity at pH values below pH 10 and above 11.5 [31]. High alkalinity is a desirable characteristic for proteases used in detergent formulation [7,32]. Studies on pH stability showed that the enzyme was very stable in the pH range of 7.0 to 12.0 with above 60% residual protease activity (Figure 4). It is quite interesting that the purified protease had residual activity of 93% at pH 10.0 and 64% at pH 12. The protease from the strain of *B. brevis* MWB-01 under study distinctively demonstrated higher stability over a broader pH range of 6.0 to 12.0 than the previously reported protease from another strain of *B. brevis* which had optimum activity at pH 10.5 [24]. This exhibited characteristic by protease from *B. brevis* MWB-01 is of crucial importance in determining the suitability of proteases for use in detergents especially in ultrafiltration membrane cleaning which requires activity of protease over a broad pH [33].

Table 4 Substrate specificity of purified protease from *Bacillus brevis* MWB-01

Substrate	Relative activity (%)
Control (Casein)	100
BSA	74.5 ± 1.5
Egg albumin	63.2 ± 1.3
Collagen	4.1 ± 0.5
Gelatin	18.4 ± 0.7

Effect of inhibitors on activity of protease

All the inhibitors tested at 5 mM inhibited the protease (Table 3). PMSF inhibited the protease activity by 91.7% which suggests that the purified protease from *B. brevis* MWB-01 is a serine protease. PMSF sulphonates the essential serine residue in the active site of the protease and has been reported to result in the loss of enzyme activity [34]. β-ME inhibited protease activity by 90.4% which shows that the purified protease is not a cysteine protease which is activated in the presence of β-ME. Inhibition of the protease by β-ME might be due to denaturation of the enzyme resulting from cleavage of at least one disulfide bond that is critical to the stability of the protein and maintenance of the enzyme in its active form [35]. The enzyme lost 92.5% of its original activity in the presence of EDTA which infers that EDTA chelates metal ions at the active site of the purified protease which possibly serve as cofactors for the protease [36] confirming the enzyme is a metalloprotease.

Effect of metal ions on activity of protease

The protease activity was enhanced in the presence of Ca^{2+} and Mn^{2+} with 110% and 103.5% residual activity

Table 5 Detergent compatibility of purified protease from *Bacillus brevis* MWB-01

Detergent	Relative residual activity (%)
Control	100
Omo	40.0 ± 1.5
Sunlight	76.0 ± 2.1
Ariel	11.0 ± 0.9

respectively above the control (Figure 5) which indicates that these metal ions play an important role in maintaining the active site conformation of the purified protease [37]. This is similar to some previous reports on alkaline protease from *Bacillus* species which were activated in the presence of Ca^{2+} [17,24,28]. However, Hg^{2+} and Zn^{2+} strongly inhibited the activity of the enzyme with 89% and 86% loss in activity when compared with the control.

Substrate specificity of protease from *B. brevis* MWB-01
The purified protease exhibited highest level of hydrolytic activity on casein which was used as control. The enzyme hydrolysed bovine serum albumin and egg albumin showing relative activity of 74.5% and 63.2%, respectively (Table 4). Gelatin and collagen were poorly hydrolysed with relative activity of 18.4% and 4.1%, respectively. Earlier reports revealed that alkaline proteases showed highest activity towards casein relative to other proteins like BSA, gelatin [36,38-40]. Hydrolysis of various substrates has been reported as one of the important criteria for selection of proteases for application in laundry detergent formulation [41].

Detergent compatibility of protease
Purified protease from *B. brevis* MWB-01 was stable in the presence of tested commercial laundry detergents, Omo and Sunlight with relative residual activity of 76% and 40%, respectively (Table 5) in the absence of $CaCl_2$ and glycine. Similarly, protease from *B. amyloliquefaciens* PFB-01 retained above 60% of its original activity in the presence of Omo [17]. The detergent compatibility of alkaline protease from *B. brevis* MWB-01 in the absence of Ca^{2+} and glycine is particularly novel considering reports on similar studies that used Ca^{2+} and glycine to enhance detergent compatibility of the alkaline proteases studied [24,36]. Purified protease from *B. brevis* MWB-01 was least stable in the presence of Ariel with relative residual activity of 11%. The demonstrated detergent stability of protease from *B. brevis* MWB-01 infers that the enzyme had good compatibility with ingredients present in detergent formulation.

Conclusion
The characteristics of purified protease from *B. brevis* MWB-01 reveal that the enzyme is a thermotolerant serine alkaline metalloprotease which is compatible with detergent formulation aids. These results suggest that protease from *B. brevis* MWB-01 is a good bioresource for industrial applications.

Competing interests
The authors declare that they have no competing interests.

Authors' contributions
FMO conceived and designed the study and analyzed and interpreted the acquired data. AMF carried out the purification and characterization studies. Both authors participated in writing the manuscript and read and approved the final manuscript.

Acknowledgements
This research was supported by the International Foundation for Science, Sweden through research grant (F/3775-2) to Dr. Folasade M. Olajuyigbe. The authors gratefully acknowledge the contribution of Mr. S.O. Olatope and other technical staff of Biotechnology Unit, Federal Institute of Industrial Research Oshodi (F. I. I. R. O), Lagos, Nigeria in the identification of bacterial isolates.

References
1. Gupta R, Beg Q, Khan S, Chauhan B (2002) An overview on fermentation, downstream processing and properties of microbial alkaline proteases. Appl Microbiol Biotechnol 60:381–395
2. Vellard M (2003) The enzyme as drug: application of enzymes as pharmaceuticals. Curr Opin Biotechnol 14:444–450
3. Pastor MD, Lorda GS, Balatti A (2001) Protease obtention using *Bacillus subtils 3411* and amaranth seed meal medium at different aeration rates. Braz J Microbiol 32:6–9
4. Nascimento WC, Martins ML (2004) Production and properties of an extracellular protease from thermophilic *Bacillus* sp. SMIA2. Brazil. J. Microbiology 35:91–96
5. Beg QK, Gupta R (2003) Purification and characterization of an oxidation stable, thiol-dependent serine alkaline protease from *Bacillus mojavensis*. Enzyme Microb Technol 32:294–304
6. Breithaupt H (2001) The hunt for living gold. EMBO Rep 2:968–971
7. Sanatan PT, Lomate PR, Giri AP, Hivrale VK (2013) Characterization of a chemostable serine alkaline protease from *Periplaneta americana*. BMC Biochem 14:32
8. Olajuyigbe FM, Kolawole AO (2011) Purification and partial characterization of a thermostable alkaline protease from *Bacillus licheniformis* LHSB-05 isolated from hot spring. Afr J Biotechnol 10:11703–11710
9. Olajuyigbe FM, Ehiosun KI (2013) Production of organic solvent-tolerant and thermostable alkaline protease from *Bacillus coagulans* PSB-07 under different submerged fermentation conditions. Afr J Biotechnol 12:3341–3350
10. Aslim B, Yuksekdag ZN, Beyatli Y (2002) Determination of PHB growth quantities of certain *Bacillus* species isolated from soil. Turkish Electronic J Biotechnol Special issue:24–32
11. Vos P, Garrity G, Jones D, Krieg NR, Ludwig W, Rainey FA, Schleifer K, Whitman WB (eds) (2009) Bergey's manual of systematic bacteriology: the firmicutes, 2nd edn. New York, Springer
12. Buchanan RE, Gibbons NE (1990) Bergey's manual of determinative bacteriology, 19th edn. Williams and Wilkinsco, Baltimore
13. Fujiwara N, Masui A, Imanaka T (1993) Purification and properties of highly thermostable alkaline protease from an alkalophilic and thermophilic *Bacillus* sp. J Biotechnol 30:245–256
14. Bradford M (1976) A rapid and sensitive method for the quantitation of microgram quantities of protein utilizing the principle of protein-dye binding. Anal Biochem 72:248–254
15. Olajuyigbe FM, Ajele JO (2005) Production dynamics of extracellular protease from *Bacillus* species. Afr J Biotechnol 4:776–779
16. Olajuyigbe FM, Ajele JO (2011) Thermostable alkaline protease from *Bacillus licheniformis* LBBL-11 isolated from traditionally fermented African locust bean (*Parkia biglobosa*). J Food Biochem 35:1–10
17. Olajuyigbe FM, Ogunyewo OA (2013) Enhanced production and physicochemical properties of a commercially viable alkaline protease from *Bacillus amyloliquefaciens* PFB-01. Curr Biotechnol 2:73–80
18. Yu J, Jin H, Choi W, Yoon M (2006) Production and characterization of an alkaline protease from *Bacillus licheniformis* MH31. Agric Chem Biotechnol 49:135–139
19. Aoyama M, Yasuda M, Nakachi K, Kobamoto N, Oku H, Kato F (2000) Soybean-milk-coagulating activity of *Bacillus pumilus* derives from a serine proteinase. Appl Microbiol Biotechnol 53:390–395

20. Yang J, Shih I, Tzeng Y, Wang S (2000) Production and purification of protease from *Bacillus subtilis* that can deproteinize crustacean wastes. Enzyme Microb Technol 26:406–413

21. Joo H-S, Kumar CG, Park G-C, Paik SR, Chang C-S (2003) Oxidant and SDS-stable alkaline protease from *Bacillus clausii* I-52: production and some properties. J Appl Microbiol 95:267

22. Feng YY, Yang WB, Ong SL, Hu JY, Ng WJ (2001) Fermentation of starch for enhanced alkaline protease production by constructing an alkalophilic *Bacillus pumilus* strain. Appl Microbiol Biotechnol 57:153–160

23. Park I, Cho J (2011) Production of an extracellular protease by an Antarctic bacterial isolate (*Bacillus sp.* JSP1) as a potential feed additive. Revista Colombiana de Ciencias Pecuarias 24:1

24. Banerjee UC, Sani RK, Azmi W, Soni R (1999) Thermostable alkaline protease from *Bacillus brevis* and its characterization as a laundry detergent additive. Process Biochem 35:213–219

25. Prescott L, Harley J, Klein D (1996) Microbiology, 3rd edn. WCB Publishers, Chicago

26. Sun B, Wang X, Wang F, Jiang Y, Zhang X (2013) Assessing the relative effects of geographic location and soil type on microbial communities associated with straw decomposition. Appl Environ Microbiol 79:3327–3335

27. Papke RT, Ward DM (2004) The importance of physical isolation to microbial diversification. FEMS Microbiol Ecol 48:293–303

28. Ghorbel B, Sellami-Kamoun A, Nasri M (2003) Stability studies of protease from *Bacillus cereus* BG1. Enzyme Microb Technol 32:513–518

29. Asker MMS, Mahmoud MG, Shebwy KE, Abdelaziz MS (2013) Purification and characterization of two thermostable protease fractions from *Bacillus megaterium*. J Genet Eng Biotechnol 11(2):103–109

30. Genkal H, Tari C (2006) Alkaline protease production from alkalophilic *Bacillus sp* isolated from natural habitats. Enzyme Microb Technol 39:703–710

31. Takeda M, Iohara K, Shinmaru S, Suzuki I, Koizumi J (2000) Purification and properties of an enzyme capable of degrading the sheath of *Sphaerotilus natans*. Appl Environ Microbiol 66:4998–5004

32. Dias DR, Vilela DM, Silvestre MPC, Schwan RF (2008) Alkaline protease from *Bacillus sp.* isolated from coffee bean grown on cheese whey. World J Microb Biotechnol 24:2027–2034

33. Kumar CG (2002) Purification and characterization of a thermostable alkaline protease from alkalophilic Bacillus pumilus. Lett Appl Microbiol 34:13–17

34. Gold AM, Fahrney D (1964) Sulfonyl fluorides as inhibitors of esterases. II. Formation and reactions of phenylmethane sulfonyl alpha-chymotrypsin. Biochemistry 3:783–791

35. Nelson DL, Cox MM (2004) The three-dimensional structure of proteins. In: Lehninger principles of biochemistry, 4th edn. WH Freeman & Co, New York

36. Adinarayana K, Ellaiah P, Prasad DS (2003) Purification and partial characterization of thermostable serine alkaline protease from a newly isolated *Bacillus subtilis* PE-11. AAPS Pharm Sci Tech 4:56

37. Shivan P, Jayaraman G (2011) Isolation and characterization of a metal ion-dependent alkaline protease from a halotolerant *Bacillus aquimaris* VITP4. Ind J Biochem Biophys 48:95–100

38. Rahman RNZA, Razak CN, Ampon K, Basri M, Yunus WMZW, Salleh AB (1994) Purification and characterization of a heat stable alkaline protease from *Bacillus stearothermophilus* F1. Appl Microbiol Biotechnol 40:822–827

39. Jung H-J, Kim H, Kim J-I (1999) Purification and characterization of Co^{2+}-activated extracellular metalloprotease from *Bacillus* sp. JH108. J Microb Biotechnol 9:861–869

40. Yossana S, Reungsang A, Yasuda M (2006) Purification and characterization of alkaline protease from *Bacillus megaterium* isolated from Thai fish sauce fermentation process. Sci Asia 32:377–383

41. Gouda MK (2006) Optimization and purification of alkaline proteases produced by marine *Bacillus sp.* MIG newly isolated from Eastern Harbour of Alexandria. Pol J Microbiol 55:119–126

12

Effect of fluid shear stress on catalytic activity of biopalladium nanoparticles produced by *Klebsiella Pneumoniae* ECU-15 on Cr(VI) reduction reaction

Bin Lei, Xu Zhang[*], Minglong Zhu and Wensong Tan

Abstract

Background: Biopalladium (bioPd(0)) nanoparticles on *Klebsiella Pneumoniae* ECU-15 were synthesized mainly on the microorganism's surface. Data suggest that the resistance of mass transfer around the cell surface region plays a critical role in bioPd(0) synthesis process. However, the mechanisms for its role remains elusive.

Results: The experimental results indicated that 1) diffusion resistance existed around the microorganism's cell in reaction vessel and 2) fluid shear stress affected the mass transfer rates differently according to its strength and thus had varying effects on the bioPd(0) synthesis. More than $97.9 \pm 1.5\%$ Chromium(VI)(Cr(VI)) (384 µM) was reduced to Cr(III) within 20 min with 5% Pd/bioPd(0) as catalyst, which was generated by the *K. Pneumoniae* ECU-15, and the catalytic performance of Pd/bioPd(0) was stable over 6 months. The optimal condition of bioreduction of Pd(II) to Pd(0) was determined at the Kolmogorov eddy length of 7.33 ± 0.5 µm and lasted for 1 h in the extended reduction process after the usual adsorption and reduction process.

Conclusions: It is concluded that a high bioPd(0) catalytic activity can be achieved by controlling the fluid shear stress intensity in an extended reduction process in the bioreactor.

Keywords: Biopalladium; Mass transfer resistance; Fluid shear stress; *K. pneumoniae*

Background

The recovery of nanopalladium particles from waste has been of great interest recently [1]. Certain physical properties of nanopalladium particles differ from those of the bulk material [2,3]. Palladium is used extensively because of its catalytic activity in some chemical reactions. Traditionally, the preparation of nanopalladium requires the rigorous experimental procedures. In contrast, the microbiological biosynthetic process of converting Pd(II) to Pd(0) works through the binding of metal Pd(II) ions onto highly reactive bacterial cell surfaces concomitant with Pd(II) ion reduction [4] at ambient temperature and pressure. This process does not require hazardous chemicals and, thus, is effective and environmentally friendly.

Many kinds of bacterial species have been used in the reduction of Pd(II) salts to their elementary metallic form [1]. Humphries et al. investigated the reduction of

Cr(chromium(VI) by immobilized cells of *Desulfovibrio vulgairs* NCIMB8303 and *Microbacterium sp.* NCIMB 13776, which showed that the best immobilization matrices were 130 (agarose) and 15 (agar) nmol h^{-1} mg dry cell wt^{-1}, with the highest Cr(VI) reducing efficiency [5]. Macaskie et al. studied palladium catalysts generated on gram negative (*Desulfovibrio*) and gram positive (*Bacillus*) bacterial surfaces. Discrete nanoparticles were located in the periplasmic space of *Desulfovibrio desulfuricans* [6] and in the peptidoglycan and proteinaceous surface layer (S-layer) of *Bacillus sphaericu* [7]. De Windt et al. found that the bioreductive deposition of Pd(0) occurred on the cell wall and in the periplasmic space of *Shewanella oneidensis* in the presence of a series of electron donors [8]. Deplanche et al. investigated the involvement of hydrogenase in the formation of Pd(0) using *Escherichia coli* mutant strains [9]. Martins et al. found that a Pd(II) resistant bacterial community could biorecover this metal from a solution, and that the bacterial consortium was closely related to several *Clostridium* species, *Bacteroides*, and *Citrobacter* [10]. Many fermentative species

* Correspondence: zhangxu@ecust.edu.cn
State Key Laboratory of Bioreactor Engineering, East China University of Science and Technology, Shanghai 200237, People's Republic of China

could produce hydrogen during fermentation and subsequently reduce Pd(II) to Pd(0). An addition of Pd(II) to the fermenting culture of *Clostridium pasteurianum* could result in the formation of Pd(0) nanoparticles on the bacterial cell wall and in the cytoplasm [11]. *Klebsiella pneumoniae* is a potential biohydrogen producer [12] and has the ability to bioreduce Pd(II) [13]. The evaluation of the capacity of *K. pneumoniae* to reduce Pd(II) to Pd(0) nanoparticles would be significant because of its respiration diversity, its wide distribution in various environments, and its potentially effective approach to the remediation of aggressive metal waste.

Biopalladium nanoparticles were synthesized mainly inside the cell surface region [14]. The resistance to mass transfer inside this region during bioPd(0) preparation should not be neglected. Han et al. found that besides the heating effect, microwaves could also enhance the mass transfer rates of active constituents. The bound potential of cell walls could be overcome by a certain microwave intensity to increase the extraction efficiency [15]. The fluid shear stress could also be used to open the periplasmic layer, reduce its mass transfer resistance and enhance the rates [16]. However, the influence of shear stress on the chemical and physical properties of biosynthetic nanopalladium particles was not fully understood. In this study, the influence of fluid shear stress during the bioPd(0) preparation process on its catalytic activity for Cr(VI) reduction to Cr(III) was investigated. Physical properties of the bioPd(0) were measured by a transmission electron microscopy (TEM), energy-dispersive X-ray spectroscopy (EDAX), and X-ray diffractometry (XRD). The results would be significant for the development of the efficient and stable bioPd(0) catalyst production biotechnology.

Methods

Preparation of the bioPd(0) catalysts

K. pneumoniae ECU-15 was isolated from the anaerobic sewage sludge [12] and cultured to a high-cell concentration in an experimental reactor. The biomass at mid-logarithmic phase cultures was harvested by a centrifuge (4000 g, 15 min, 4°C). The pellets were washed three times by 100 mL of MOPS-NaOH buffer (20 mM, pH 7.2) and resuspended in 50 mL of the same buffer. And then it was stored at 4°C under N_2 for no more than 24 h until use. Cell concentration (mg/mL) was determined by a correlation to a predetermined OD600 to dry weight conversion.

A known volume of the concentrated resting cell suspension was transferred anaerobically into a 100-mL serum bottle, which contained an appropriate volume of degassed 2-mM $PdCl_2$ solution to make the final weight ratio of Pd to dry cells equal to 1:19. That means a final loading of 5% Pd on biomass could be obtained. A series of steps was scheduled as follows: firstly, the static adsorption process between the microorganism and the Pd(II) lasted for 30 min, then the solution was sparged by H_2 for 20 min in the reduction process. The H_2 was retained in the serum bottle till the end of the preparation process. The degree of the Pd(II) reduction from the solution was confirmed by assaying the residual Pd(II) ion concentration by the $SnCl_2$ method in cell/Pd mixture supernatant [17]. The Pd(0)-coated biomass was harvested by centrifugation(4,000 g, 15 min, 25°C) and rinsed three times by a distilled water (dH_2O). The black precipitate was washed once by acetone and left to dry in air for overnight to a constant weight and finely ground in a mortar. The physical properties of the Pd-loaded bacterial cells were measured using TEM (JEM-2100, JEOL, Akishima-shi, Japan), EDAX (DX-4, EDAX, Mahwah, NJ, USA), and the XRD (D/max2550B/PC, Rigaku, Shibuya-ku, Japan). The catalytic activity of the bioPd(0) was tested without further processing.

Transmission electron microscopy

The bacteria loaded with the palladium were rinsed twice with the distilled water and then fixed in the 2.5% glutaraldehyde. After being centrifuged at 10,000 rpm, the pellets of palladium-loaded bacteria were resuspended in 15 mL of 0.1 M Na-cacodylate buffer and then stained for an hour in the 1% osmium tetroxide solution in 0.1 M phosphate buffer at pH 7 for the measurement of the TEM. Cells were dehydrated for 15 min by the 70%, 90%, and 100% ethanol solution, respectively. Then, it was washed twice in the propylene oxide for 15 min. The cells were embedded in the epoxy resin and then the mixture was left to be polymerized at 60°C for 24 h. A section (100-nm thick) was cut from the resin block, placed onto a copper grid, and then viewed with TEM (JEM-2100, JEOL, Akishima-shi, Japan).

EDAX and XRD measurements

For the EDAX and the XRD analysis, the parallel samples of Pd(0)-loaded biomass were washed for three times in the distilled water and once in acetone. After the centrifugation process, the Pd(0)-loaded biomass were resuspended in a small volume of acetone and then air-dried to a constant weight. Finally, it was grounded to a fine powder by an agate mortar. The powder was analyzed using the EDAX (DX-4, EDAX, Mahwah, NJ, USA) and the XRD (D/max2550B/PC, Rigaku, Shibuya-ku, Japan). The powder pattern was compared to the references in the Joint Committee for Powder Diffraction Studies (JCPDS) database.

Evaluation of catalytic activity via Cr(VI) reduction to Cr(III)

The bioPd(0) could be used as the catalyst for the reduction of Cr(VI) to Cr(III). In this paper, a comparative study of the catalytic activity of bioPd(0) under various condition of preparation was carried out. During the experimental process, 10 mg bioPd(0) were accurately weighed and then put into a 25-mL serum bottle sealed with butyl rubber stopper, containing 10 mL, 384 μM Cr (VI). The mixture was made anaerobic by sparging the solution with oxygen free nitrogen for 10 min. The bottle was stalled in a 30°C water bath for 30 min before the reaction. The reaction was initiated when 1 mL of sodium formate as the electron donor was added with the final concentration to 25 mM. The samples were periodically withdrawn from the bottle via the rubber septa and centrifuged at 4,000 rpm for 2 min to remove the bioPd(0) catalyst. The supernatant was measured to determine the concentration of Cr(VI) and Cr(III).

Effect of different conditions on catalytic activity in the extended reduction of Cr(VI) to Cr(III)

Comparative studies of the catalytic activity of bioPd(0) for Cr(VI) reduction to Cr(III) were carried out under different bioPd(0) preparation conditions. The preparation of bioPd(0) nanoparticles through the reduction of Pd(II) by *K. pneumoniae* ECU-15 comprised three steps, as represented schematically in Figure 1. In the adsorption step, the Pd(II) ions in the solution were adsorbed on the cell surface. In the reduction step, the adsorbed Pd(II) ions were reduced to Pd(0) at the cell surface with hydrogen sparging. And in the third step, the reduction process was extended through the introduction of stationary or shaking factors, which represented the new findings of this paper.

Stationary operation in the extended reduction process referred to the cell/Pd(0)/Pd(II) mixture solution being allowed to stand from 0 to 13 h. BioPd(0) was harvested and used as Cr(VI) reduction catalyst after several hours of standing time. The variation in shaking factors referred to a variation in shaker speed from 50 to 200 rpm and time from 1 to 3 h only after hydrogen sparging. The reduction of Cr(VI) was estimated as the percent of Cr(VI) which were reduced to Cr(III) at the set time. The reduction ratio could be calculated from the following formulas:

$$\eta = \frac{C_0 - C_t}{C_0} \tag{1}$$

Where C_0 was the initial concentration of Cr(VI), and C_t was the concentration of Cr(VI) at time t.

Results and discussion

Examination of bioPd(0) samples by electron microscopy, energy dispersive X-ray spectroscopy, and X-ray diffraction

As shown in Figure 2, the color of the solution in the reduction system changed from milk white to black, which indicated that the Pd(0) might be generated through the reduction of Pd(II) ions by the *K. pneumoniae* ECU-15. After the centrifugation and rinse treatment, the black precipitate was dried to a constant weight and finely ground in a mortar, which was finally stored in the vacuum desiccator.

The TEM photographs of the Pd-located cells of the *K. pneumoniae* ECU-15 were shown in Figure 3a,b,c,d. The nanoPd(0) deposits were almost the round black particles with similar morphology. It was also shown that the Pd particles were distributed on the periplasmic space and the cytoderm of the *K. pneumoniae* ECU-15, which was consistent with that of the previous research [18]. It seems that there was no bioPd(0) appeared in the intercellular space, which indicated that the nano bioPd(0) might be generated by some catalytic reduction reaction on the cell surface of *K. pneumoniae* ECU-15. According to the various magnifications of the TEM images of the bioPd(0) nanoparticals, it seems that the particles size ranged from 5.0 to 26.5 nm, with the average article size of 15.8 nm. The identity of the nanoparticle deposits as bioPd(0) was confirmed by the EDAX analysis shown in Figure 3e, where the Pd components was obvious. The C, O, and P components shown in Figure 3e were the cellular material. The characteristic diffraction peak of the crystalline Pd(0) particles at $2\theta = 40°$, $47°$, and $68°$ were also shown by the XRD diffractograms, as reported by Hennebel et al. [13]. It was also confirmed that the cell-mediated reduction of the Pd(II) to the metallic Pd(0), with a face-centered cubic structure [19,20] through the XRD patterns of the Pd(II)-challenged cells, were shown in Figure 3f. All these results proved that the generation of the bioPd(0) was through the reduction of the Pd(II) to Pd(0) on the surface of the *K. pneumoniae* ECU-15 cells. The location of the bioPd(0) nanoparticles generated might be caused by the distribution and properties of hydrogenase and other enzymes involved in Pd(II) reduction [11]. However, Rotaru found that the cell surface functional groups really played a

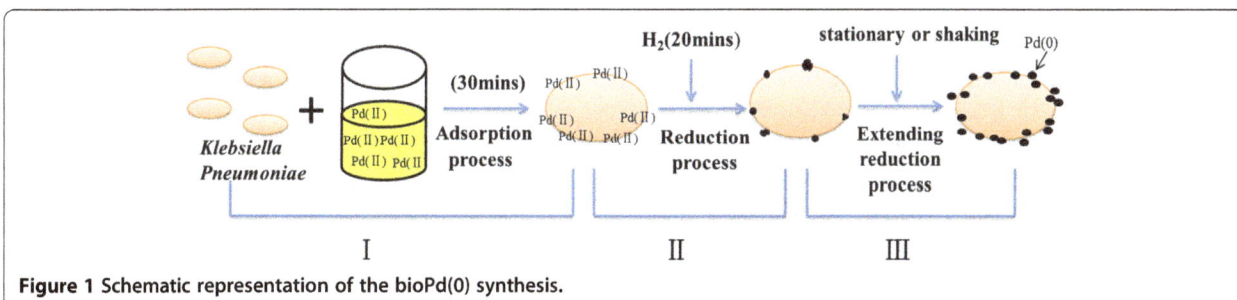

Figure 1 Schematic representation of the bioPd(0) synthesis.

Figure 2 Photographs of color during the process of Pd(II) reduction to Pd(0) with *K. pneumoniae* ECU-15. **(a)** Pd(II) solution, **(b)** Pd(II) and the *K. pneumoniae* suspension, and **(c)** bioPd(0) and the *K. pneumoniae* solution.

catalytic function role for the Pd(II) reduction to the bioPd(0) [3]. These mechanisms were not yet unified and needed to be explored carefully. The accepted view was that the reactant should firstly transfer from the solution to the function groups or the enzymes on the cell surface, and thus, the resistance of mass transfer inside the cell surface scaffold should not be neglected.

Effect of extended reduction standing time during bioPd (0) preparation on catalytic activity for Cr(VI) reduction to Cr(III)

The cell wall of *K. pneumoniae* ECU-15 is composed mainly of peptidoglycan, which contributes to the mechanical strength of the bacterial cell structure and prevents the intrusion of various ions as a barrier to mass transfer [21]. As discussed above, the synthesis process of bioPd(0) nanoparticles by the reduction of Pd(II) on *K. pneumoniae* ECU-15 included an adsorption process and a reduction process [20]. In order to confirm the time of the adsorption process, the adsorption efficiency of Pd(II) in the solution could be determined by detecting the residual Pd(II) concentration through the $SnCl_2$ method. The preliminary experiments showed that the residual Pd(II) concentration gradually decreased within the first 30 min, and the adsorption equilibration of Pd (II) would be reached at 30 min. The influence of standing time on the extended reduction of Cr(VI) to Cr(III) by bioPd(0) catalytic activity was given in Figure 4. The catalytic activity was improved by extending the reduction time. An increase in standing time from 0 to 6 h resulted in the complete reduction of Cr(VI) to Cr(III) in 1 h instead of 5 h. Especially, the Cr(VI) reduction ratio (η) has increased exponentially in the first 1 h with the standing treatment. With a further increase in standing time, the improvement in catalytic activity of the bioPd (0) was not obvious. Extending the reduction time would influence the equilibrium time for bioPd(0) generation, which influence the bioPd(0) catalytic activity indirectly. The optimal standing time was found to be 6 h. The TEM micrograph in Figure 5 showed that the nanoparticles

were distributed homogeneously with higher density in the periplasmic space and cytoderm with the longer standing time (Figure 5a,c), while in Figure 5b,d, the palladium particles demonstrated with the properties of uneven size and inhomogeneity distribution. This observation further confirmed the results in Figure 4.

During the Pd(II) reduction process, the Pd(II) ions and molecular hydrogen in the solution would pass through several layers, such as the gas-liquid, liquid-liquid, and solid–liquid layers before arriving at the hydrogenase and functional organic groups on the membrane and periplasmic space of the cells [22]. It was followed by the nucleation of Pd(0) and further cluster growth. The three-dimensional structure of the periplasmic space around the bacterial cell had a certain resistance to the ions [23] and the molecular diffusion process and more time was required for the delivery of Pd(II) ions and molecular hydrogen to the microorganism cell surface to complete the reduction of Pd(II) to Pd(0). Thus, it's necessary to overcome the resistance to mass transfer around the microgram cell for the efficient preparation of stable and high efficiency bioPd(0).

Effect of fluid shear stress during Pd(II) adsorption and reduction processes on the catalytic activity of bioPd(0) for Cr(VI) reduction to Cr(III)

It is critical that the Pd(II) ion and molecular hydrogen were transferred efficiently to the synthesis site of the bioPd(0) on the cell surface of *K. pneumoniae* ECU-15. Therefore, mass transfer problems in the preparation of bioPd(0) should be investigated. The fluid shear stress, which was generated by various flask shaking intensities, could reduce the mass transfer resistance in the periplasmic layer [23,24]. The movement of the turbulent eddies, which were produced by fluid turbulence intensities, would generate fluid shear stress [25]. The turbulence intensity could be demonstrated by the Kolmogorov eddy length of the fluid [26]. For an internal fluid in the Erlenmeyer flask at different shaking speeds, the Kolmogorov eddy length of the fluid turbulence could be calculated from Equation (2) [27]. The estimated values for the energy dissipation rate and smallest eddy size at different shaking speeds were given in Table 1. When the shaking speeds increased from 50 to 200 rpm, the Kolmogorov eddy length would decrease from 160.69 to 3.26 μm. And thus, the impact of fluid shear force could be expressed quantitatively by the Kolmogorov eddy length.

$$\lambda = \left(\frac{v^3}{\varepsilon}\right)^{0.25}, \quad \varepsilon = k\rho^{-1}V_t^{-0.25}N^{2.81} \qquad (2)$$

where λ is the Kolmogorov eddy length (μm), v is the fluid kinematic viscosity (m²/s), ε is the unit average

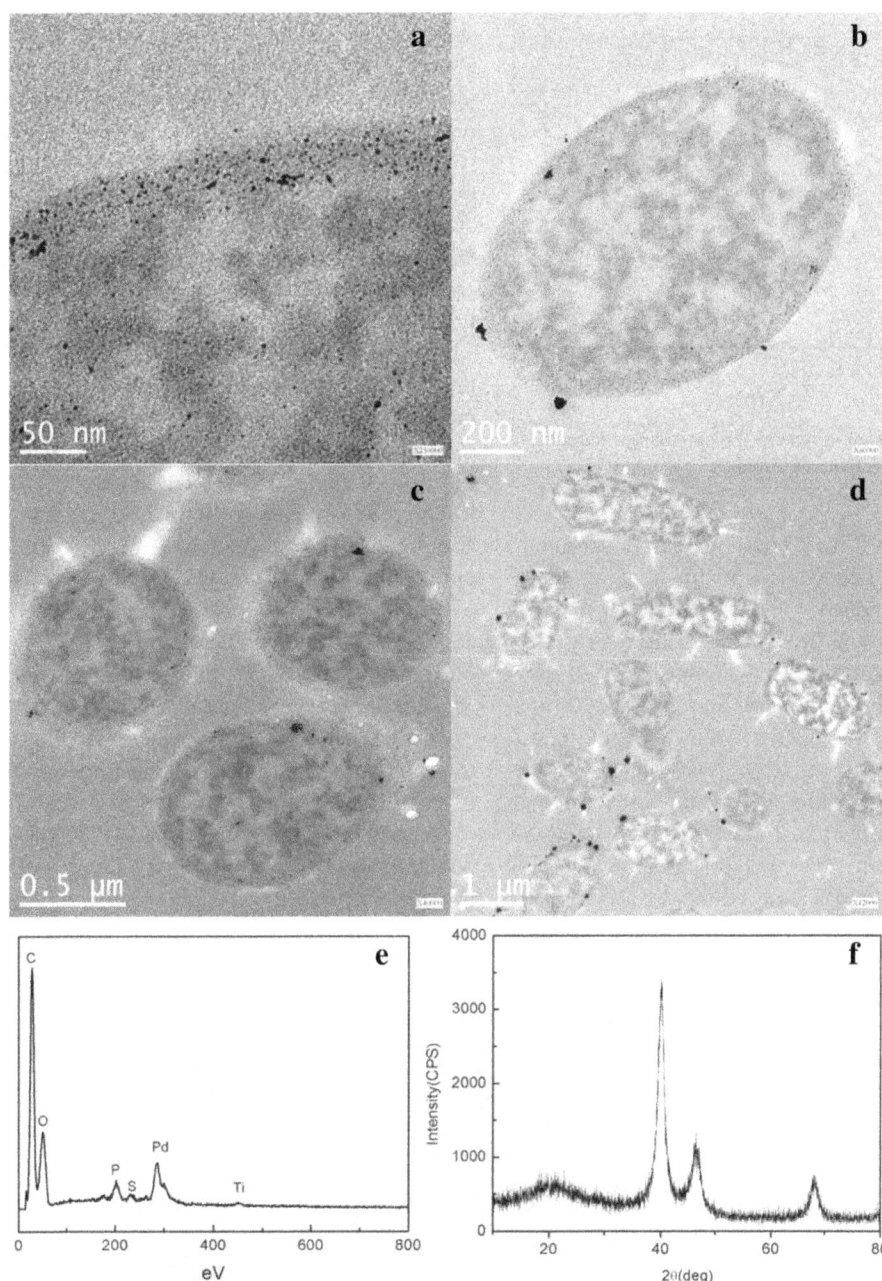

Figure 3 Properties of bioPd(0) generated by Pd(II) reduction on the cell surface of the *K. pneumoniae* ECU-15. **(a-d)** The TEM of nanoparticles of Pd(0) at different scale; **(e)** The EDAX results of Pd(0); **(f)** The XRD patterns of Pd(0).

energy consumption power of the Erlenmeyer flask, k is the constant (1.09×10^{-3}), ρ is the liquid density (kg/m^3), V_t is the working volume in the Erlenmeyer flask (m^3), and N is the shaking speed of the Erlenmeyer flask (s^{-1}).

Additional file 1: Figure S1 showed the recovery ratios of Pd(0) at different shaking speed. It seems that all the recovery of Pd(0) exceeded 90% with little difference shown among the various shaking speeds. The possible reason was that the Pd(II) was mainly reduced by H$_2$, but the other factors, such as the fluid shear stress, the

hydrogenase, etc., in the bioreduction process predominantly regulated and controled the physical properties of the Pd(0), which fundamentality influenced the catalytic activity of Pd(0). The effects of shaking speed during the adsorption and the reduction process on the catalytic activity of bioPd(0) for Cr(VI) reduction to Cr(III) were studied in this study with results shown in Figure 6a. The η of the Cr(VI) reduction to Cr(III) increased as the reaction time increased, and gradually added to a certain value. At the initial time of the reduction process, the

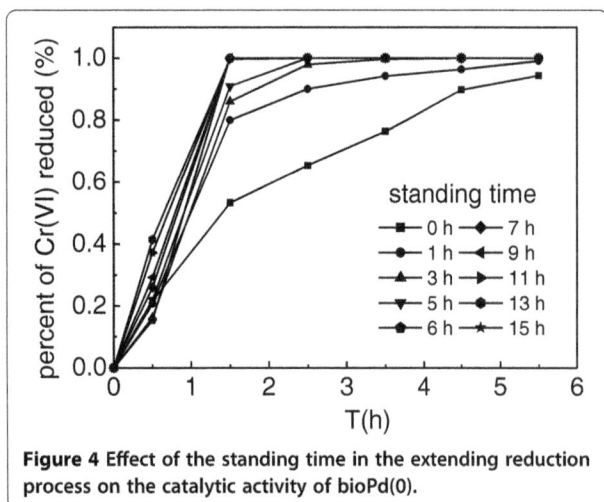

Figure 4 Effect of the standing time in the extending reduction process on the catalytic activity of bioPd(0).

reduced rates of Cr(VI) to Cr(III) increased exponentially by the bioPd(0) generated with the shaking treatment, which was obviously higher than those without shaking. After 3.5 h, 84.6% Cr(VI) was reduced to Cr(III) by the bioPd(0) generated at the shaking speed of 150 rpm. It reached the 95% with the bioPd(0) prepared at 200 rpm, while only 60% Cr(VI) was reduced with 300 rpm. It was indicated that the higher shear stress was beneficial to the synthesis of bioPd(0) with higher catalytic activity. However, too high shear stress, such as 300 rpm, would be harmful, which adversely affect the preparation of bioPd (0). Approximately 5.5 h was required for the accomplishment of the reduction of Cr(VI) to Cr(III) by the bioPd(0)

generated at a shaking speed of 150 or 200 rpm in the generated process. The data suggested that the shaking during adsorption or reduction process did not obviously impact on the improvements in the catalytic activity of the bioPd (0). It was estimated that this process was reaction controlled, and the increase of the mass transfer rates would not visibly influence the bioPd(0) catalytic activity.

Effect of fluid shear stress during the extended Pd(II) reduction process on catalytic activity of bioPd(0) for Cr(VI) reduction to Cr(III)

The cell-Pd(II) mixed with liquor after 30 min adsorption was sparged with hydrogen for 20 min under static conditions before it was placed on a shaking table. As seen in Figure 6b, the η was studied as functions of the different shaking speeds. The catalytic activity of bioPd (0) on the η could be improved severalfold when the shaking speeds were increased from 50 to 200 rpm. The optimum reduction condition was determined at 75 min, 150 rpm. When the shaking speeds increased from 50 to 150 rpm, the catalytic activity of bioPd(0) increased proportionately. However, the catalytic activity of the bioPd (0) decreased at 200 rpm. The catalytic activity of bioPd (0) with shear stress was higher than that without. And then the influence of the shaking time on the catalytic activity in the extended reduction process was investigated. As seen in Additional file 1: Figure S2, there was no obvious improvement in the catalytic activity of bioPd(0) with the time extended. The effect of fluid shear stress on bioPd(0) catalytic activity varied among the adsorption,

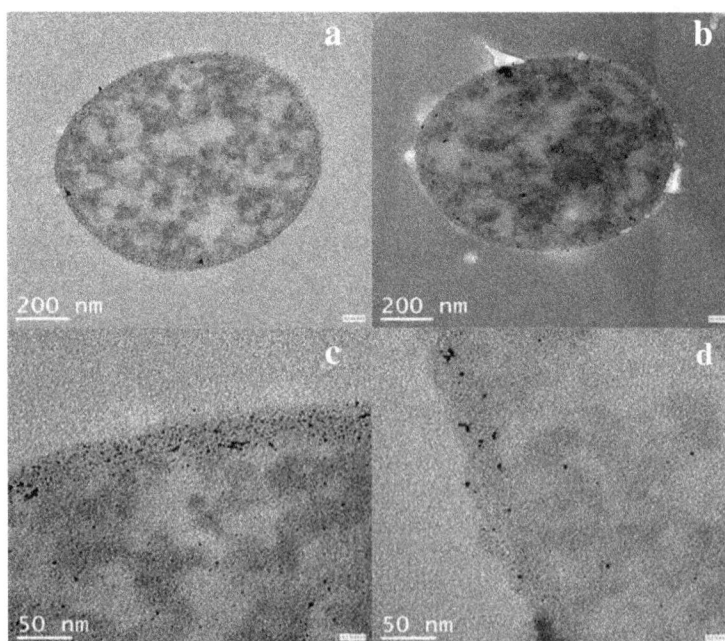

Figure 5 TEM photo of bioPd(0) at different standing time in the extending reduction process. TEM photo of bioPd(0) collected at a standing time of 6 h **(a,c)** and collected immediately **(b,d)** in the extending reduction process.

Table 1 Estimated values for energy dissipation rate and smallest eddy size of the reactor fitted at different speeds

Shaking speeds(rpm)	$\varepsilon(W/Kg$ or $m^2/s^3)$	λ (μm)
50	0.006	160.69
100	0.044	22.91
150	0.138	7.33
200	0.312	3.26

reduction, and extended reduction process. In the adsorption and reduction processes, more time was required for the dissolved hydrogen molecules to reach a certain concentration around the cell surface before the beginning of the bioPd(0) synthesized process; whereas in the extended reduction process, when the dissolved hydrogen molecule and Pd(II) concentrations in the periplasmic region reached a certain level, strengthening the mass transfer for generating biological palladium would have a significant effect on the catalytic activity of bioPd(0). Hence, it was speculated that the shaking might assist in the even dispersion of Pd(II) on or into the location of Pd(0) generated sites. The fluid shear stress in the extended reduction process could affect the catalytic activity of bioPd(0) significantly. A possible explanation was that this stage was mass transfer controlled, and strengthening the mass transfer would be beneficial to generate bioPd(0) with high catalytic activities [28]. It required the further research.

The diffusion rates of Pd(II) ions and hydrogen molecules from the solution to the liquid boundary layer surrounding the bacterial cell increased with fluid turbulence intensities. The thickness of the liquid boundary layer and the external diffusion resistance decreased with turbulence intensity and fluid shear stress [29]. The too high shear stress might reduce the enzyme activity and destroy the cellular structure [24,30]. Small eddies, similar in size to those of the cells, could cause the higher shear stresses on the cells and lead to the physical damage [31]. As shown in Table 1, when the shaking speeds increased from 50 to 150 rpm, the smallest eddy sizes decreased from 160.69 to 7.33 μm, which was greater than the size of the *K. pneumoniae* ECU-15 cells (1–1.5 μm). However, when the shaking speed was 200 rpm, the smallest eddy size (3.26 μm) was close to that of the *K. pneumoniae* ECU-15 cells with the catalytic activity of the bioPd(0) for Cr(VI) reduction to Cr (III) decreased. A suitable fluid shear stress would probably decrease the mass transfer resistance with more bioPd(0) being nucleated and evenly distributed on the cells. Considering the time consumption and cost of resources, the optimal conditions for the preparation of high catalytic activity bioPd(0) were to shake the mixed liquor at 150 rpm during the extended reduction process.

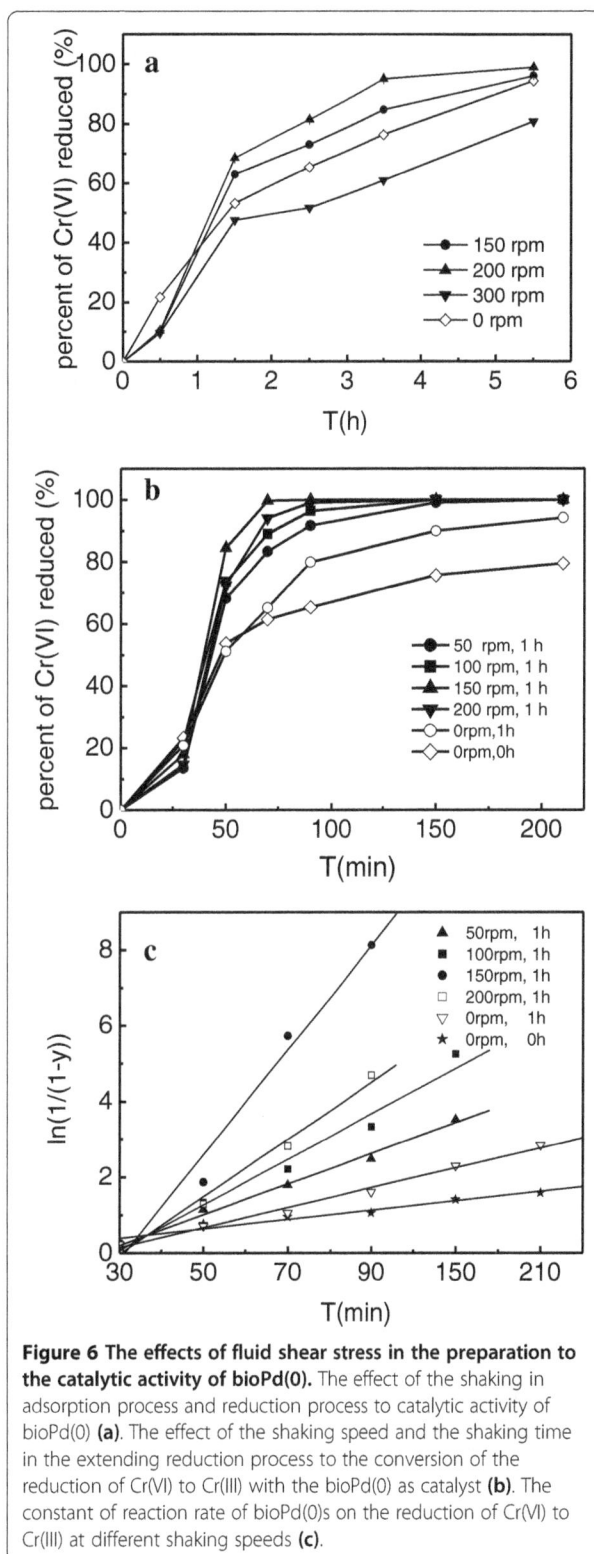

Figure 6 The effects of fluid shear stress in the preparation to the catalytic activity of bioPd(0). The effect of the shaking in adsorption process and reduction process to catalytic activity of bioPd(0) **(a)**. The effect of the shaking speed and the shaking time in the extending reduction process to the conversion of the reduction of Cr(VI) to Cr(III) with the bioPd(0) as catalyst **(b)**. The constant of reaction rate of bioPd(0)s on the reduction of Cr(VI) to Cr(III) at different shaking speeds **(c)**.

More than $97.9 \pm 1.5\%$ Cr(VI) (384 μM) was reduced to Cr(III) within 20 min with 5% Pd/bioPd(0) as the catalyst. The catalytic performance of the bioPd(0) was stable over 6 months. The 5% Pd/carbon commercial catalyst could reduce $96 \pm 1.3\%$ Cr(VI)(500 µM) in 60 min [9]. It seems that the bioPd(0) generated from the fluid shear stress regulation process could be a competitive biocatalyst.

The Cr(VI) reaction ratio changed with different types of bioPd(0) catalyst and was solely attributed to the different shaking speeds in the extended reduction process. The reaction ratio constants (k) of Cr(VI) to Cr(III) with different types of bioPd(0) as catalysts were calculated from Equation (3). The slope of the lines in Figure 6c was the reaction ratio constant (k). A maximum k of 1.199 min^{-1} was obtained at 150 rpm. The difference in maximum k was mainly because of the maximum catalytic activity of the bioPd(0), which resulted from the different shaking speeds in the extended reduction process. A suitable shaking speed would reduce the reaction time of Cr(VI) to Cr(III) reduction process.

$$In\left(\frac{1}{1-y}\right) = k\,t \tag{3}$$

where y is the percentage Cr(VI) consumption, t is the test time (s), and k is the reaction rate constant.

Conclusions

BioPd(0) was produced inside the periplasm of *K. pneumoniae* ECU-15 cells and could catalyze the reduction of Cr(VI) to Cr(III). Extending the standing time in the reduction process would be benefical for the biosynthesis of the bioPd(0) with high catalytic activity. Increasing the mass transfer rates through regulating the fluid shear stress for the extended reduction process during the bioPd(0) generation process would improve the catalytic activity of bioPd(0) for the reduction of Cr(VI) to Cr(III). The optimal condition of the bioreduction of Pd(II) to Pd(0) was at the Kolmogorov eddy length of 7.33 ± 0.5 µm, 1 h in the extended reduction process after the usual adsorption and reduction process. Increasing the mass transfer rate by the regulation of the fluid shear stress could provide certain advantages and possibilities for the development of bioreactor technologies for the noble metal recovery.

Additional file

Additional file 1: The supplementary information. The bioreduced of Pd(II) as the function of the shaking speed **(Figure S1)**. The effect of rotating speed with different time to the catalytic activity of BioPd(0) **(Figure S2)**.

Competing interests

The authors declare that they have no competing interests.

Authors' contributions

BL, XZ, MLZ, and WST have made substantive intellectual contributions to this study, substantial contributions to the conception and design of it as well as to the acquisition, analysis, and interpretation of data. All of them have been also involved in the drafting and revision of the manuscript. All authors have contributed to and seen the manuscript. All authors read and approved the final manuscript.

Acknowledgements

This study was financially supported by the open Project Funding of State Key Laboratory of Bioreactor Engineering of China and the National High Technology Research and Development Program of China (No. 2007AA060904 and No. 2012AA061503).

References

1. Yong P, Rowson NA, Farr JPG, Harris IR, Macaskie LE (2002) Bioreduction and biocrystallization of palladium by *Desulfovibrio desulfuricans* NCIMB 8307. Biotechnol Bioeng 80:369–379
2. Hennebel T, De Gusseme B, Boon N, Verstraete W (2009) Biogenic metals in advanced water treatment. Trends Biotechnol 27:90–98
3. Rotaru AE, Jiang W, Finster K, Skrydstrup T, Meyer RL (2012) Non-enzymatic palladium recovery on microbial and synthetic surfaces. Biotechnol Bioeng 109:1889–1897
4. Karthikeya S, Beveridge TJ (2002) Pseudomonas aeruginosa biofilms react with and precipitate toxic soluble gold. Environ Microbiol 4:667–675
5. Humphries AC, Nott KP, Hall LD, Macaskie LE (2005) Reduction of Cr(VI) by immobilized cells of *Desulfovibrio vulgaris* NCIMB 8303 and *Microbacterium sp.* NCIMB 13776. Environ Microbiol 90:589–596
6. Lloyd JR, Yong P, Macaskie LE (1998) Enzymatic recovery of elemental palladium by using sulfate-reducing bacteria. Appl Environ Microbiol 64:4607–4609
7. Creamer NJ, Mikheenko IP, Yong P, Deplanche K, Sanyahumbi D, Wood J, Pollmann K, Merroun M, Selenska-Pobell S, Macaskie L (2007) Novel supported Pd hydrogenation bionanocatalyst for hybrid homogeneous/heterogeneous catalysis. Catal Today 128:80–87
8. De Windt W, Aelterman P, Verstraete W (2005) Bioreductive deposition of palladium (0) nanoparticles on Shewanella oneidensis with catalytic activity towards reductive dechlorination of polychlorinated biphenyls. Environ Microbiol 7:314–325
9. Deplanche K, Caldelari I, Mikheenko IP, Sargent F, Macaskie LE (2010) Involvement of hydrogenases in the formation of highly catalytic Pd (0) nanoparticles by bioreduction of Pd (II) using Escherichia coli mutant strains. Microbiol 156:2630–2640
10. Martins M, Assuncao A, Martins H, Matos AP, Costa MC (2013) Palladium recovery as nanoparticles by an anaerobic bacterial community. J Chem Technol Biotechnol 88:2039–2045
11. Chidambaram D, Hennebel T, Taghavi S, Mast J, Boon N, Verstraete W, van der Lelie D, Fitts JP (2010) Concomitant microbial generation of palladium nanoparticles and hydrogen to immobilize chromate. Environ Sci Technol 44:7635–7640
12. Niu K, Zhang X, Tan WS, Zhu ML (2010) Characteristics of fermentative hydrogen production with *Klebsiella pneumoniae* ECU-15 isolated from anaerobic sewage sludge. Int J Hydrogen Energy 35:71–80
13. Hennebel T, Van Nevel S, Verschuere S, De Corte S, De Gusseme B, Cuvelier C, Fitts JP, Van der Lelie D, Boon N, Verstraete W (2011) Palladium nanoparticles produced by fermentatively cultivated bacteria as catalyst for diatrizoate removal with biogenic hydrogen. Appl Microbiol Biotechnol 91:1435–1445
14. Deplanche K, Merroun ML, Casadesus M, Tran DT, Mikheenko IP, Bennett JA, Zhu J, Jones IP, Attard GA, Swlenska-Pobell S, Macaskie LE (2012) Microbial synthesis of core/shell gold/palladium nanoparticles for applications in green chemistry. J R Soc Interface 9:1705–1712
15. Han GZ, Chen MD (2008) Microwave peak absorption frequency of liquid. S Sci China Ser G-Phys Mech Astron 51:1254–1263
16. Tzima E, Irani-Tehrani M, Kiosses WB, Dejana E, Schultz DA, Engelhardt B, Cao G, DeLisser H, Schwartz MA (2005) A mechanosensory complex that mediates the endothelial cell response to fluid shear stress. Nature 437:426–431
17. Deplanche K, Mikheenko I, Bennett J, Merroun M, Mounzer H, Wood J, Macaskie L (2011) Selective oxidation of benzyl-alcohol over biomass-supported Au/Pd bioinorganic catalysts. Top Catal 54:1110–1114
18. Bunge M, Sobjerg LS, Rotaru AE, Gauthier D, Lindhardt AT, Hause G, Finster K, Kingshott P, Skrydstrup T, Meyer RL (2010) Formation of palladium (0) nanoparticles at microbial surfaces. Biotechnol Bioeng 107:206–215
19. Wang RF, Wang H, Feng HQ, Ji S (2013) Palladium decorated nickel nanoparticles supported on carbon for formic acid oxidation. Int J Electrochem Sci 8:6068–6076

20. Tobin JM, White C, Gadd GM (1994) Metal accumulation by fungi: applications in environmental biotechnology. J Ind Microbiol 13:126–130
21. Gumbart JC, Beeby M, Jensen GJ, Roux B (2014) *Escherichia coli* peptidoglycan structure and mechanics as predicted by atomic-scale simulations. Plos Comput Biol 10:1003475
22. Nies DH (2003) Efflux-mediated heavy metal resistance in prokaryotes. FEMS Microbiol Rev 27:313–339
23. Hodgson L, Tarbell JM (2002) Solute transport to the endothelial intercellular cleft: the effect of wall shear stress. Ann Biomedl Eng 30:936–945
24. Silva-Santisteban BOY, Filho FM (2005) Agitation, aeration and shear stress as key factors in inulinase production by *Kluyveromyces marxianus*. Enzyme Microb Technol 36:717–724
25. Lantz J, Gardhagen R, Karlsson M (2012) Quantifying turbulent wall shear stress in a subject specific human aorta using large eddy simulation. Med Eng Phys 34:1139–1148
26. Allen JJ, Shockling MA, Kunkel GJ, Smits AJ (2007) Turbulent flow in smooth and rough pipes. Phil Trans R Soc A 365:699–714
27. Onishi R, Matsuda K, Takahashi K, Kurose R, Komori S (2008) Retrieval of collision kernels from the change of droplet size distributions with linear inversion. Phys Scripta 2008:014050
28. Klaewkla R, Arend M, Hoelderich WF (2011) A review of mass transfer controlling the reaction rate in heterogeneous catalytic systems. In: de Mass Transfer-Advanced Aspects. InTech, Germany, pp 668–684
29. Evans JR, Davids WG, MacRae JD, Amirbahman A (2002) Kinetics of cadmium uptake by chitosan-based crab shells. Water Res 36:3219–3226
30. Ghadge RS, Patwardhan AW, Joshi JB (2006) Combined effect of hydrodynamic and interfacial flow parameters on lysozyme deactivation in a stirred tank bioreactor. Biotechnol Prog 22:660–672
31. Cherry RS, Papoutsakis ET (1986) Hydrodynamic effects on cells in agitated tissue culture reactors. Bioproc Eng 1:29–41

13

Engineering of *Corynebacterium glutamicum* for growth and production of L-ornithine, L-lysine, and lycopene from hexuronic acids

Atika Hadiati, Irene Krahn, Steffen N Lindner and Volker F Wendisch[*]

Abstract

Background: Second-generation feedstocks such as lignocellulosic hydrolysates are more and more in the focus of sustainable biotechnological processes. *Corynebacterium glutamicum*, which is used in industrial amino acid production at a million-ton scale, has been engineered towards utilization of alternative carbon sources. As for other microorganisms, the focus has been set on the pentose sugars present in lignocellulosic hydrolysates. Utilization of the hexuronic acids D-galacturonic acid (abundant in pectin-rich waste streams such as peels and pulps) and D-glucuronic acid (a component of the side-chains of plant xylans) for growth and production with *C. glutamicum* has not yet been studied.

Results: Neither aldohexuronic acid supported growth of *C. glutamicum* as sole or combined carbon source, although its genome encodes a putative uronate isomerase sharing 28% identical amino acids with UxaC from *Escherichia coli*. Heterologous expression of the genes for both uptake and catabolism of D-galacturonic acid and D-glucuronic acid was required to enable growth of *C. glutamicum* with either aldohexuronic acid as the sole carbon source. When present in mixtures with glucose, the recombinant *C. glutamicum* strains co-utilized D-galacturonate with glucose and D-glucuronate with glucose, respectively. When transformed with the plasmid for uptake and catabolism of the aldohexuronates, model producer strains were able to grow with and produce from D-galacturonate or D-glucuronate as sole carbon source.

Conclusions: An easily transferable metabolic engineering strategy for access of *C. glutamicum* to aldohexuronates was developed and applied to growth and production of the amino acids L-lysine and L-ornithine as well as the terpene lycopene from D-galacturonate or D-glucuronate.

Background

Corynebacterium glutamicum is a rod-shaped Gram-positive aerobic bacterium, which can be found in soil, sewages, vegetables, and fruits [1]. This bacterium is capable of utilizing various sugars as well as organic acids [2]. Among others, *C. glutamicum* has the ability to metabolize glucose, fructose, and sucrose as well as lactate, pyruvate, and acetate [2-4]. Characteristic of *C. glutamicum* is the capability of growing on mixtures of different carbon sources with a monoauxic growth [5,6] as opposed to diauxic growth observed for many other microorganisms such as *Escherichia coli* and *Bacillus subtilis* [7]. Only a few exceptions have been reported as

in the case of glucose-ethanol or acetate-ethanol mixtures, where preferential substrate utilization was observed [8].

Since its discovery, *C. glutamicum* has become an indispensable microorganism for the biotechnological industry [9]. From its initial use as a natural L-glutamate producer [10], it is currently used for production of other amino acids such as L-lysine, L-ornithine, L-methionine, and L-aspartate [11-14]. However, its importance has further increased as it was for production of of non-natural products [9] such as isobutanol [15], ethanol [16], putrescine [14,17,18], cadaverine [19], carotenoids and terpenoids [20-25], and xylitol [26].

Recently, efforts with *C. glutamicum* have shifted from optimizing production processes to also include access to alternative carbon sources. As yet, feed in the industry

---

* Correspondence: volker.wendisch@uni-bielefeld.de
Genetics of Prokaryotes, Center for Biotechnology, Faculty of Biology, Bielefeld University, Universitaetsstraße 25, Bielefeld 33615, Germany

relies mainly on glucose and fructose [8], which also have competing uses in the food industry. So far, alternative carbon source utilization in *C. glutamicum* has been successfully established, among others, for xylose [27,28], galactose [29], arabinose [14,30], glucosamine [31], N-acetyl-glucosamine [32], and glycerol [33]. Plant cell wall materials such as lignocellulose and pectin are promising alternatives as carbon source. These materials are readily and abundantly available as agricultural waste or forestry residues [34]. Among the sugar constituents of plant cell wall are the hexuronic acids D-galacturonate and D-

glucuronate found in pectin. These acidic sugars are naturally consumed by most plant pathogenic bacteria such as *Erwinia carotovora*, *Pseudomonas syringae*, and *Agrobacterium tumefaciens*, but also by *E. coli*.

There are three pathways for utilization of hexuronic acids, namely via isomerization, oxidation, and reduction [35]. The isomerization pathway in *E. coli* consists of seven reactions, which yield the central intermediates D-glyceraldehyde-3-phosphate and pyruvate (Figure 1). Both *uxaC-uxaA* and *uxuA-uxuB* are located within operons. However, these operons and the remaining genes

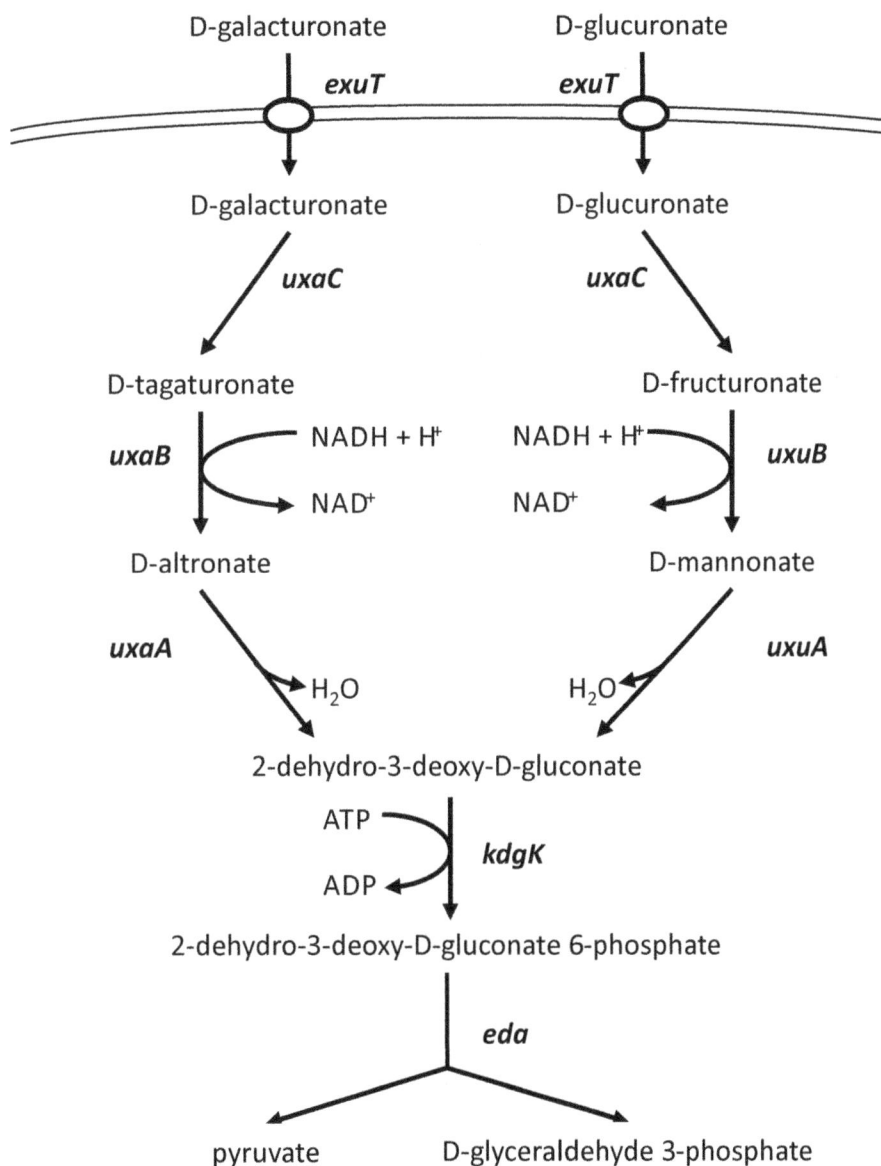

Figure 1 Schematic representation of the D-galacturonate and D-glucuronate catabolic pathways of *E. coli*. Gene names are given adjacent to the reactions: *eda*, 2-keto-3-deoxygluconate-6-phosphate aldolase; *exuT*, uptake system for hexuronic acids; *kdgK*, 2-keto-3-deoxygluconokinase; *uxaA*, altronate dehydratase; *uxaB*, altronate oxidoreductase; *uxaC*, uronate isomerase; *uxuA*, mannonate dehydratase; *uxuB*, mannonate oxidoreductase.

are not clustered but spread across the genome of *E. coli*. The genes *kdgK* and *eda* belong to the modified Entner-Doudoroff pathway, which is present in many Gram-negative bacteria, pseudomonads, and Archaea [36]. Bioinformatic analysis revealed an *uxaC* ortholog in the genome of *C. glutamicum* with 28% protein similarity to that of *E. coli*. However, homologs of *uxaB*, *uxuA*, *uxuB*, *kdgK*, and *eda* appear to be absent. Therefore, in this work, *C. glutamicum* was engineered for the uptake and utilization of D-galacturonate and D-glucuronate as alternative carbon sources. The potential of this synthetic pathway was then analyzed with respect to the production of the amino acids L-lysine and L-ornithine as well as the carotenoid lycopene.

Methods

Microorganisms, plasmids, and cultivation conditions

The wild-type strain *C. glutamicum* ATCC 13032 used in this study was obtained from the American Type Culture Collection (ATCC). Other strains include *C. glutamicum* ORN1 [14], *C. glutamicum* DM1933 [37], and *C. glutamicum* ΔcrtYEB [23] that are derived from the wild-type strain (Table 1). The hexuronic acid utilization and transporter genes originated from *E. coli* MG1655, whereas the strain *E. coli* DH5α [38] was used for plasmid construction. Both *E. coli* strains were obtained from the Coli Genetic Stock Center (CGSC). For cultivations, the Luria broth (LB) complex medium and CGXII minimal medium [39] were used and contained glucose,

Table 1 Strains, plasmids, and oligonucleotides used in this study

Name	Relevant genotype/information	Reference
***E. coli* strain**		
DH5α	(φ80lacZΔM15) Δ(lacZYA-argF)U169 deoR recA1 endA1 hsdR17(rK−, mK−) supE44 thi1 gyrA96 relA1	[38]
***C. glutamicum* strains**		
WT	Wild-type strain ATCC 13032, auxotrophic for biotin	[40]
ORN1	In-frame deletion of *argR* and *argF*, L-ornithine overproducing strain derived from *C. glutamicum* WT ATCC 13032, auxotrophic for L-arginine	[14]
DM1933	Δpck deletion mutant of *C. glutamicum* WT also carrying the chromosomal changes pyc(P458S), hom(V59A), two copies of lysC(T311I), two copies of asd, two copies of dapA, two copies of dapB, two copies of ddh, two copies of lysA, two copies of lysE	[37]
ΔcrtYEb	crtYeYfEb deletion mutant of *C. glutamicum* MB001	[23]
Plasmids		
pEKEx3	Spec^R; *E. coli*/*C. glutamicum* shuttle vector for regulated gene expression (P_tac, lacI^q, pBL1 oriV_Cg)	[41]
pHexA	pEKEx3 derivative for IPTG-inducible expression of *uxaCAB*, *uxuAB*, *kdgK*, and *eda* from *E. coli* containing artificial ribosome binding sites each	This work
pVWEx1	Km^R; *E. coli*/*C. glutamicum* shuttle vector for regulated gene expression (P_tac, lacI^q, pCG1 oriV_Cg)	[42]
pVWEx1-*exuT*	pVWEx1 derivative for IPTG-inducible expression of *exuT* from *E. coli* containing artificial ribosome binding site	This work
Primers		
G1a	GCAGGTCGACTCTAGAGGATCCCCGAAAGGAGGCCCTTCAGATGACTCCGTTTATGACTGAAGATTTC	
G1b	GTACTAGCTAATGCAATCAGTGATGTTATAGCGTTACGCCGCTTTTG	
G2d	CATCACTGATTGCATTAGCTAGTACGAAAGGAGGCCCTTCAGATGAAAACACTAAATCGTCGCGAT	
G2c	GCTAATGGTGCTATCTGGTACGATCTTAGCACAACGGACGTACAG	
G3d	GATCGTACCAGATAGCACCATTAGCGAAAGGAGGCCCTTCAGATGGAACAGACCTGGCGC	
G3f	GCAGGTCGACTCTAGAGGATCCCCATGGAACAGACCTGGCGCTGGTACGGCC	
G3c	CGTTCTAGTTACTTTGGAACGTACCTTACAGCGCAGCCACACA	
G4d	GGTACGTTCCAAAGTAACTAGAACGGAAAGGAGGCCCTTCAGATGTCCAAAAAGATTGCCGTGAT	
G4c	GTCAATCCATGGCATTCTAGCCAAGTTACGCTGGCATCGCCTC	
G5d	CTTGGCTAGAATGCCATGGATTGACGAAAGGAGGCCCTTCAGATGAAAAACTGGAAAACAAGTGCAG	
G5c	GCCAGTGAATTCGAGCTCGGTACCCTTACAGCTTAGCGCCTTCTACAG	
ExuT-fw	CTGCAGGTCGACTCTAGAGGAAAGGAGGCCCTTCAGA	
ExuT-rv	CGGTACCCGGGGGATCTTAATGTTGCGGTGCGGGATC	

D-galacturonate, or D-glucuronate in concentrations as indicated in the 'Results' section. Kanamycin (25 µg ml^{-1}), spectinomycin (100 µg ml^{-1}), and/or isopropyl-β-D-thio-galactopyranoside (IPTG; 20 µM) were added to the medium when necessary. Cultivations were carried out in 50-ml cultures in 500-ml baffled Erlenmeyer flasks on a rotary shaker at 120 rpm and 30°C for C. glutamicum or 37°C for E. coli. In both cases, the growth in liquid cultures was followed by measuring the optical density at 600 nm (OD$_{600}$).

DNA preparation, modification, and transformation

Standard procedures were used for plasmid and chromosomal DNA isolation, molecular cloning, and transformation of E. coli. Plasmid isolation for C. glutamicum was carried out as described previously [43]. Transformation of C. glutamicum by electroporation was carried out as described [39]. PCR experiments were carried out in a thermocycler (Analytik Jena AG, Jena, Germany) with KOD Hot Start DNA Polymerase (Novagen, Merck KGaA, Darmstadt, Germany) and with oligonucleotides obtained from Eurofins MWG Operon (Eurofins Genomics, Ebersberg, Germany) as listed in Table 2. Restriction enzymes, T4 DNA Ligase, and alkaline phosphatase were obtained from New England BioLabs, Inc. (Ipswich, MA, USA) and used according to the manufacturer's protocol.

Construction of plasmids and strains

For the construction of pHexA, the uxaCA and uxuAB operons as well as genes uxaB, kdgK, and eda were amplified by PCR from E. coli MG1655. The uxaCA operon was amplified with primers G1a + G1b resulting in a 2,981-bp product, whereas the uxuAB operon was amplified with primers G3f + G3c resulting in a 2,775-bp product. Gene uxaB was amplified with primers G2d + G2c, gene kdgK with primers G4d + G4c, and eda was amplified with primers G5d + G5c resulting in 1,519-, 997-, and 709-bp PCR products, respectively. Through these primers, appropriate linker sequences and a ribosomal binding site (RBS) sequence were attached to each gene or operon to facilitate the Gibson assembly [44]. The genetic load was first divided due to the insert size. Therefore, genes uxuAB, kdgK, and eda were cloned into the SmaI-digested pEKEx3 resulting in pEKEx3-uxuAB-kdgK-eda, designated as pAB5. The insert of pAB5 was amplified via PCR with primers G3d and G5d with a 4481-bp PCR product. The pAB5 amplicon, uxaB, and uxaCA were then used for Gibson assembly into the SmaI-digested pEKEx3 and yielded the final vector pHexA. The aldohexuronate transporter gene exuT was amplified via PCR with primers ExuT-fw and ExuT-rv from E. coli MG1655 and used for Gibson assembly into the BamHI-digested pVWEx1. The plasmid inserts were verified by sequencing (Sequencing Core Facility CeBiTec, Bielefeld, Germany), and the plasmids were used to transform C. glutamicum ATCC 13032.

DNA microarray analysis

C. glutamicum ATCC 13032 was cultivated in CGXII medium with 50 mM D-galacturonate plus 50 mM glucose, 50 mM D-glucuronate plus 50 mM glucose, or 50 mM glucose as carbon source. Exponentially growing cells were harvested after 4 h. RNA isolation, cDNA synthesis, and microarray hybridization were performed according to previous protocols [45]. Microarray images were analyzed with ImaGene software (BioDiscovery, Inc., Hawthorne, CA, USA), whereas the EMMA platform was used for data evaluation.

Quantification of amino acids and carbohydrates

To evaluate the amino acid and carbohydrate production, culture samples were taken and centrifuged (13.000×g, 10 min), and the supernatant analyzed by high-pressure liquid chromatography (HPLC, 1200 series, Agilent Technologies Inc., Santa Clara, CA, USA) as described previously [14,21].

Computational analysis

Protein alignments were carried out via the BLASTP algorithm [46] of NCBI (Bethesda, MD, USA). The GenBank accession number for the annotated genome sequence of C. glutamicum ATCC 13032 and E. coli MG1655 is NC_006958 [47] and NC_000913 [48], respectively.

Results

Response of C. glutamicum WT to D-galacturonate and D-glucuronate

D-galacturonate and D-glucuronate were tested as potential carbon sources of C. glutamicum. However, although its genome encodes a putative uronate isomerase sharing 28% identical amino acids with UxaC from E. coli, both compounds did not support growth as sole carbon sources at 50 mM (data not shown). When present in addition to 50 mM glucose, C. glutamicum wild type (WT) grew to comparable maximal OD$_{600}$ values of about 18 irrespective of the presence or absence of D-galacturonate or D-glucuronate (data not shown). Surprisingly, the presence of 50 mM D-galacturonate in minimal medium with 50 mM glucose accelerated growth slightly (0.24 h^{-1} as compared to 0.17 h^{-1}), while the addition of 50 mM D-glucuronate to glucose minimal medium slowed growth of WT (0.13 h^{-1}). These observations prompted us to carry out DNA microarray experiments to study global gene expression under these conditions. Genes differentially expressed in cells growing exponentially with 50 mM D-galacturonate plus 50 mM glucose, 50 mM D-glucuronate plus 50 mM glucose, and 50 mM glucose alone are listed in Table 2. As expected, the presence of

Table 2 Gene expression analysis of *C. glutamicum* WT in CGXII minimal medium with 50 mM glucose[a]

	Gene ID	Description	M
Differentially expressed genes in the presence of D-glucuronate	cg3219	*ldh*, L-lactate dehydrogenase	1.7
	cg3303	Transcriptional regulator PadR family	1.6
	cg0580	Hypothetical protein	1.5
	cg2789	*nrdH*, glutaredoxin-like protein NrdH	−1.5
	cg2182	ABC-type peptide transport system, permease component	−1.5
	cg3300	Cation transport ATPase	−1.5
	cg2477	Hypothetical protein	−1.6
	cg1809	DNA-directed RNA polymerase subunit omega	−1.7
	cg0935	Hypothetical protein	−1.8
	cg1286	Hypothetical protein	−1.8
Differentially expressed genes in the presence of D-galacturonate	cg2313	*idhA3*, myo-inositol 2-dehydrogenase	2.0
	cg1118	Pyrimidine reductase, riboflavin biosynthesis	1.9
	cg0687	*gcp*, putative O-sialoglycoprotein endopeptidase	1.9
	cg1116	*tdcB*, threonine dehydratase	1.9
	cg1784	*ocd*, putative ornithine cyclodeaminase	1.9
	cg3096	*ald*, aldehyde dehydrogenase	1.9
	cg0792	Thioredoxin domain-containing protein	1.8
	cg0682	ATPase or kinase	1.7
	cg1003	*fthC*, 5-formyltetrahydrofolate cycloligase	1.7
	cg1134	*pabAB*, para-aminobenzoate synthase components I and II	1.7
	cg1438	ABC-type transport system, ATPase component (C-terminal fragment)	1.7
	cg2430	Hypothetical protein	1.7
	cg1560	*uvrA*, excinuclease ATPase subunit	1.7
	cg1014	*pmt*, protein O-mannosyltransferase	1.6
	cg1668	Putative membrane protein	1.6
	cg2625	*pcaF*, β-ketoadipyl CoA thiolase	1.6
	cg2094	Hypothetical protein	1.6
	cg1241	Hypothetical protein	1.5
	cg1876	Glycosyltransferase	1.5

Table 2 Gene expression analysis of *C. glutamicum* WT in CGXII minimal medium with 50 mM glucose[a] *(Continued)*

Gene ID	Description	M
cg2417	Short-chain-type oxidoreductase	1.5
cg3118	*cysI*, ferredoxin-sulfite reductase	−1.5
cg0504	*qsuD*, shikimate 5-dehydrogenase	−1.5
cg1740	Putative nucleoside-diphosphate-sugar epimerase	−1.5
cg3225	Putative serine/threonine-specific protein phosphatase	−1.5
cg2945	*ispD*, 2-C-methyl-D-erythritol 4-phosphate cytidylyltransferase	−1.5
cg1614	Hypothetical protein	−1.5
cg3427	*parA1*, ATPase involved in chromosome partitioning	−1.5
cg1252	*fdxC*, ferredoxin	−1.5
cg0518	*hemL*, glutamate-1-semialdehyde 2,1-aminomutase	−1.5
cg2587	Phosphoglycerate dehydrogenase or related dehydrogenase	−1.6
cg1551	*uspA1*, universal stress protein UspA	−1.6
cg0059	*pknA*, serine/threonine protein kinase	−1.7
cg0156	*cysR*, transcriptional activator of assimilatory sulfate reduction	−1.7
cg0966	*thyA*, thymidylate synthase	−1.7
cg0060	*pbpA*, D-alanyl-D-alanine carboxypeptidase	−1.7
cg1045	Hypothetical protein	−1.7
cg3119	*fpr2*, probable sulfite reductase (flavoprotein)	−1.8
cg1253	Succinyldiaminopimelate aminotransferase	−1.8
cg0045	Putative ABC-type transporter membrane protein	−1.8
cg3117	*cysX*, ferredoxin-like protein	−1.8
cg3430	Hypothetical protein	−1.9
cg3115	*cysD*, sulfate adenylyltransferase subunit 2	−1.9
cg1037	*rfp2*, RPF2 precursor	−1.9

[a]Statistical evaluation was carried out with the *t*-test, where $p \leq 0.05$, log expression ratio $M \geq 1.5$ or ≤ -1.5, and signal intensity $A \geq 10$. Values are averages of three independent cultivations.

D-galacturonate elicited different gene expression changes than D-glucuronate. Since neither D-galacturonate nor D-glucuronate was metabolized, these gene expression changes are likely due to regulatory or secondary effects. To elicit such regulatory changes, transport of minute amounts of D-galacturonic or D-glucuronic acid might be sufficient.

Increased expression of fermentative lactate dehydrogenase gene *ldhA* in the presence of D-glucuronate might have slowed growth with glucose since lactate is known to accumulate transiently and since high *ldhA* levels have been implied to slow growth of *C. glutamicum* with sugars [49]. Furthermore, expression of the gene for subunit omega of RNA polymerase was reduced, thus transcription might have been negatively affected more in general. The gene expression changes due to the presence of D-galacturonate did not allow deriving a potential explanation for faster growth with glucose. However, decreased expression of *cg1551* encoding putative universal stress protein UspA is in line with faster growth of *C. glutamicum* in the presence of D-galacturonate.

Expression of genes for catabolism of D-galacturonate and D-glucuronate in *C. glutamicum* WT

Plasmid pHexA was constructed for heterologous expression of the *E. coli* genes for degradation of D-galacturonate and D-glucuronate to the glycolytic intermediates pyruvate and glyceraldehyde-3-phosphate. To this end, the operons *uxaCA* and *uxuAB* as well as genes *uxaB*, *kdgK*, and *eda* were cloned with attached RBS sequences as synthetic operon into IPTG-inducible gene expression vector pEKEx3. The resulting plasmid pHexA was used to transform *C. glutamicum* WT. However, the transformants were unable to grow with either D-galacturonate or D-glucuronate as sole carbon sources (data not shown).

Co-expression of *exuT* from *E. coli* was required for uptake of hexuronic acids

Since endowing *C. glutamicum* with D-galacturonate or D-glucuronate catabolism proved insufficient for utilization of these substrates, the gene for the respective uptake system *exuT* was co-expressed from a compatible plasmid. In preliminary experiments, *C. glutamicum* WT (pHexA)(pVWEx1-*exuT*) did indeed grow with either D-galacturonate or D-glucuronate as sole carbon source, however, very slowly. Based on the assumption that overproduction of transmembrane protein ExuT perturbed growth, the concentration of the inducer IPTG was titrated. Moreover, it was required to pre-cultivate the strain in minimal medium with a mixture of 50 mM glucose and 50 mM of either D-galacturonate or D-glucuronate as carbon source. In the main culture with 50 mM of either D-galacturonate or D-glucuronate, no growth was observed for *C. glutamicum* WT(pHexA)(pVWEx1). By contrast, *C. glutamicum* WT(pHexA)(pVWEx1-*exuT*) was able to grow with D-galacturonate and D-glucuronate, respectively, with growth rates of 0.06 ± 0.01 and 0.05 ± 0.01 h^{-1}, respectively (data not shown). Thus, these results revealed that *C. glutamicum* WT lacks the ability for uptake and

catabolism of hexuronic acids and that heterologous expression of the genes from *E. coli* for uptake and catabolism of hexuronic acids enabled access of *C. glutamicum* to D-galacturonate and D-glucuronate as growth substrates.

Co-utilization of hexuronic acids with glucose expression by *C. glutamicum* WT(pHexA)(pVWEx1)

A hallmark of *C. glutamicum* is its ability to co-utilize various carbon sources when these are added as carbon source mixtures. To assay if hexuronic acids are utilized simultaneously with glucose, the preferred carbon source of *C. glutamicum*, the growth and substrate consumption of *C. glutamicum* WT(pHexA)(pVWEx1) in minimal medium containing either 100 mM D-galacturonate plus 100 mM glucose or 100 mM D-glucuronate plus 100 mM glucose were determined (Figure 2). In minimal medium with the mixture of D-galacturonate plus glucose, *C. glutamicum* WT(pHexA)(pVWEx1) grew with a growth rate of 0.25 ± 0.02 h^{-1} and co-utilized glucose with D-galacturonate (Figure 2). Specific uptake rates of 28 ± 3 and 39 ± 4 nmol (mg cell dry weight (CDW))$^{-1}$ min^{-1} were derived for utilization of D-galacturonate and glucose, respectively. In minimal medium with a blend of 100 mM D-glucuronate plus 100 mM glucose, both carbon sources were utilized simultaneously and support a growth rate of 0.25 ± 0.02 h^{-1} for *C. glutamicum* WT(pHexA)(pVWEx1) (Figure 2). The specific uptake rates were 21 ± 2 nmol (mg CDW)$^{-1}$ min^{-1} for glucose and 18 ± 2 nmol (mg CDW)$^{-1}$ min^{-1} for D-glucuronate.

Production of L-lysine, L-ornithine, and lycopene by recombinant *C. glutamicum* strains from D-galacturonate and D-glucuronate

The natural substrate spectrum of *C. glutamicum* has been broadened to realize a flexible feedstock concept for production processes using this bacterium [27-33]. To test if recombinant *C. glutamicum* strains engineered to accept D-galacturonate and D-glucuronate as growth substrates are able to produce, e.g., amino acids from these substrates, model L-lysine, L-ornithine, and lycopene producer strains were transformed with plasmids pHexA and pVWEx1-*exuT*. These strains were cultivated in CGXII minimal medium with 20 μM IPTG and either 100 mM D-galacturonate or 100 mM D-glucuronate as sole carbon source.

The lysine-producing strain *C. glutamicum* DM1933 carries a number of chromosomal changes known to be beneficial for L-lysine production [37]. DM1933(pHexA)(pVWEX1-*exuT*) hardly grew with either D-galacturonate or D-glucuronate (Table 3). However, DM1933(pHexA)(pVWEX1-*exuT*) produced 6.5 ± 0.2 mM L-lysine from 100 mM D-galacturonate and 9.3 ± 1.1 mM L-lysine from 100 mM D-glucuronate (Table 3).

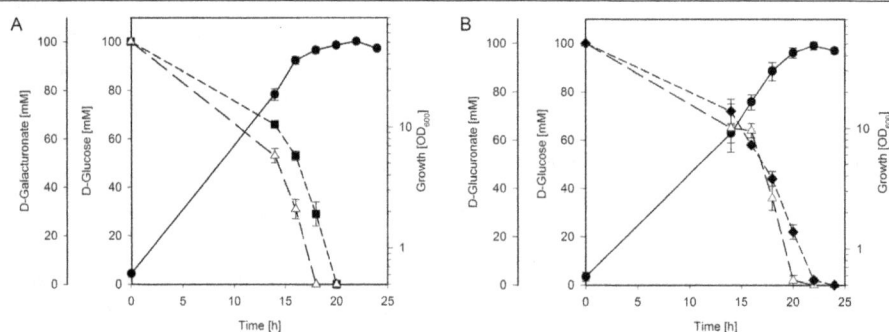

Figure 2 Growth of *C. glutamicum* strains WT(pHexA)(pVWEx1-*exuT*) with blends of D-glucose and D-galacturonate (A) or D-glucose and D-glucuronate (B). Residual concentrations of D-galacturonate (closed squares), D-glucuronate (closed diamonds), and D-glucose (open triangles) and OD600 values (closed circles) are given as means and standard deviations of triplicate cultivations in CGXII minimal medium, pH 6.3, with 20 μM IPTG, and either a mixture of 100 mM D-galacturonate and 100 mM D-glucose **(A)** or a mixture of 100 mM D-glucuronate plus 100 mM D-glucose **(B)** as carbon sources.

Lycopene accumulates in *C. glutamicum* ΔcrtYEb due to disruption of the pathway for biosynthesis of the endogenous carotenoid decaprenoxanthin [20]. *C. glutamicum* ΔcrtYEb(pHexA)(pVWEX1-*exuT*) grew with D-galacturonate $(0.02 \pm 0.01 \ h^{-1})$ and D-glucuronate $(0.04 \pm 0.01 \ h^{-1})$, respectively, as sole carbon source (Table 3). The strain produced $0.7 \pm 0.1 \ \text{mg} \ (\text{g CDW})^{-1}$

Table 3 Batch fermentations of L-lysine-, L-ornithine-, and lycopene-producing strains in minimal medium with D-galacturonate or D-glucuronate

		D-galacturonate	D-glucuronate
L-lysine by DM1933(pHexA) (pVWEx1-*exuT*)			
	C_{Lys} (mM)	6.5 ± 0.2	9.3 ± 1.1
	μ (h^{-1})	0.01 ± 0.01	0.02 ± 0.01
	$Y_{p/s}$ (g$_{Lys}$ g$_{substrate}$$^{-1}$)	0.04 ± 0.01	0.07 ± 0.01
Lycopene by ΔcrtYEb(pHexA) (pVWEx1-*exuT*)			
	C_{Lyc} (mg/g)	0.7 ± 0.1	0.8 ± 0.3
	μ (h^{-1})	0.02 ± 0.01	0.04 ± 0.01
	$Y_{p/s}$ (g$_{Lyc}$ g$_{substrate}$$^{-1}$)	0.09 ± 0.01	0.08 ± 0.02
L-ornithine by ORN1(pHexA) (pVWEx1-*exuT*)			
With 0.75 mM L-arginine	C_{Orn} (mM)	2.4 ± 0.2	<0.5
	μ (h^{-1})	0.03 ± 0.01	0.03 ± 0.01
	$Y_{p/s}$ (g$_{Orn}$ g$_{substrate}$$^{-1}$)	0.01 ± 0.01	<0.01
With 0.125 mM L-arginine	C_{Orn} (mM)	1.7 ± 0.1	0.6 ± 0.2
	μ (h^{-1})	0.04 ± 0.01	0.03 ± 0.01
	$Y_{p/s}$ (g$_{Orn}$ g$_{substrate}$$^{-1}$)	0.01 ± 0.01	<0.01

CGXII minimal medium was used with 100 mM of the indicated carbon source.

lycopene in minimal medium with D-galacturonate and $0.8 \pm 0.3 \ \text{mg} \ (\text{g CDW})^{-1}$ lycopene with D-glucuronate (Table 3).

C. glutamicum ORN1 is an L-arginine auxotrophic derivative of *C. glutamicum* WT that secretes L-ornithine due to deletions of the L-ornithine carbamoyltransferase gene *argF* and the L-arginine biosynthesis repressor gene *argR* [14]. When supplemented with either 0.75 mM or 0.125 mM L-arginine, *C. glutamicum* ORN1(pHexA) (pVWEX1-*exuT*) grew in D-galacturonate and D-glucuronate minimal medium with growth rates of about $0.04 \pm 0.01 \ h^{-1}$ (Table 3). However, L-ornithine accumulated only to low concentrations corresponding to yields of about 1 to 2 mol% (Table 3).

Discussion

C. glutamicum WT is not capable of utilizing hexuronic acids. Heterologous expression of gene for catabolism and uptake of the hexuronic acid pathway from *E. coli* in *C. glutamicum* enabled utilization of both D-galacturonate and D-glucuronate as sole carbon sources in minimal medium. Moreover, both hexuronates were co-utilized with glucose by the recombinant *C. glutamicum* strains developed here. Simultaneous utilization of several carbon sources as required for efficient utilization of substrate mixtures such as in lignocellulosic hydrolysates is a hallmark of *C. glutamicum* [6,8] and also pertains to co-utilization of non-native substrates by the respective recombinants [14,27,28,30-33].

Notably, the aldohexuronate transporter ExuT was strictly required, indicating that *C. glutamicum* lacks the capacity to import sugar acids. ExuT belongs to the major facilitator superfamily (MFS) class of transporters, more specifically to the anion:cation symporter (ACS) family [50]. This class of symporters transfers organic/inorganic anions while simultaneously co-transporting H$^+$/Na$^+$, respectively. ExuT has not been reported to transport other substrates than the aldohexuronic acids. Inspection of the genome of *C. glutamicum* WT showed only one protein with 22% similarity to ExuT, namely putative lincomycin resistance protein LMRB (YP 226924.1). Engineering *C. glutamicum* for growth with other non-native carbon sources does not necessarily require heterologous expression of a gene encoding a transport system. Introduction of catabolic genes for conversion of glycerol, arabinose, and xylose was sufficient to enable utilization of these carbon sources by these recombinant *C. glutamicum* strains, while additional introduction of the respective uptake system accelerated carbon source utilization [27,28,33,51,52]. Transport engineering was not required for the amino sugar glucosamine, which is a substrate of the endogenous glucose-specific PTS [31], whereas the amino sugar N-acetylglucosamine could only be utilized if NagE from the related *Corynebacterium glycinophilum* was introduced [32]. In the latter case as well as in the present study, it was necessary to adjust the concentration of the inducer IPTG. It is often observed that too high levels of a transmembrane protein such as a transport protein results in growth perturbation [53,54]. In addition, expression levels of several genes of a pathway may need to be tuned to avoid accumulation of potentially inhibitory intermediates as demonstrated for *C. glutamicum* engineered for decaprenoxanthin overproduction [23].

It is not known if the recombinant *C. glutamicum* strain WT(pHexA)(pVWEX1-*exuT*) would be able to grow with sugar acids related to the aldohexuronates D-galacturonate and D-glucuronate since ExuT is specific for aldohexuronate uptake. In *E. coli*, the intermediate of aldohexuronate catabolism D-fructuronate serves as carbon source and its utilization requires uptake by GntP [55]. *C. glutamicum* possesses GntP for gluconate uptake [56], but it is unknown whether GntP from *C. glutamicum* accepts D-fructuronate or the related intermediates of aldohexuronate catabolism D-tagaturonate, D-altronate, or D-mannonate as substrates. Recently, it was shown that *E. coli* may grow with L-galactonate and L-gulonate as sole carbon sources with L-galactonate-5-dehydrogenase YjjN being required for their conversion to D-tagaturonate and D-fructuronate, respectively [57]. Under osmotic stress conditions, *E. coli* may use a different pathway, i.e., 5-keto 4-deoxyuronate isomerase KduI and 2-deoxy-D gluconate 3-dehydrogenase KduD may compensate for reduced levels of UxaC, UxaB, and UxuB under osmotic stress conditions [58]. Since ExuT from *E. coli* was required for aldohexuronate utilization by recombinant *C. glutamicum*, it is likely that introduction of the respective uptake systems for the related sugar acids described above is a prerequisite for their use as carbon sources.

Degradation of aldohexuronate to pyruvate yields one mole of ATP per mole of aldohexuronate by substrate-level phosphorylation as compared to two moles of ATP

per glucose in the Embden-Meyerhof-Parnas (EMP) pathway of glycolysis as present in *C. glutamicum*. In the EMP pathway of glycolysis, two moles of nicotinamide adenine dinucleotide (NADH) are generated from glucose, while no net formation of NADH occurs in aldohexuronate conversion to pyruvate (Figure 1). The maximal OD_{600} values reflecting the maximal biomass concentration with 50 mM D-galacturonate consequently were lower (OD_{600} of about 6.5) than with 50 mM glucose (OD_{600} of about 18). The maximal OD_{600} (about 4.5) with 50 mM D-glucuronate was even lower (data not shown). The reduced biomass yields for aerobic growth with the aldohexuronates can be explained at least in part by their lower ATP yields as compared to glucose catabolism. The growth rates in minimal medium with D-galacturonate ($0.06\ h^{-1}$) and D-glucuronate ($0.05\ h^{-1}$), respectively, obtained with *C. glutamicum* WT(pHexA)(pVWEx1-*exuT*) are five to six times lower than with glucose. Aldohexuronate utilization may be accelerated, e.g., by improving heterologous gene expression or by using catabolic enzymes of other microorganisms as in the case of xylose [28]. Instability of the plasmids pHexA and pVWEX1-*exuT* were not observed in the experiments described here; however, it might be possible that plasmid instability poses a challenge when using these strains in large fermentation vessels.

The low biomass yields and slow growth rates observed with *C. glutamicum* WT(pHexA)(pVWEX1-*exuT*) were also found when the respective plasmids were transformed into model producer strains. The product yields observed were low, e.g., about 6 to 9 mol% for L-lysine (Table 3). Thus, the aldohexuronates are not good substrates for growth and production by *C. glutamicum*. However, endowing *C. glutamicum* strains with aldohexuronate catabolism may be a requirement for complete and efficient utilization of second-generation feedstocks ensuring that not only the major sugar fractions are converted to value-added products.

Conclusions

Access of *C. glutamicum* to the aldohexuronates D-galacturonate and D-glucuronate was established by heterologous expression of genes for catabolism and uptake of the aldohexuronates from *E. coli* in *C. glutamicum*. This metabolic engineering strategy could be applied to D-galacturonate- or D-glucuronate-based growth and production of the amino acids L-lysine and L-ornithine as well as the terpene lycopene.

Competing interests
The authors declare that they have no competing interests.

Authors' contributions
AH planned and performed the experiments, analyzed the data, and drafted the paper. IK performed the experiments and analyzed the data. SNL planned the experiments and analyzed the data. VFW designed and

coordinated the study, analyzed the data, and wrote the paper. All authors read and approved the final manuscript.

Acknowledgements

We thank Petra Peters-Wendisch for help with artwork. Work in the lab of the authors was funded in part by grants 0315589G from BMBF in the CRP 'Corynebacterium: improving flexibility and fitness for industrial production'.

References

1. Eggeling L, Bott M (2005) Handbook of Corynebacterium glutamicum. CRC, Boca Raton
2. Blombach B, Seibold GM (2010) Carbohydrate metabolism in Corynebacterium glutamicum and applications for the metabolic engineering of L-lysine production strains. Appl Microbiol Biotechnol 86(5):1313–1322, doi:10.1007/s00253-010-2537-z
3. Cocaign M, Monnet C, Lindley ND (1993) Batch kinetics of Corynebacterium glutamicum during growth on various carbon sources: use of substrate mixtures to localise metabolic bottlenecks. Appl Microbiol Biotechnol 40:526–530
4. Dominguez H, Cocaign-Bousquet M, Lindley ND (1997) Simultaneous consumption of glucose and fructose from sugar mixtures during batch growth of Corynebacterium glutamicum. Appl Microbiol Biotechnol 47(5):600–603
5. Dominguez H, Nezondet C, Lindley ND, Cocaign M (1993) Modified carbon flux during oxygen limited growth of Corynebacterium glutamicum and the consequences for amino acid overproduction. Biotech Lett 15:449–454
6. Wendisch VF, de Graaf AA, Sahm H, Eikmanns BJ (2000) Quantitative determination of metabolic fluxes during coutilization of two carbon sources: comparative analyses with Corynebacterium glutamicum during growth on acetate and/or glucose. J Bacteriol 182(11):3088–3096
7. Monod J (1949) The growth of bacterial cultures. Ann Rev Microbiol 3:371–394
8. Zahoor A, Lindner SN, Wendisch VF (2012) Metabolic engineering of Corynebacterium glutamicum aimed at alternative carbon sources and new products. Comput Struct Biotechnol J 3:e201210004, doi:10.5936/csbj.201210004
9. Wendisch VF (2014) Microbial production of amino acids and derived chemicals: synthetic biology approaches to strain development. Curr Opin Biotechnol 30C:51–58, doi:10.1016/j.copbio.2014.05.004
10. Ikeda M (2003) Amino acid production processes. Adv Biochem Eng Biotechnol 79:1–35
11. Wendisch VF (2007) Amino acid biosynthesis – pathways, regulation and metabolic engineering. Springer, Heidelberg
12. Hermann T (2003) Industrial production of amino acids by coryneform bacteria. J Biotechnol 104(1–3):155–172
13. Leuchtenberger W, Huthmacher K, Drauz K (2005) Biotechnological production of amino acids and derivatives: current status and prospects. Appl Microbiol Biotechnol 69(1):1–8
14. Schneider J, Niermann K, Wendisch VF (2011) Production of the amino acids L-glutamate, L-lysine, L-ornithine and L-arginine from arabinose by recombinant Corynebacterium glutamicum. J Biotechnol 154(2–3):191–198, doi:10.1016/j.jbiotec.2010.07.009
15. Blombach B, Riester T, Wieschalka S, Ziert C, Youn JW, Wendisch VF, Eikmanns BJ (2011) Corynebacterium glutamicum tailored for efficient isobutanol production. Appl Environ Microbiol 77(10):3300–3310, doi:10.1128/AEM. 02972-10
16. Inui M, Kawaguchi H, Murakami S, Vertes AA, Yukawa H (2004) Metabolic engineering of Corynebacterium glutamicum for fuel ethanol production under oxygen-deprivation conditions. J Mol Microbiol Biotechnol 8(4):243–254, doi:10.1159/000086705
17. Schneider J, Wendisch VF (2010) Putrescine production by engineered Corynebacterium glutamicum. Appl Microbiol Biotechnol 88(4):859–868, doi:10.1007/s00253-010-2778-x
18. Schneider J, Eberhardt D, Wendisch VF (2012) Improving putrescine production by Corynebacterium glutamicum by fine-tuning ornithine transcarbamoylase activity using a plasmid addiction system. Appl Microbiol Biotechnol 95(1):169–178, doi:10.1007/s00253-012-3956-9
19. Mimitsuka T, Sawai H, Hatsu M, Yamada K (2007) Metabolic engineering of Corynebacterium glutamicum for cadaverine fermentation. Biosci Biotechnol Biochem 71(9):2130–2135, doi:10.1271/bbb.60699
20. Heider SA, Peters-Wendisch P, Netzer R, Stafnes M, Brautaset T, Wendisch VF (2014) Production and glucosylation of C50 and C 40 carotenoids by metabolically engineered Corynebacterium glutamicum. Appl Microbiol Biotechnol 98(3):1223–1235, doi:10.1007/s00253-013-5359-y
21. Heider SA, Peters-Wendisch P, Wendisch VF (2012) Carotenoid biosynthesis and overproduction in Corynebacterium glutamicum. BMC Microbiol 12:198, doi:10.1186/1471-2180-12-198
22. Heider SA, Peters-Wendisch P, Wendisch VF, Beekwilder J, Brautaset T (2014) Metabolic engineering for the microbial production of carotenoids and related products with a focus on the rare C50 carotenoids. Appl Microbiol Biotechnol 98(10):4355–4368, doi:10.1007/s00253-014-5693-8
23. Heider SA, Wolf N, Hofemeier A, Peters-Wendisch P, Wendisch VF (2014) Optimization of the IPP precursor supply for the production of lycopene, decaprenoxanthin and astaxanthin by Corynebacterium glutamicum. Front Bioeng Biotechnol 2:28, doi:10.3389/fbioe.2014.00028
24. Frohwitter J, Heider SA, Peters-Wendisch P, Beekwilder J, Wendisch VF (2014) Production of the sesquiterpene (+)-valencene by metabolically engineered Corynebacterium glutamicum. J Biotechnol, doi:10.1016/j.jbiotec.2014.05.032
25. Kang MK, Eom JH, Kim Y, Um Y, Woo HM (2014) Biosynthesis of pinene from glucose using metabolically-engineered Corynebacterium glutamicum. Biotechnol Lett 36(10):2069–2077, doi:10.1007/s10529-014-1578-2
26. Sasaki M, Jojima T, Inui M, Yukawa H (2009) Xylitol production by recombinant Corynebacterium glutamicum under oxygen deprivation. Appl Microbiol Biotechnol 86:1057–1066, doi:10.1007/s00253-009-2372-2
27. Gopinath V, Meiswinkel TM, Wendisch VF, Nampoothiri KM (2011) Amino acid production from rice straw and wheat bran hydrolysates by recombinant pentose-utilizing Corynebacterium glutamicum. Appl Microbiol Biotechnol 92(5):985–996, doi:10.1007/s00253-011-3478-x
28. Meiswinkel TM, Gopinath V, Lindner SN, Nampoothiri KM, Wendisch VF (2013) Accelerated pentose utilization by Corynebacterium glutamicum for accelerated production of lysine, glutamate, ornithine and putrescine. Microb Biotechnol 6(2):131–140, doi:10.1111/1751-7915.12001
29. Barrett E, Stanton C, Zelder O, Fitzgerald G, Ross RP (2004) Heterologous expression of lactose- and galactose-utilizing pathways from lactic acid bacteria in Corynebacterium glutamicum for production of lysine in whey. Appl Environ Microbiol 70(5):2861–2866
30. Kawaguchi H, Sasaki M, Vertes AA, Inui M, Yukawa H (2009) Identification and functional analysis of the gene cluster for L-arabinose utilization in Corynebacterium glutamicum. Appl Environ Microbiol 75(11):3419–3429, doi:10.1128/AEM.02912-08
31. Uhde A, Youn JW, Maeda T, Clermont L, Matano C, Kramer R, Wendisch VF, Seibold GM, Marin K (2013) Glucosamine as carbon source for amino acid-producing Corynebacterium glutamicum. Appl Microbiol Biotechnol 97(4):1679–1687, doi:10.1007/s00253-012-4313-8
32. Matano C, Uhde A, Youn JW, Maeda T, Clermont L, Marin K, Kramer R, Wendisch VF, Seibold GM (2014) Engineering of Corynebacterium glutamicum for growth and L-lysine and lycopene production from N-acetyl-glucosamine. Appl Microbiol Biotechnol 98(12):5633–5643, doi:10.1007/s00253-014-5676-9
33. Rittmann D, Lindner SN, Wendisch VF (2008) Engineering of a glycerol utilization pathway for amino acid production by Corynebacterium glutamicum. Appl Environ Microbiol 74(20):6216–6222, doi:10.1128/AEM. 00963-08
34. Zaldivar J, Nielsen J, Olsson L (2001) Fuel ethanol production from lignocellulose: a challenge for metabolic engineering and process integration. Appl Microbiol Biotechnol 56(1–2):17–34
35. Boer H, Maaheimo H, Koivula A, Penttila M, Richard P (2010) Identification in Agrobacterium tumefaciens of the D-galacturonic acid dehydrogenase gene. Appl Microbiol Biotechnol 86(3):901–909, doi:10.1007/s00253-009-2333-9
36. Conway T (1992) The Entner-Doudoroff pathway: history, physiology and molecular biology. FEMS Microbiol Rev 9(1):1–27
37. Blombach B, Hans S, Bathe B, Eikmanns BJ (2009) Acetohydroxyacid synthase, a novel target for improvement of L-lysine production by Corynebacterium glutamicum. Appl Environ Microbiol 75(2):419–427, doi:10.1128/AEM. 01844-08
38. Hanahan D (1983) Studies on transformation of Escherichia coli with plasmids. J Mol Biol 166(4):557–580
39. Eggeling L, Reyes O (2005) Experiments. In: Eggeling L, Bott M (eds) Handbook of Corynebacterium glutamicum. CRC, Boca Raton, pp 3535–3566
40. Abe S, Takayarna K, Kinoshita S (1967) Taxonomical studies on glutamic acid-producing bacteria. J Gen Appl Microbiol 13:279–301

41. Stansen C, Uy D, Delaunay S, Eggeling L, Goergen JL, Wendisch VF (2005) Characterization of a *Corynebacterium glutamicum* lactate utilization operon induced during temperature-triggered glutamate production. Appl Environ Microbiol 71(10):5920–5928, doi:10.1128/AEM. 71.10.5920-5928.2005

42. Peters-Wendisch PG, Schiel B, Wendisch VF, Katsoulidis E, Mockel B, Sahm H, Eikmanns BJ (2001) Pyruvate carboxylase is a major bottleneck for glutamate and lysine production by *Corynebacterium glutamicum*. J Mol Microbiol Biotechnol 3(2):295–300

43. Eikmanns BJ, Thum-Schmitz N, Eggeling L, Lüdtke KU, Sahm H (1994) Nucleotide sequence, expression and transcriptional analysis of the *Corynebacterium glutamicum gltA* gene encoding citrate synthase. Microbiology 140(Pt 8):1817–1828

44. Gibson DG, Young L, Chuang RY, Venter JC, Hutchison CA, 3rd, Smith HO (2009) Enzymatic assembly of DNA molecules up to several hundred kilobases. Nat Methods 6(5):343–345, doi:10.1038/nmeth.1318

45. Wendisch VF (2003) Genome-wide expression analysis in *Corynebacterium glutamicum* using DNA microarrays. J Biotechnol 104(1–3):273–285

46. Altschul SF1, Gish W, Miller W, Myers EW, Lipman DJ (1990) Basic local alignment search tool. J Mol Biol 215:403–410

47. Kalinowski J, Bathe B, Bartels D, Bischoff N, Bott M, Burkovski A, Dusch N, Eggeling L, Eikmanns BJ, Gaigalat L, Goesmann A, Hartmann M, Huthmacher K, Kramer R, Linke B, McHardy AC, Meyer F, Mockel B, Pfefferle W, Puhler A, Rey DA, Ruckert C, Rupp O, Sahm H, Wendisch VF, Wiegrabe I, Tauch A (2003) The complete *Corynebacterium glutamicum* ATCC 13032 genome sequence and its impact on the production of L-aspartate-derived amino acids and vitamins. J Biotechnol 104(1–3):5–25

48. Blattner FR, Plunkett G 3rd, Bloch CA, Perna NT, Burland V, Riley M, Collado-Vides J, Glasner JD, Rode CK, Mayhew GF, Gregor J, Davis NW, Kirkpatrick HA, Goeden MA, Rose DJ, Mau B, Shao Y (1997) The complete genome sequence of *Escherichia coli* K-12. Science 277(5331):1453–1474

49. Engels V, Lindner SN, Wendisch VF (2008) The global repressor SugR controls expression of genes of glycolysis and of the L-lactate dehydrogenase LdhA in *Corynebacterium glutamicum*. J Bacteriol 190(24):8033–8044, doi:10.1128/JB.00705-08

50. Pao SS, Paulsen IT, Saier MH Jr (1998) Major facilitator superfamily. Microbiol Mol Biol Rev 62(1):1–34

51. Meiswinkel TM, Rittmann D, Lindner SN, Wendisch VF (2013) Crude glycerol-based production of amino acids and putrescine by *Corynebacterium glutamicum*. Bioresour Technol 145:254–258, doi:10.1016/j. biortech.2013.02.053

52. Sasaki M, Jojima T, Kawaguchi H, Inui M, Yukawa H (2009) Engineering of pentose transport in *Corynebacterium glutamicum* to improve simultaneous utilization of mixed sugars. Appl Microbiol Biotechnol 85(1):105–115, doi:10.1007/s00253-009-2065-x

53. Youn JW, Jolkver E, Kramer R, Marin K, Wendisch VF (2008) Identification and characterization of the dicarboxylate uptake system DccT in *Corynebacterium glutamicum*. J Bacteriol 190(19):6458–6466, doi:10.1128/JB.00780-08

54. Youn JW, Jolkver E, Kramer R, Marin K, Wendisch VF (2009) Characterization of the dicarboxylate transporter DctA in *Corynebacterium glutamicum*. J Bacteriol 191(17):5480–5488, doi:10.1128/JB.00640-09

55. Bates Utz C, Nguyen AB, Smalley DJ, Anderson AB, Conway T (2004) GntP is the *Escherichia coli* fructuronic acid transporter and belongs to the UxuR regulon. J Bacteriol 186(22):7690–7696, doi:10.1128/JB.186.22.7690-7696.2004

56. Frunzke J, Engels V, Hasenbein S, Gatgens C, Bott M (2008) Co-ordinated regulation of gluconate catabolism and glucose uptake in *Corynebacterium glutamicum* by two functionally equivalent transcriptional regulators, GntR1 and GntR2. Mol Microbiol 67(2):305–322, doi:10.1111/j.1365-2958.2007.06020.x

57. Kuivanen J, Richard P (2014) The *yjjN* of *E. coli* codes for an L-galactonate dehydrogenase and can be used for quantification of L-galactonate and L-gulonate. Appl Biochem Biotechnol 173(7):1829–1835, doi:10.1007/s12010-014-0969-0

58. Rothe M, Alpert C, Loh G, Blaut M (2013) Novel insights into *E. coli*'s hexuronate metabolism: KduI facilitates the conversion of galacturonate and glucuronate under osmotic stress conditions. PLoS One 8(2):e56906, doi:10.1371/journal.pone.0056906

14

Programming the group behaviors of bacterial communities with synthetic cellular communication

Wentao Kong[1,2†], Venhar Celik[1,2†], Chen Liao[1,2], Qiang Hua[3] and Ting Lu[1,2,4*]

Abstract

Synthetic biology is a newly emerged research discipline that focuses on the engineering of novel cellular behaviors and functionalities through the creation of artificial gene circuits. One important class of synthetic circuits currently under active development concerns the programming of bacterial cellular communication and collective population-scale behaviors. Because of the ubiquity of cell-cell interactions within bacterial communities, having an ability of engineering these circuits is vital to programming robust cellular behaviors. Here, we highlight recent advances in communication-based synthetic gene circuits by first discussing natural communication systems and then surveying various functional engineered circuits, including those for population density control, temporal synchronization, spatial organization, and ecosystem formation. We conclude by summarizing recent advances, outlining existing challenges, and discussing potential applications and future opportunities.

Keywords: Synthetic biology; Gene circuits; Bacterial communities; Cellular communication; Collective behaviors; Dynamics

Background

Synthetic biology is a newly emerged research discipline that focuses on the engineering of novel cellular behaviors and functionalities. Since the launch of the field in 2000 [1,2], a wide range of synthetic gene devices have been created, including switches [3-9], oscillators [10-13], memory elements [7,14,15], and communication modules [13,16-18], as well as other electronics-inspired genetic devices, such as digital logic gates [19-22], pulse generators [23], and filters [24,25]. With designed cellular behaviors and functionalities, engineered circuits have been exploited to understand biological questions and to address various real-world problems [26]. The field has shown tremendous potential for biomedical, environmental, and energy-related applications [27]. For example, towards biomedical applications, engineered genetic circuits contribute to the understanding of disease mechanisms, provide novel diagnostic tools, enable economic production of therapeutics, and enable the design of novel treatment strategies for various diseases including cancer, metabolic disorders, and infectious diseases [28,29].

In the last few years, the advances of synthetic circuits have been further expedited, empowered by recent breakthroughs in genetic engineering techniques such as novel DNA assembly [30-33] and genome editing tools [34-37], advances in methodologies including those for rational circuit design and optimization [38-40], and quick enrichment of parts and elements [41,42]. As a result, synthetic biologists are now in a position to engineer desired cellular phenotypes in a larger, faster, and cheaper fashion.

One important class of synthetic circuits that are under active development concerns the programming of bacterial cell-cell communication and the group behaviors of communities [43-48]. Successful examples include gene constructs responsible for cellular density control [18], spatiotemporal patterning [13,16,49,50], and ecosystem formation [51,52]. The engineering of community-based circuits is essential and invaluable towards the implementation of complex but robust cellular functionality because of the following reasons: First, although microbes are

* Correspondence: luting@illinois.edu
†Equal contributors
[1]Department of Bioengineering, University of Illinois at Urbana-Champaign, 1304 W Springfield Avenue, Urbana, IL 61801, USA
[2]Institute for Genomic Biology, University of Illinois at Urbana-Champaign, 1206 W Gregory Drive, Urbana, IL 61801, USA
Full list of author information is available at the end of the article

single cell organisms, they are present dominantly in the form of communities in nature and in live bodies, such as biofilms [53,54] and the human microbiome [55,56]. Second, microbial physiology and functionality are strongly correlated with their forms - for instance, bacterial antibiotic resistance is distinct when cells are in planktonic forms and biofilm forms [57,58]. Third, recent advances in the biotechnological industry have clearly shown that microbial consortia may provide many compelling advantages in producing products of interest and controlling fermentation processes [59,60].

We are thus motivated in this article to overview the advances of synthetic gene circuits towards the programming of bacterial cellular communication and community behaviors. We will first discuss basic communication modules that confer cell-cell coordination in communities. We will then overview various functional gene circuits that enable the implementation of desired dynamic group behaviors, including those for population density control, temporal synchronization, spatial organization, and ecosystem formation. We will conclude by summarizing recent advances and discussing existing challenges, potential applications, and future opportunities.

Although not discussed here, it is important to note that there has been considerable progress in developing synthetic cellular communication in eukaryotes such as mammalian cells and yeast, which has been surveyed in the literature [51,61,62].

Review
Basic communication modules
Despite their species diversity, bacteria often utilize similar signaling systems for the implementation of their group behaviors [63,64]. For instance, quorum sensing (QS) is prevalent in bacteria for coordinating their group behaviors such as bioluminescence [65], biofilm formation [66], pathogenesis [67] and antibiotic synthesis [68-70].

Bacterial communication via nonvolatile signaling molecules
In Gram-negative bacteria, acyl-homoserine lactones (AHLs) are commonly used as QS molecules for intra-species communication. These molecules are composed of a homoserine lactone ring with an acyl chain of C4 to C18 in length [71,72]. AHL molecules are synthesized by the LuxI family synthases and detected by the corresponding LuxR-type receptors [73]. One canonical example of this class of communication is the QS system discovered in the bioluminescent marine bacterium *Vibrio fischeri* [74]. As shown in Figure 1A, LuxI, the autoinducer synthase, produces the AHL molecule $3OC_6HSL$ that can diffuse freely across the cell membrane and accumulate with the increase of cell density. Once the AHLs reach a threshold concentration, they form a complex with the LuxR receptor and activate the transcription of the

downstream genes (*luxI* and *luxR* in this case). A positive feedback regulatory architecture arises here from the self-activation of LuxI synthesis to facilitate the synchronization of the cellular population. Similar to *V. fischeri*, many other Gram-negative bacteria also possess QS systems, including the LasI/LasR and RhlI/RhlR systems in *Pseudomonas aeruginosa* [75], the CarI/CarR system in *Erwinia carotovora* [76], and the EsaI/EsaR system in *Pantoea stewartii* [77]. From an engineering perspective, these QS systems can be decomposed into two separate modules with one for signal production and the other for signal detection and response - when engineered in different cells, the two functional modules will confer communications between the two cells as shown in Figure 1C.

In Gram-positive bacteria, modified oligopeptides often serve as the signaling molecules for cellular communication with the cooperation of two-component systems. One classic example of this type of system is the Agr system in *Staphylococcus aureus* (Figure 1B) [78]. Here, the auto-inducing peptide (AIP) precursor, encoded by the gene *agrD*, is modified on its thiolactone ring and exported by AgrB protein. Upon the binding of AIP with the transmembrane protein AgrC, the transcriptional factor AgrA inside the cell is phosphorylated and then activated, which leads to the induction of the transcription of the downstream genes (*agrB/D/C/A* here). In addition to the Agr system, there are many communication systems based on auto-inducing peptides, such as the fsr system in *Enterococcus faecalis* [79], the Com system of *Streptococcus pneumonia* [80], and the nisRK system in *Lactococcus* [70]. To program collective behaviors in Gram-positive bacteria, a modular partition of those AI systems can thus been exploited (Figure 1D).

Other than the QS and AIP systems that are primarily present in intra-species communication, there are inter-species communication systems that coordinate cellular behaviors over multiple bacterial species. One such example is the communication systems mediated by the universal signaling molecule autoinducer-2 (AI-2), a furanosyl borate diester synthesized by LuxS from *S*-adenosyl-methionine and present in roughly half of all sequenced bacterial genome [81,82]. Towards programmable behaviors in multiple bacterial species, AI-2 is hence an ideal candidate for exploitation.

Bacterial communication via volatile and gas molecules
The adoption of nonvolatile molecules, such as AHLs and AIs, as the broadcast signal enables cellular coordination across various species. However, communications via those molecules require the presence of the both sender and receiver species in the same liquid environments or in gel-like setting within a short distance to allow for diffusion of signaling molecules. Volatile molecules, in contrast, can diffuse through air and circumvent the need of

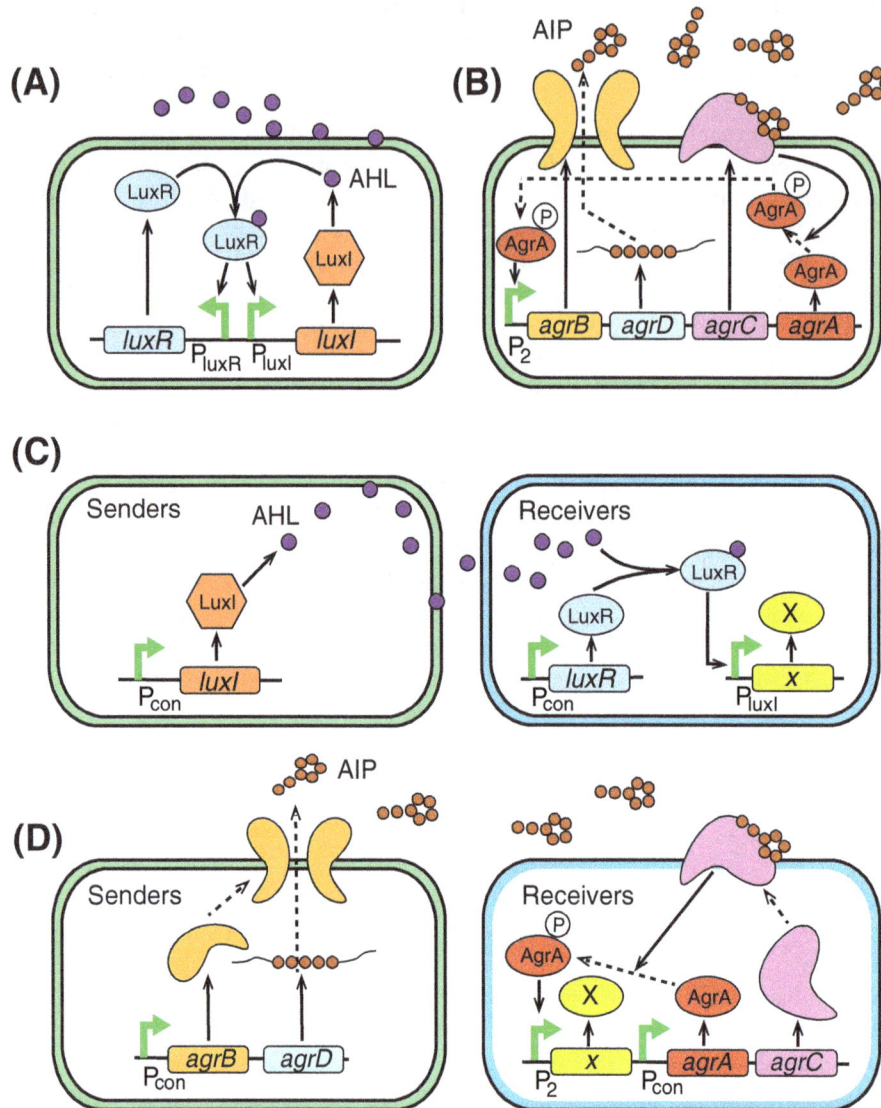

Figure 1 Cellular communication in bacteria. (A) The LuxI/LuxR quorum sensing (QS) system in the Gram-negative bacterium *Vibrio fischeri*. The system consists of the genes *luxI* and *luxR* and the cognate promoter P_{luxI} and P_{luxR}. Its signaling molecule is the acyl-homoserine lactone (AHL) 3OC$_6$HSL. **(B)** The Agr QS system in the Gram-positive bacterium *Staphylococcus aureus*. It consists of the genes, *agrD*, *agrB*, *agrC*, and *agrA* and the cognate promoter P_2. The auto-inducing peptide (AIP) is the signaling molecule of the system. **(C)** An engineered communication module adapted from the wild-type LuxI/LuxR system in **(A)**. By expressing *luxI*, the sender cell (left) produces the signal AHL that diffuses to the extracellular milieu and further into the receiver cell (right) to alter the expression of the downstream genes *X*. **(D)** A synthetic communication module built from the Agr system in **(B)**. The sender cell (left) produces and secretes the signaling molecule AIP that is sensed by the receiver cell (right), resulting in the expression shift of the gene *X* in the receiver cell.

physical mediating settings for signaling, allowing for more versatile, rapid, and large-scale communications of communities.

Weber et al. recently established a communication system that utilizes acetaldehyde as signaling molecules [51]. In their study, a bacterial strain (sender) was engineered to constitutively express alcohol dehydrogenase (ADH), an enzyme that converts ethanol in the medium to acetaldehyde. Due to its low boiling point (21°C), acetaldehyde volatized and was broadcast to neighboring

cells (receiver) via air to trigger the expression of genes controlled by the cognate acetaldehyde-inducible promoters. Therefore, the sender cells produced a concentric gradient of acetaldehyde that induced the dose-dependent gene expression of the receiver cells with the expression level defined by the distance between the sender and receiver cells.

In another example, Hasty and colleagues constructed *ndh-2*, a gene encoding NADH dehydrogenase II (membrane-bound respiratory enzyme), into an *Escherichia coli* strain to confer the production of hydrogen peroxide (H_2O_2) [83]. H_2O_2 is a thermodynamically unstable chemical compound and is able to enter neighboring cells quickly to alter their redox state and inactivate ArcAB, resulting in the shift of the activity of the corresponding downstream genes. Through the exploitation of H_2O_2, a novel route of airborne signaling molecule was created for fast and large-scale colony coordination.

Other communication mechanisms

In addition to the common signaling mechanisms discussed above, bacteria also exploit a wide range of alternative approaches for communications, such as quinolone signal [84], diffusible signal factor [85], cyclic dipeptide [86], diketopiperazines [86,87], and others [88,89]. One such representative mode of signaling is the use of indole, an aromatic heterocyclic organic compound that is produced by over 85 species of Gram-positive and Gram-negative bacteria and used as an extracellular signal for global coordination of various bacterial species [90]. Although little of those mechanisms have been explored for synthetic biology applications, the broad spectrum of signaling systems provides a rich reservoir for engineering multicellular functionality.

Dynamic group behaviors of bacterial communities via engineered communications

Cellular communications enable the coordination of single cells by sending and sensing the states of individuals. Inspired by this natural capability of bacteria, synthetic biologists have developed a set of engineered bacterial populations with their group behaviors programmed from designed artificial cell-cell communications.

Population density control

The first communication-based synthetic circuit was built by You et al. in 2004 with the goal of creating a dynamic, autonomous regulation of the cell density of an *E. coli* population [18]. As illustrated in Figure 2A, the Lux system from *V. fischeri* was introduced to construct cell-cell communication and was coupled to cell survival and killing via the CcdA/B toxin system. Here, the LuxI protein catalyzes the synthesis of a small, diffusible AHL signaling molecule, $3OC_6HSL$, which accumulates in the extracellular milieu and the intracellular environment as the cell density increases. When cells reach a sufficient density, the AHL binds to LuxR and forms the LuxR/AHL complex that activates the expression of the killer protein LacZα-CcdB, leading to cell death. On the other hand, cell death can cause a reduction of total population density and hence the level of AHL production, which in turn allows the population to recover after killing. The continuous production and degradation of AHL make the cell density approach a steady

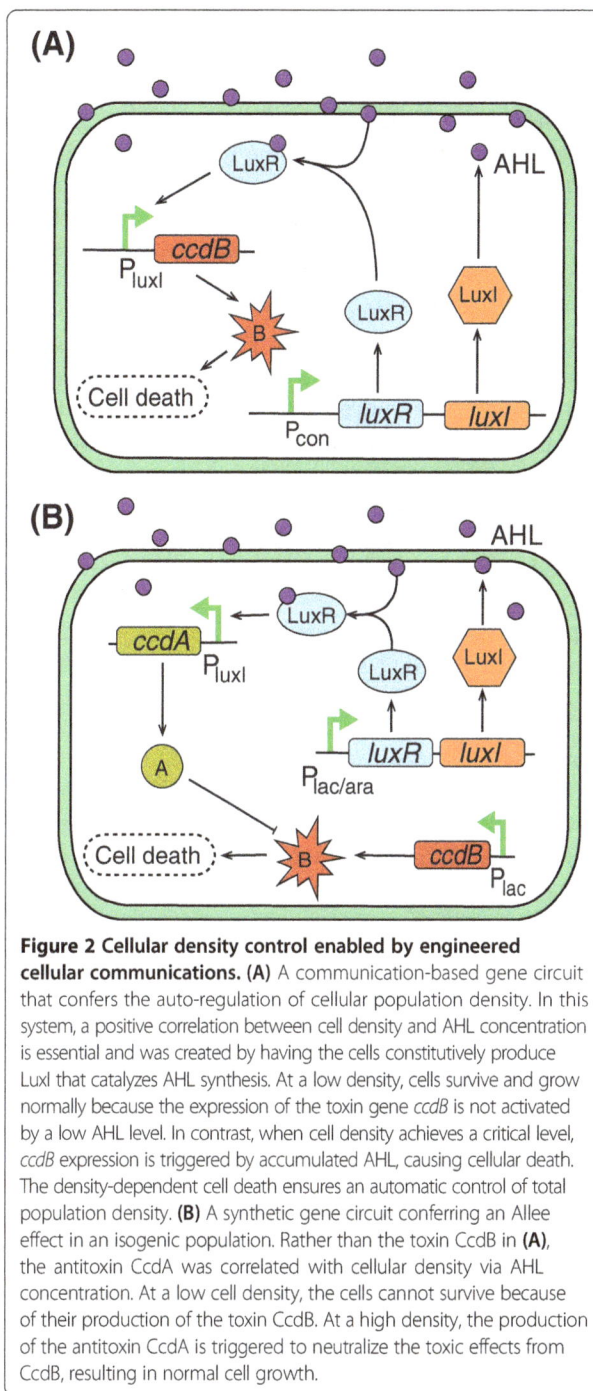

Figure 2 Cellular density control enabled by engineered cellular communications. (A) A communication-based gene circuit that confers the auto-regulation of cellular population density. In this system, a positive correlation between cell density and AHL concentration is essential and was created by having the cells constitutively produce LuxI that catalyzes AHL synthesis. At a low density, cells survive and grow normally because the expression of the toxin gene *ccdB* is not activated by a low AHL level. In contrast, when cell density achieves a critical level, *ccdB* expression is triggered by accumulated AHL, causing cellular death. The density-dependent cell death ensures an automatic control of total population density. **(B)** A synthetic gene circuit conferring an Allee effect in an isogenic population. Rather than the toxin CcdB in **(A)**, the antitoxin CcdA was correlated with cellular density via AHL concentration. At a low cell density, the cells cannot survive because of their production of the toxin CcdB. At a high density, the production of the antitoxin CcdA is triggered to neutralize the toxic effects from CcdB, resulting in normal cell growth.

state. Indeed, a stable cell density was maintained for more than 30 h with the variation within less than 5% in the study. This density control circuit laid a foundation for using cellular communications to program bacterial communities, allowing the extension of the control of population dynamics to the engineering of more sophisticated synthetic ecosystems.

In a recent work, Smith et al. utilized the density control circuit constructed above to create an artificial Allee effect in *E. coli* populations [91]. The Allee effect is a biological phenomenon characterized by a correlation between population density and the mean individual fitness of a population [92]. To create such an effect, a synthetic gene circuit was constructed to contain the LuxI/LuxR system and the CcdA/B toxin-antitoxin system (Figure 2B). In this setting, the expression of LuxR/LuxI and CcdB (killer) is under the control of $P_{lac/ara}$ promoter, while CcdA (rescue) was regulated by the cell density-dependent P_{lux} promoter. When IPTG induction is on, the cellular population growth rate is negative if the initial cell density is less than the critical value (C_{crit}) at which CcdA expression is not activated. However, if the initial cell density is above C_{crit}, AHL activates the production of LuxR and further drives the production of CcdA which rescues the population by inhibiting the toxicity of CcdB. An Allee effect population was thus established to have a negative fitness below a threshold of cell density but a positive fitness when the density is beyond the threshold. This study provided new implications of engineered cellular communication for controlling invasive species and the spread of infectious diseases.

Temporal synchronization

Complex cellular behaviors, such as biofilm formation and host invasion, often require the temporal coordination and collective action of cellular populations [93,94]. Towards this need, engineered communications offer a powerful solution.

In a recent study, Hasty and colleagues reported the development of an artificial gene circuit that synchronizes the oscillation of gene expression in individual cells [13]. Figure 3A shows their circuit design based on the QS elements of *V. fischeri* (*luxI*, *luxR*) and *Bacillus thurigensis* (*aiiA*). The AHL 3OC$_6$HSL, synthesized by LuxI, binds to transcriptional factor LuxR to form a complex (LuxR-AHL) that activates the expression of *luxI*, which leads to a positive feedback loop in regulation. At the same time, the LuxR-AHL complex also activates the expression of *aiiA*, a gene encoding the AHL degradation enzyme, which leads to a negative feedback loop in regulation. The dual positive and negative feedback loops drive the sustained oscillation of gene expression of individual cells, and in the meantime, the signaling molecule AHL confers the synchronization of individual oscillations. Using a custom-tailored microfluidic device, the authors were, for the first time, able to establish and tune synchronized oscillations of an entire cellular population (thousands of cells). Compared with the single cell oscillators developed by the same group [11] and other researchers [1,10], the engineered cellular communication indeed conferred the synchrony of cellular gene expression dynamics at a robust and yet tunable fashion.

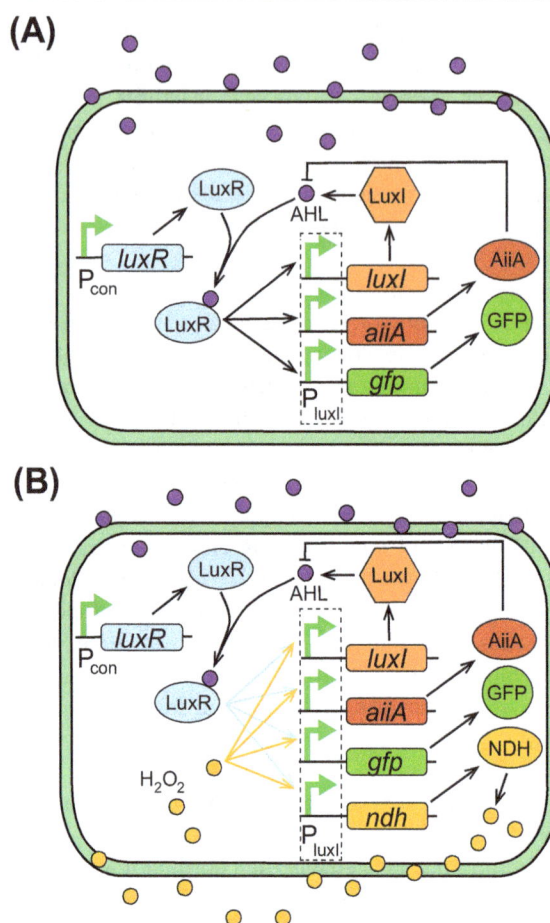

Figure 3 Synchronization of genetic oscillations by communication-based circuits. (A) A QS-based gene circuit that synchronizes the oscillation of gene expression in an isogenic bacterial population. The oscillation is enabled by positive feedback, arising from the self-activation of AHL synthesis, and negative feedback, mediated by the AHL-degrading gene *aiiA*. The coupling of the two feedback loops results in robust oscillations of gene expression of an entire bacterial population (thousands of cells). **(B)** An advanced gene circuit modified from **(A)** that enables large-scale synchronization of oscillatory gene expression. In addition to the coupled positive and negative feedback in **(A)**, an additional positive feedback loop is introduced by coupling the production of thermally unstable H_2O_2 by NADH dehydrogenase (*ndh*) with AHL biosynthesis, leading to global oscillation synchronization of millions of cells.

Building on their success of the synchronized oscillator, the same group further advanced to create a more sophisticated genetic network that is capable of synchronizing oscillatory gene expression of populations across multiple spatial scales [83]. As illustrated in Figure 3B, the researchers placed a copy of the *ndh-2* gene, which encodes NADH dehydrogenase II, under the control of an additional copy of P_{lux} promoter (compared with Figure 3A). The NDH-2 produces a low level of H_2O_2 that vapors and passes through the walls of the oxygen-permeable polydimethylsiloxane (PDMS) chips. Driven by the oscillation of gene expression mediated by the AHL-based synchronized oscillation circuit, H_2O_2 was periodically produced and exchanged between the cells within individual chambers. When entering cells, H_2O_2 changes the redox state of the cells and inactivates their lux promoter binding protein ArcAB, causing the global activation of the lux promoter of the cells in different chambers. As a result, thousands of oscillating colony 'biopixels' (approximately 2.5 million cells) were synchronized over centimeter-length scales through the use of synergistic intercellular coupling involving both quorum sensing within a colony and gas-phase redox signaling between colonies. As a proof-of-concept application, this system was further employed to sense arsenic in environments via differential modulations of the period of the oscillatory cells that resemble a liquid crystal display (LCD)-like macroscopic clock.

Spatial organization

One of the most fascinating aspects of biological systems is their ability to generate complex but highly reproducible organisms through differential spatial patterning of morphogens across isogenic cells [95]. Towards the ultimate goal of biological engineering for creating desired tissues,

organs, or even entire organisms, one critical step is to develop an engineering strategy that enables robust spatiotemporal pattern formation of living cells. Engineered cellular communications hold a great promise towards this goal, in addition to their roles in conferring temporal coordination of cellular behaviors.

In fact, synthetic biologists have already made several interesting attempts through the exploitation of artificial communication-based gene circuits. For instance, Sohka et al. constructed a circuit implementing Wolpert's French flag model [96], enabling the determination of cell fates in a concentration-dependent manner [25]; Payne et al. created a circuit that allows self-organized pattern formation without morphogen gradients in bacteria [97]; Basu et al. engineered a band detector that allows for differential response of gene expression according to the local concentration of AHL, creating a bull's eye-like spatiotemporal pattern [16].

One elegant example for this line of applications is the programming of bacterial stripe patterns by Liu and co-workers [49]. As shown in Figure 4A, the gene circuit consists of two functional parts: density-sensing module and motility-control module. The density-sensing module centers on the LuxI/LuxR QS system that enables the synthesis and excretion of the AHL and the activation of the

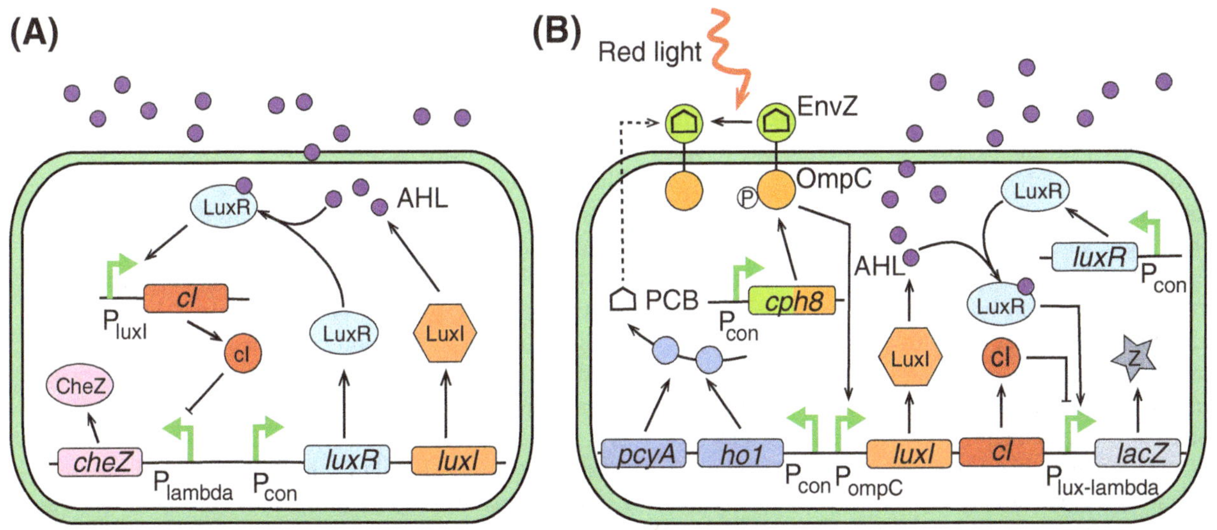

Figure 4 Spatial organization of cellular populations via engineered communication circuits. (A) A genetic circuit that generates periodic stripes in space. The LuxI/LuxR QS system is coupled to cellular motility via the transcriptional repressor gene *cI*, which is induced by AHL via the promoter P$_{luxI}$ and inhibits the expression of *cheZ*, one of the essential genes in bacteria motility. At a low cell density, AHL concentration remains low and no CI is produced, leading to a constant production of CheZ and hence a high cell motility. In a high cell density, sufficient accumulation of the AHL induces CI production which in turn suppresses *cheZ* expression, resulting in a deficiency in cell motility. The density-dependent motility of the population generates periodic stripe patterns in an expanding cell population. **(B)** A multi-module, communication-based synthetic circuit that allows accurate edge detection. Three functional modules are involved, including a light sensor, a cell communication module, and an X AND (NOT Y) gate. Upon exposure to red light, the light-sensing protein Cph8 induces the expression of *cI* and *luxI*: CI represses the expression of *lacZ* in the same cell regardless of AHL concentration while LuxI triggers the production of AHL that can diffuse to neighboring cells in the dark region to induce the production of LacZ. On the other hand, cells far from the light cannot produce LacZ because no AHL is available to trigger the transcription. As a result, only the cells near the edge of the light-exposed area actively express *lacZ*, which results in a dark pigment due to the enzymatic cleavage of a substrate in the plate by LacZ.

downstream gene *cI* when cell density is sufficiently high. The motility-control module is based on the bacterial motility system that is regulated by the transcription of *cheZ*. Upon the replacement of the wild-type *cheZ* with an inducible version (*cheZ* is under the control of the cI-repressed P_{lambda} promoter), cellular motility becomes regulated by the expression of cI. With the coupling of the two modules, engineered *E. coli* populations were able to form robust but tunable periodic stripes of high and low cell densities sequentially and autonomously. These results established cellular motility as a simple route to create recurrent spatial structures without the need for an extrinsic pacemaker. As a novel mechanism, it offered an alternative solution for the formation of biological spatial patterns that is distinct from the well-acknowledged Turing mechanism [98].

In addition to autonomous pattern formation, the QS-based communication mechanism can also be applied to detect complex spatial signals. Tabor et al. recently developed a multi-module gene circuit system for edge detection, a signal processing algorithm common in artificial intelligence and image recognition [99]. As illustrated in Figure 4B, the biological edge detection algorithm is composed of three modules: a dark sensor (NOT light), cell-cell communication cassette, and an X AND (NOT Y) genetic logic. The darker sensor was engineered based on the light-sensitive protein Cph8, a chimeric sensor kinase. With the covalent association of chromophore phycocyanobilin produced from heme via ho1 and pcyA [100,101], Cph8 is able to activate the ompC promoter (P_{ompC}) by transferring a phosphoryl group to the response regulator OmpR. However, in the presence of red light, the kinase activity of Cph8 is inhibited, which precludes the transcription from P_{ompC} and causes a NOT light transcriptional logic gate. The cell-cell communication was implemented through the Lux QS system and was used to convert light information into spatial distribution of AHL. With the incorporation of the converter cI and the hybrid promoter $P_{lux-lambda}$, the state of P_{ompC} is converted via an X AND (NOT Y) logical operation into the state of the promoter $P_{lux-lambda}$, which is displayed via the production of LacZ that produces black pigment. Upon the loading of the programs, a lawn of isogenic *E. coli* populations was able to sense an image of light, communicate to identify the light-dark edges, and visually present the result of the computation.

Ecosystem formation

Artificial cellular communications can enable not only the coordination of isogenic cell populations but also heterogeneous ecosystems that are composed of multiple species. You and co-works recently developed two gene circuits into a predator-prey ecosystem that consists of two *E. coli* populations [52].

Figure 5A shows the design of the ecosystem that involves two QS modules, LuxI/LuxR from *V. fischeri* and LasI/LasR from *P. aeruginosa*, for two-way communications. The predator cell (top) produces and secretes the AHL $3OC_{12}HSL$ that induces the expression of the toxin gene *ccdB* in the prey cell (bottom), leading to the death of the prey. In the meantime, the prey produces another AHL molecule, $3OC_6HSL$, which rescues the predator by inducing the production of antitoxin CcdA that neutralizes the toxin from CcdB. With appropriate modulations of the system parameters, the researchers were able to create a bacterial version of predation with different population dynamics generated, including extinction, coexistence, and oscillation. Similar to this work, another bidirectional intercellular communication network was also engineered by Brenner et al. [50], in which the LasI/LasR and RhlI/RhlR QS systems from *P. aeruginosa* were adopted to create a two-species microbial consensus consortium. In that ecosystem, the gene expression of any of the two species mutually depends on the presence of the other.

Beyond predation and consensus, designer cellular communications can be used to create a wide spectrum of inter-species interactions. As revealed by metagenomics and 16S pyrosequencing, microbial interactions in nature such as biofilms and the microbiome are extremely complicated and diverse - for instance, there can be parasitism, predation, commensalism, mutualism, competition, and amensalism within a single pair of species [102]. As one of the earliest efforts towards the programming of complicated cellular consortia, Weber and Fussenegger developed a set of pairwise interactions between *E. coli* and Chinese hamster ovary (CHO) cells [51].

As illustrated in Figure 5B, the designs of the ecosystems center on an airborne transmission of the transcription system that allows one species (*E. coli*) to convert ethanol into volatile acetaldehyde and broadcast this airborne signal (boiling point: 21°C) to another species (CHO-K1 cell line) for the activation of functionally specific, rationally engineered genes. The commensal ecosystem (top) was created by constructing an *E. coli* strain capable of converting ethanol into acetaldehyde for air broadcast and placing a neomycin resistance gene (*neo*) under the control of an acetaldehyde-induced promoter (P_{air}) in a CHO-K1 cell line. In addition, secreted alkaline phosphatase (SEAP) was used as a reporter of the CHO-K1 cells. When cultivated proximate to synthetic CHO-K1, the engineered *E. coli* cells confer survival of the mammalian cells while keeping their own growth unaffected by the mammalian cells cultured in a separate dish. The amensal ecosystem (middle) was synthesized by cultivating an acetaldehyde-broadcasting *E. coli* strain in close proximity to a CHO-K1 cell line that was engineered to have acetaldehyde-controlled expression of RipDD, a gene that encodes an

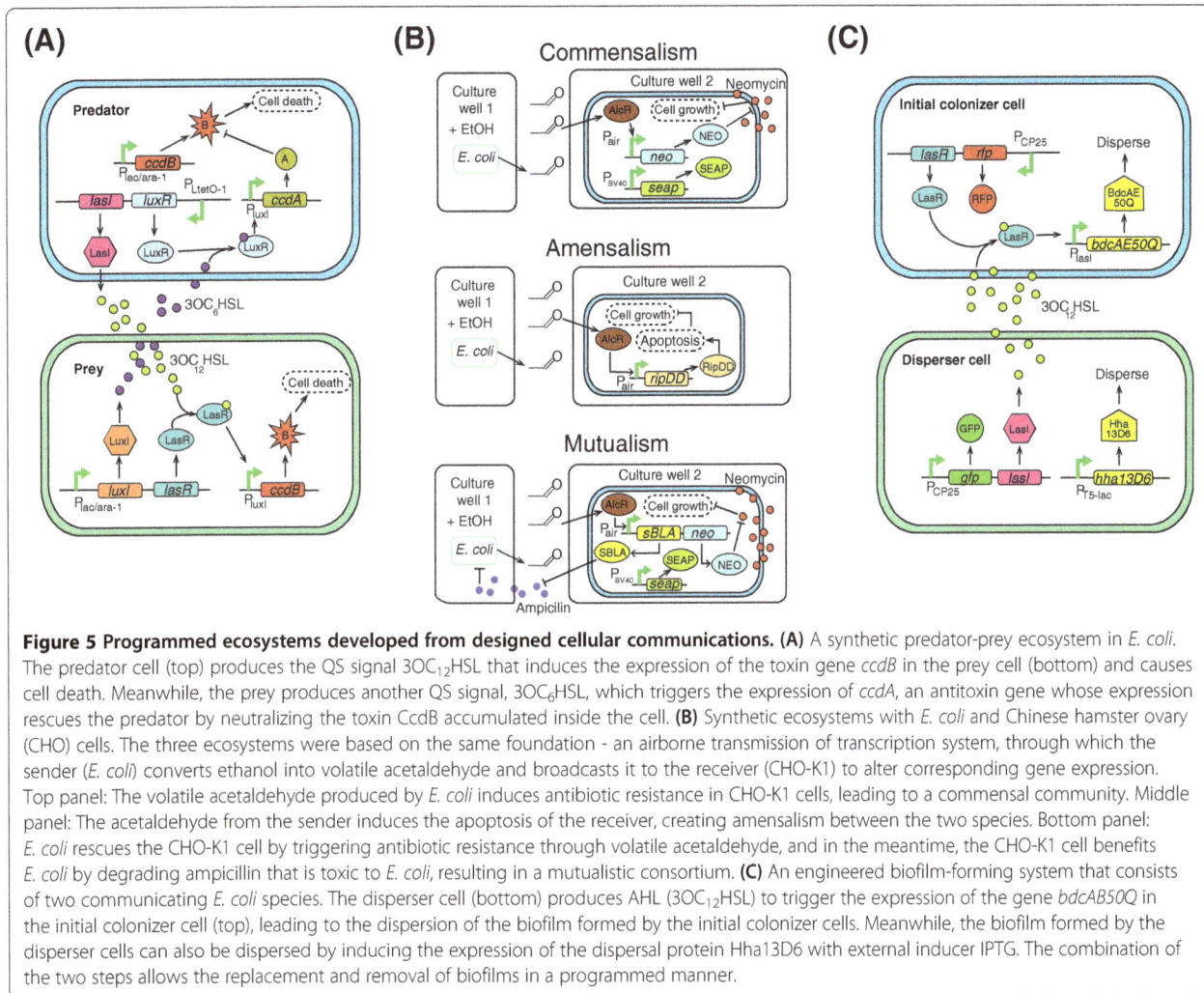

Figure 5 Programmed ecosystems developed from designed cellular communications. (A) A synthetic predator-prey ecosystem in *E. coli*. The predator cell (top) produces the QS signal 3OC$_{12}$HSL that induces the expression of the toxin gene *ccdB* in the prey cell (bottom) and causes cell death. Meanwhile, the prey produces another QS signal, 3OC$_6$HSL, which triggers the expression of *ccdA*, an antitoxin gene whose expression rescues the predator by neutralizing the toxin CcdB accumulated inside the cell. **(B)** Synthetic ecosystems with *E. coli* and Chinese hamster ovary (CHO) cells. The three ecosystems were based on the same foundation - an airborne transmission of transcription system, through which the sender (*E. coli*) converts ethanol into volatile acetaldehyde and broadcasts it to the receiver (CHO-K1) to alter corresponding gene expression. Top panel: The volatile acetaldehyde produced by *E. coli* induces antibiotic resistance in CHO-K1 cells, leading to a commensal community. Middle panel: The acetaldehyde from the sender induces the apoptosis of the receiver, creating amensalism between the two species. Bottom panel: *E. coli* rescues the CHO-K1 cell by triggering antibiotic resistance through volatile acetaldehyde, and in the meantime, the CHO-K1 cell benefits *E. coli* by degrading ampicillin that is toxic to *E. coli*, resulting in a mutualistic consortium. **(C)** An engineered biofilm-forming system that consists of two communicating *E. coli* species. The disperser cell (bottom) produces AHL (3OC$_{12}$HSL) to trigger the expression of the gene *bdcAB50Q* in the initial colonizer cell (top), leading to the dispersion of the biofilm formed by the initial colonizer cells. Meanwhile, the biofilm formed by the disperser cells can also be dispersed by inducing the expression of the dispersal protein Hha13D6 with external inducer IPTG. The combination of the two steps allows the replacement and removal of biofilms in a programmed manner.

apoptosis-inducing human receptor interacting protein. As a result, the CHO-K1 cells survive only in the absence of the *E. coli* cells because, otherwise, they induce the death of the CHO-K1 cells by producing acetaldehyde. To create a mutualistic interaction between *E. coli* and CHO-K1 cells (bottom), the commensal ecosystem developed earlier (top) was modified to incorporate a mammalian beta-lactamase gene *sBLA* under the control of the acetaldehyde-inducible promoter (P$_{air}$). Here, sBLA can be secreted to the extracellular milieu to hydrolyze the bacterial antibiotic ampicillin in the culture medium to promote the survival of co-cultured *E. coli*, resulting in bidirectional benefits between the two cell species. Following a similar idea, three additional types of ecosystem interactions were created, including parasitism, third party-inducible parasitism, and predator-prey interaction (not shown in Figure 5). This example demonstrated the ability of programming microbial consortia via rational design of cellular interactions by rewiring cellular communication systems, providing novel insights in

understanding and programming microbial community patterns that orchestrate the complex coexistence of living systems.

In addition to programming planktonic bacterial populations, synthetic communication circuits have also been exploited in controlling complex communities such as biofilms. Hong et al. recently developed quorum-sensing circuits to program the formation and dispersal of artificial *E. coli* biofilms [103]. As shown in Figure 5C, the circuits have two functional parts with one belonging to the initial colonizer cell (top) and the other belonging to the disperser cell (bottom). The initial colonizer part consists of the constitutively expressed repressor gene *lasR* and its cognate promoter P$_{lasI}$ that drives the expression of the biofilm dispersion gene *bdcAB50Q*; the disperser part is composed of the AHL-producing gene *lasI* that is constitutively expressed and another biofilm dispersion gene, *hha13D6*, controlled by external inducer IPTG. Such a design allows the disperser cell to trigger the expression of the gene *bdcAB50Q* in the initial colonizer cell by producing AHL ($3OC_{12}HSL$), leading to the dispersion and replacement of the biofilm formed by the initial colonizer cells. Meanwhile, the circuit in the disperser enables the biofilm formed by the dispersers to be removed with the external signal inducer IPTG. These types of functional circuits can be powerful in creating designer biofilms and enabling precise manipulation of community composition in the fields of biorefinery, medicine, and bioproduction.

Conclusions

With the advances of synthetic biology technologies and a consensus on the need for community-based functionality engineering, synthetic microbial consortia have undergone a rapid development in the past few years. This review has surveyed recent advances of engineered biological systems that utilize cell-cell communication to program bacterial group behaviors, covering both the basic communication modules and functional gene circuits that confer desired community-based dynamic behaviors.

Although there has been significant progress, the engineering of microbial communities is still in its infancy and is subject to a set of challenges. In fact, almost all synthetic circuits to date have involved many rounds of trial and error before achieving the desired functionality. Difficulties in the efficient construction of engineered circuits often stem from a lack of biological knowledge. Specifically, to facilitate gene circuit engineering, it is needed to have a deep understanding of stochasticity in gene expression [104-106], the inherent interplay between a synthetic circuit and the host organism [1], and issues related to multicellular physiology and metabolism [107]. Another big challenge arises from the technical side of synthetic biology, which includes the lack of powerful rational design platforms, limited availability of parts and modules, efficient systematic optimization strategies and toolkits, and high-throughput assays for circuit validation. Addressing the above challenges will foster our engineering capability and help to achieve the ultimate goal of efficient and reliable development of synthetic circuits with defined functionality.

Despite the challenges, the future of engineered microbial communities is bright. In fact, synthetic consortia have already started to show tremendous potential in both understanding biological questions and addressing real-world concerns. For example, extended from the programming of cellular dynamics, synthetic bacterial systems have been applied to understand ecological and evolutionary questions that are difficult to address with natural communities [108]. Towards real-world applications, bacterial consortia synthesized with designer communication modules have been used for information processing [109,110], bio-computation [111], and therapeutics [112-114], as well as material and chemical productions [115-117]. There are a variety of research fields where synthetic bacterial consortia have started to play an important role: In metabolic engineering, cellular communication can be used to implement self-regulated control between cellular growth and product manufacturing in bioreactors for autonomous bioproduction. In biomedical applications, custom-tailored probiotic bacteria can be introduced into the human body to alter the composition and hence the function of the gut microbiota for disease treatment. In areas relating to the environment, biofilms and microbial consortia in soil and other natural settings can be perturbed and even reprogrammed with engineered microbes for desired purposes. We thus expect that microbial communities programmed via engineered cellular communication will become a versatile strategy in addressing both scientific and practical challenges in the near future.

Abbreviations
ADH: alcohol dehydrogenase; AHL: acyl-homoserine lactones; AI-2: autoinducer-2; AIP: auto-inducing peptide; CHO cells: Chinese hamster ovary cells; H_2O_2: hydrogen peroxide; Ndh-2: NADH dehydrogenase II; QS: quorum sensing; SEAP: secreted alkaline phosphatase.

Competing interests
The authors declare that they have no competing interests.

Authors' contributions
QH and TL conceived the study and designed the project. WT, VC, and TL drafted the manuscript. CL analyzed the data. All authors read and approved the final manuscript.

Acknowledgements
We thank Andrew Blanchard for commenting and editing the manuscript. This work was supported by the American Heart Association (Grant No. 12SDG12090025), the Network for Computational Nanotechnology at UIUC sponsored by National Science Foundation (Grant No. 1227034), and the UIUC Research Board.

Author details
[1]Department of Bioengineering, University of Illinois at Urbana-Champaign, 1304 W Springfield Avenue, Urbana, IL 61801, USA. [2]Institute for Genomic Biology, University of Illinois at Urbana-Champaign, 1206 W Gregory Drive, Urbana, IL 61801, USA. [3]State Key Laboratory of Bioreactor Engineering, East China University of Science and Technology, 130 Meilong Road, Shanghai 200237, People's Republic of China. [4]Department of Physics, University of Illinois at Urbana-Champaign, 1110 W Green Street, Urbana, IL 61801, USA.

References
1. Elowitz MB, Leibler S (2000) A synthetic oscillatory network of transcriptional regulators. Nature 403(6767):335–338
2. Gardner TS, Cantor CR, Collins JJ (2000) Construction of a genetic toggle switch in Escherichia coli. Nature 403(6767):339–342
3. Atkinson MR, Savageau MA, Myers JT, Ninfa AJ (2003) Development of genetic circuitry exhibiting toggle switch or oscillatory behavior in Escherichia coli. Cell 113(5):597–607
4. Kramer BP, Viretta AU, Daoud-El Baba M, Aubel D, Weber W, Fussenegger M (2004) An engineered epigenetic transgene switch in mammalian cells. Nat Biotechnol 22(7):867–870
5. Bayer TS, Smolke CD (2005) Programmable ligand-controlled riboregulators of eukaryotic gene expression. Nat Biotechnol 123(3):337–343

6. Deans TL, Cantor CR, Collins JJ (2007) A tunable genetic switch based on RNAi and repressor proteins for regulating gene expression in mammalian cells. Cell 130(2):363–372

7. Friedland AE, Lu TK, Wang X, Shi D, Church G, Collins JJ (2009) Synthetic gene networks that count. Science 324(5931):1199–1202

8. Ellis T, Wang X, Collins JJ (2009) Diversity-based, model-guided construction of synthetic gene networks with predicted functions. Nat Biotechnol 27(5):465–471

9. Wu M, Su R-Q, Li X, Ellis T, Lai Y-C, Wang X (2013) Engineering of regulated stochastic cell fate determination. Proc Natl Acad Sci U S A 110(26):10610–10615

10. Fung E, Wong WW, Suen JK, Bulter T, S-g L, Liao JC (2005) A synthetic gene–metabolic oscillator. Nature 435(7038):118–122

11. Stricker J, Cookson S, Bennett MR, Mather WH, Tsimring LS, Hasty J (2008) A fast, robust and tunable synthetic gene oscillator. Nature 456(7221):516–519

12. Tigges M, Marquez-Lago TT, Stelling J, Fussenegger M (2009) A tunable synthetic mammalian oscillator. Nature 457(7227):309–312

13. Danino T, Mondragón-Palomino O, Tsimring L, Hasty J (2010) A synchronized quorum of genetic clocks. Nature 463(7279):326–330

14. Ham TS, Lee SK, Keasling JD, Arkin AP (2008) Design and construction of a double inversion recombination switch for heritable sequential genetic memory. PLoS One 3(7):e2815

15. Ajo-Franklin CM, Drubin DA, Eskin JA, Gee EP, Landgraf D, Phillips I, Silver PA (2007) Rational design of memory in eukaryotic cells. Genes Dev 21(18):2271–2276

16. Basu S, Gerchman Y, Collins CH, Arnold FH, Weiss R (2005) A synthetic multicellular system for programmed pattern formation. Nature 434(7037):1130–1134

17. Kobayashi H, Kærn M, Araki M, Chung K, Gardner TS, Cantor CR, Collins JJ (2004) Programmable cells: interfacing natural and engineered gene networks. Proc Natl Acad Sci U S A 101(22):8414–8419

18. You L, Cox RS, Weiss R, Arnold FH (2004) Programmed population control by cell–cell communication and regulated killing. Nature 428(6985):868–871

19. Guet CC, Elowitz MB, Hsing W, Leibler S (2002) Combinatorial synthesis of genetic networks. Science 296(5572):1466–1470

20. Rackham O, Chin JW (2005) Cellular logic with orthogonal ribosomes. J Am Chem Soc 127(50):17584–17585

21. Anderson JC, Voigt CA, Arkin AP (2007) Environmental signal integration by a modular AND gate. Mol Syst Biol 3:133

22. Win MN, Smolke CD (2008) Higher-order cellular information processing with synthetic RNA devices. Science 322(5900):456–460

23. Basu S, Mehreja R, Thiberge S, Chen M-T, Weiss R (2004) Spatiotemporal control of gene expression with pulse-generating networks. Proc Natl Acad Sci U S A 101(17):6355–6360

24. Hooshangi S, Thiberge S, Weiss R (2005) Ultrasensitivity and noise propagation in a synthetic transcriptional cascade. Proc Natl Acad Sci U S A 102(10):3581–3586

25. Sohka T, Heins RA, Phelan RM, Greisler JM, Townsend CA, Ostermeier M (2009) An externally tunable bacterial band-pass filter. Proc Natl Acad Sci U S A 106(25):10135–10140

26. Church GM, Elowitz MB, Smolke CD, Voigt CA, Weiss R (2014) Realizing the potential of synthetic biology. Nat Rev Mol Cell Biol 15(4):289–94

27. Khalil AS, Collins JJ (2010) Synthetic biology: applications come of age. Nat Rev Genet 11(5):367–379

28. Ruder WC, Lu T, Collins JJ (2011) Synthetic biology moving into the clinic. Science 333(6047):1248–1252

29. Weber W, Fussenegger M (2011) Emerging biomedical applications of synthetic biology. Nat Rev Genet 13(1):21–35

30. Gibson DG, Young L, Chuang R-Y, Venter JC, Hutchison CA, Smith HO (2009) Enzymatic assembly of DNA molecules up to several hundred kilobases. Nat Methods 6(5):343–345

31. Engler C, Kandzia R, Marillonnet S (2008) A one pot, one step, precision cloning method with high throughput capability. PLoS One 3(11):e3647

32. Zhang Y, Werling U, Edelmann W (2012) SLiCE: a novel bacterial cell extract-based DNA cloning method. Nucleic Acids Res 40(8):e55–e55

33. Quan J, Tian J (2011) Circular polymerase extension cloning for high-throughput cloning of complex and combinatorial DNA libraries. Nat Protoc 6(2):242–251

34. Wang HH, Isaacs FJ, Carr PA, Sun ZZ, Xu G, Forest CR, Church GM (2009) Programming cells by multiplex genome engineering and accelerated evolution. Nature 460(7257):894–898

35. Ellis HM, Yu D, DiTizio T (2001) High efficiency mutagenesis, repair, and engineering of chromosomal DNA using single-stranded oligonucleotides. Proc Natl Acad Sci U S A 98(12):6742–6746

36. Jiang W, Bikard D, Cox D, Zhang F, Marraffini LA (2013) RNA-guided editing of bacterial genomes using CRISPR-Cas systems. Nat Biotechnol 31(3):233–239

37. Gaj T, Gersbach CA, Barbas CF III (2013) ZFN, TALEN, and CRISPR/Cas-based methods for genome engineering. Trends Biotechnol 31(7):397–405

38. Slusarczyk AL, Lin A, Weiss R (2012) Foundations for the design and implementation of synthetic genetic circuits. Nat Rev Genet 13(6):406–420

39. Salis HM, Mirsky EA, Voigt CA (2009) Automated design of synthetic ribosome binding sites to control protein expression. Nat Biotechnol 27(10):946–950

40. Hillson NJ, Rosengarten RD, Keasling JD (2011) j5 DNA assembly design automation software. ACS Synth Biol 1(1):14–21

41. Lou C, Stanton B, Chen Y-J, Munsky B, Voigt CA (2012) Ribozyme-based insulator parts buffer synthetic circuits from genetic context. Nat Biotechnol 30(11):1137–1142

42. Smolke CD (2009) Building outside of the box: iGEM and the BioBricks Foundation. Nat Biotechnol 27(12):1099–1102

43. Chuang JS (2012) Engineering multicellular traits in synthetic microbial populations. Curr Opin Chem Biol 16(3):370–378

44. Brenner K, You L, Arnold FH (2008) Engineering microbial consortia: a new frontier in synthetic biology. Trends Biotechnol 26(9):483–489

45. Pai A, Tanouchi Y, Collins CH, You L (2009) Engineering multicellular systems by cell–cell communication. Curr Opin Biotechnol 20(4):461–470

46. Payne S, You L (2013) Engineered cell–cell communication and its applications. Adv Biochem Eng Biotechnol 146:97-121

47. Tsao C-Y, Quan DN, Bentley WE (2012) Development of the quorum sensing biotechnological toolbox. Curr Opin Chem Eng 1(4):396–402

48. Teuscher C, Grecu C, Lu T, Weiss R (2011) Challenges and promises of nano and bio communication networks. In: Networks on Chip (NoCS), 2011 Fifth IEEE/ACM International Symposium on. IEEE, Pittsburgh, pp 247–254

49. Liu C, Fu X, Liu L, Ren X, Chau CK, Li S, Xiang L, Zeng H, Chen G, Tang L-H (2011) Sequential establishment of stripe patterns in an expanding cell population. Science 334(6053):238–241

50. Brenner K, Karig DK, Weiss R, Arnold FH (2007) Engineered bidirectional communication mediates a consensus in a microbial biofilm consortium. Proc Natl Acad Sci U S A 104(44):17300–17304

51. Weber W, Daoud-El Baba M, Fussenegger M (2007) Synthetic ecosystems based on airborne inter- and intra-kingdom communication. Proc Natl Acad Sci U S A 104(25):10435–10440

52. Balagaddé FK, Song H, Ozaki J, Collins CH, Barnet M, Arnold FH, Quake SR, You L (2008) A synthetic Escherichia coli predator–prey ecosystem. Mol Syst Biol 4:187

53. O'Toole G, Kaplan HB, Kolter R (2000) Biofilm formation as microbial development. Annu Rev Microbiol 54(1):49–79

54. Stoodley P, Sauer K, Davies D, Costerton JW (2002) Biofilms as complex differentiated communities. Annu Rev Microbiol 56(1):187–209

55. Consortium HMP (2012) Structure, function and diversity of the healthy human microbiome. Nature 486(7402):207–214

56. Cho I, Blaser MJ (2012) The human microbiome: at the interface of health and disease. Nat Rev Genet 13(4):260–270

57. Costerton J, Stewart PS, Greenberg E (1999) Bacterial biofilms: a common cause of persistent infections. Science 284(5418):1318–1322

58. Stewart PS, William Costerton J (2001) Antibiotic resistance of bacteria in biofilms. Lancet 358(9276):135–138

59. Koizumi S, Endo T, Tabata K, Ozaki A (1998) Large-scale production of UDP-galactose and globotriose by coupling metabolically engineered bacteria. Nat Biotechnol 16(9):847–850

60. Minty JJ, Singer ME, Scholz SA, Bae C-H, Ahn J-H, Foster CE, Liao JC, Lin XN (2013) Design and characterization of synthetic fungal-bacterial consortia for direct production of isobutanol from cellulosic biomass. Proc Natl Acad Sci U S A 110(36):14592–14597

61. Chen M-T, Weiss R (2005) Artificial cell-cell communication in yeast Saccharomyces cerevisiae using signaling elements from Arabidopsis thaliana. Nat Biotechnol 23(12):1551–1555

62. Williams TC, Nielsen LK, Vickers CE (2013) Engineered quorum sensing using pheromone-mediated cell-to-cell communication in Saccharomyces cerevisiae. ACS Synth Biol 2(3):136–149

63. Bassler BL, Losick R (2006) Bacterially speaking. Cell 125(2):237–246

64. Waters CM, Bassler BL (2005) Quorum sensing: cell-to-cell communication in bacteria. Annu Rev Cell Dev Biol 21:319–346

65. Engebrecht J, Nealson K, Silverman M (1983) Bacterial bioluminescence: isolation and genetic analysis of functions from Vibrio fischeri. Cell 32(3):773–781

66. Davies DG, Parsek MR, Pearson JP, Iglewski BH, Costerton J, Greenberg E (1998) The involvement of cell-to-cell signals in the development of a bacterial biofilm. Science 280(5361):295–298

67. De Kievit TR, Iglewski BH (2000) Bacterial quorum sensing in pathogenic relationships. Infect Immun 68(9):4839–4849

68. Latifi A, Winson MK, Foglino M, Bycroft BW, Stewart GS, Lazdunski A, Williams P (1995) Multiple homologues of LuxR and LuxI control expression of virulence determinants and secondary metabolites through quorum sensing in *Pseudomonas aeruginosa* PAO1. Mol Microbiol 17(2):333–343

69. Thomson N, Crow M, McGowan S, Cox A, Salmond G (2000) Biosynthesis of carbapenem antibiotic and prodigiosin pigment in *Serratia* is under quorum sensing control. Mol Microbiol 36(3):539–556

70. Kleerebezem M (2004) Quorum sensing control of lantibiotic production; nisin and subtilin autoregulate their own biosynthesis. Peptides 25(9):1405–1414

71. Fuqua C, Parsek MR, Greenberg EP (2001) Regulation of gene expression by cell-to-cell communication: acyl-homoserine lactone quorum sensing. Annu Rev Genet 35(1):439–468

72. Thiel V, Kunze B, Verma P, Wagner-Döbler I, Schulz S (2009) New structural variants of homoserine lactones in bacteria. Chem Bio Chem 10(11):1861–1868

73. Fuqua WC, Winans SC, Greenberg EP (1994) Quorum sensing in bacteria: the LuxR-LuxI family of cell density-responsive transcriptional regulators. J Bacteriol 176(2):269

74. Nealson K, Hastings JW (1979) Bacterial bioluminescence: its control and ecological significance. Microbiol Rev 43(4):496

75. Pesci EC, Pearson JP, Seed PC, Iglewski BH (1997) Regulation of las and rhl quorum sensing in *Pseudomonas aeruginosa*. J Bacteriol 179(10):3127–3132

76. Welch M, Todd DE, Whitehead NA, McGowan SJ, Bycroft BW, Salmond GP (2000) N-acyl homoserine lactone binding to the CarR receptor determines quorum-sensing specificity in *Erwinia*. EMBO J 19(4):631–641

77. Minogue TD, Trebra MW, Bernhard F, Bodman SB (2002) The autoregulatory role of EsaR, a quorum-sensing regulator in *Pantoea stewartii* ssp. *stewartii*: evidence for a repressor function. Mol Microbiol 44(6):1625–1635

78. Novick RP, Geisinger E (2008) Quorum sensing in *Staphylococci*. Annu Rev Genet 42:541–564

79. Hancock LE, Perego M (2004) The *Enterococcus faecalis* fsr two-component system controls biofilm development through production of gelatinase. J Bacteriol 186(17):5629–5639

80. Pestova E, Håvarstein L, Morrison D (1996) Regulation of competence for genetic transformation in *Streptococcus pneumoniae* by an auto-induced peptide pheromone and a two-component regulatory system. Mol Microbiol 21(4):853–862

81. Winzer K, HARDIE K, Williams P (2003) LuxS and autoinducer-2: their contribution to quorum. Adv Appl Microbiol 53:291

82. Pereira CS, Thompson JA, Xavier KB (2013) AI-2-mediated signalling in bacteria. FEMS Microbiol Rev 37(2):156–181

83. Prindle A, Samayoa P, Razinkov I, Danino T, Tsimring LS, Hasty J (2012) A sensing array of radically coupled genetic 'biopixels'. Nature 481(7379):39–44

84. Pesci EC, Milbank JB, Pearson JP, McKnight S, Kende AS, Greenberg EP, Iglewski BH (1999) Quinolone signaling in the cell-to-cell communication system of *Pseudomonas aeruginosa*. Proc Natl Acad Sci U S A 96(20):11229–11234

85. Ryan RP, Dow JM (2011) Communication with a growing family: diffusible signal factor (DSF) signaling in bacteria. Trends Microbiol 19(3):145–152

86. Holden MT, Ram Chhabra S, De Nys R, Stead P, Bainton NJ, Hill PJ, Manefield M, Kumar N, Labatte M, England D (1999) Quorum-sensing cross talk: isolation and chemical characterization of cyclic dipeptides from *Pseudomonas aeruginosa* and other gram-negative bacteria. Mol Microbiol 33(6):1254–1266

87. Tommonaro G, Abbamondi GR, Iodice C, Tait K, De Rosa S (2012) Diketopiperazines produced by the halophilic archaeon, *Haloterrigena hispanica*, activate AHL bioreporters. Microb Ecol 63(3):490–495

88. Flavier AB, Clough SJ, Schell MA, Denny TP (1997) Identification of 3-hydroxypalmitic acid methyl ester as a novel autoregulator controlling virulence in *Ralstonia solanacearum*. Mol Microbiol 26(2):251–259

89. Chen H, Fujita M, Feng Q, Clardy J, Fink GR (2004) Tyrosol is a quorum-sensing molecule in *Candida albicans*. Proc Natl Acad Sci U S A 101(14):5048–5052

90. Lee JH, Lee J (2010) Indole as an intercellular signal in microbial communities. FEMS Microbiol Rev 34(4):426–444

91. Smith R, Tan C, Srimani JK, Pai A, Riccione KA, Song H, You L (2014) Programmed Allee effect in bacteria causes a tradeoff between population spread and survival. Proc Natl Acad Sci U S A 111(5):1969–1974

92. Stephens PA, Sutherland WJ, Freckleton RP (1999) What is the Allee effect? Oikos 87:185-190

93. Shapiro JA (1998) Thinking about bacterial populations as multicellular organisms. Annu Rev Microbiol 52(1):81–104

94. Parsek MR, Greenberg E (2005) Sociomicrobiology: the connections between quorum sensing and biofilms. Trends Microbiol 13(1):27–33

95. Gurdon J, Bourillot P-Y (2001) Morphogen gradient interpretation. Nature 413(6858):797–803

96. Wolpert L (1969) Positional information and the spatial pattern of cellular differentiation. J Theor Biol 25(1):1–47

97. Payne S, Li B, Cao Y, Schaeffer D, Ryser MD, You L (2014) Temporal control of self-organized pattern formation without morphogen gradients in bacteria. Mol Syst Biol 9:697

98. Turing AM (1952) The chemical basis of morphogenesis. Phil Trans R Soc Lond B 237(641):37–72

99. Tabor JJ, Salis HM, Simpson ZB, Chevalier AA, Levskaya A, Marcotte EM, Voigt CA, Ellington AD (2009) A synthetic genetic edge detection program. Cell 137(7):1272–1281

100. Gambetta GA, Lagarias JC (2001) Genetic engineering of phytochrome biosynthesis in bacteria. Proc Natl Acad Sci U S A 98(19):10566–10571

101. Levskaya A, Chevalier AA, Tabor JJ, Simpson ZB, Lavery LA, Levy M, Davidson EA, Scouras A, Ellington AD, Marcotte EM (2005) Synthetic biology: engineering *Escherichia coli* to see light. Nature 438(7067):441–442

102. Faust K, Raes J (2012) Microbial interactions: from networks to models. Nat Rev Microbiol 10(8):538–550

103. Hong SH, Hegde M, Kim J, Wang X, Jayaraman A, Wood TK (2012) Synthetic quorum-sensing circuit to control consortial biofilm formation and dispersal in a microfluidic device. Nat Commun 3:613

104. Lu T, Hasty J, Wolynes PG (2006) Effective temperature in stochastic kinetics and gene networks. Biophys J 91(1):84–94

105. Lu T, Shen T, Zong C, Hasty J, Wolynes PG (2006) Statistics of cellular signal transduction as a race to the nucleus by multiple random walkers in compartment/phosphorylation space. Proc Natl Acad Sci U S A 103(45):16752–16757

106. Lu T, Shen T, Bennett MR, Wolynes PG, Hasty J (2007) Phenotypic variability of growing cellular populations. Proc Natl Acad Sci U S A 104(48):18982–18987

107. Mao J, Blanchard AE, Lu T (2014) Slow and steady wins the race: a bacterial exploitative competition strategy in fluctuating environments. ACS Synth Biol doi:10.1021/sb4002008

108. Song H, Payne S, Gray M, You L (2009) Spatiotemporal modulation of biodiversity in a synthetic chemical-mediated ecosystem. Nat Chem Biol 5(12):929–935

109. Prindle A, Selimkhanov J, Li H, Razinkov I, Tsimring LS, Hasty J (2014) Rapid and tunable post-translational coupling of genetic circuits. Nature 508(7496):387–91

110. Qi H, Blanchard A, Lu T (2013) Engineered genetic information processing circuits. Wiley Interdiscip Rev Syst Biol Med 5(3):273–287

111. Tamsir A, Tabor JJ, Voigt CA (2011) Robust multicellular computing using genetically encoded NOR gates and chemical 'wires'. Nature 469(7329):212–215

112. Saeidi N, Wong CK, Lo TM, Nguyen HX, Ling H, Leong SSJ, Poh CL, Chang MW (2011) Engineering microbes to sense and eradicate *Pseudomonas aeruginosa*, a human pathogen. Mol Syst Biol 7:521

113. Hwang IY, Tan MH, Koh E, Ho CL, Poh CL, Chang MW (2013) Reprogramming microbes to be pathogen-seeking killers. ACS Synth Biol 3(4):228–237

114. Duan F, March JC (2010) Engineered bacterial communication prevents *Vibrio cholerae* virulence in an infant mouse model. Proc Natl Acad Sci U S A 107(25):11260–11264

115. Chen AY, Deng Z, Billings AN, Seker UO, Lu MY, Citorik RJ, Zakeri B, Lu TK (2014) Synthesis and patterning of tunable multiscale materials with engineered cells. Nat Mater 13:515–523

116. Nocadello S, Swennen EF (2012) The new pLAI (lux regulon based auto-inducible) expression system for recombinant protein production in *Escherichia coli*. Microb Cell Fact 11(1):1–10

117. Tsao C-Y, Hooshangi S, Wu H-C, Valdes JJ, Bentley WE (2010) Autonomous induction of recombinant proteins by minimally rewiring native quorum sensing regulon of *E. coli*. Metab Eng 12(3):291–297

15

Sustaining ethanol production from lime pretreated water hyacinth biomass using mono and co-cultures of isolated fungal strains with *Pichia stipitis*

Chinnathambi Pothiraj[1][*], Ramasubramanian Arumugam[1] and Muthukrishnan Gobinath[2]

Abstract

Background: The high rate of propagation and easy availability of water hyacinth has made it a renewable carbon source for biofuel production. The present study was undertaken to screen the feasibility of using water hyacinth's hemicelluloses as a substrate for alcohol production by microbial fermentation using mono and co-cultures of *Trichoderma reesei* and *Fusarium oxysporum* with *Pichia stipitis*.

Results: In separate hydrolysis and fermentation (SHF), the alkali pretreated water hyacinth biomass was saccharified by crude fungal enzymes of *T. reesei, F. oxysporum* and then fermented by *P. stipitis*. In simultaneous saccharification and fermentation (SSF), the saccharification and fermentation was carried out simultaneously at optimized conditions using mono and co-cultures of selected fungal strains. Finally, the ethanol production kinetics were analyzed by appropriate methods. The higher crystalline index (66.7%) and the Fourier transform infrared (FTIR) spectra showed that the lime pretreatment possibly increased the availability of cellulose and hemicelluloses for enzymatic conversion. In SSF, the co-culture fermentation using *T. reesei* and *P. stipitis* was found to be promising with a higher yield of ethanol (0.411 g g^{-1}) at 60 h. The additional yield comparable with the monocultures was due to the xylanolytic activity of *P. stipitis* which ferments pentose sugars into ethanol. In SHF, the pretreatment followed by crude enzymatic hydrolysis and fermentation resulted in a significantly lesser yield of ethanol (0.344 g g^{-1}) at 96 h.

Conclusions: It is evident from the study that the higher ethanol production was attained in a shorter period in the co-culture system containing *T. reesei* and the xylose fermenting yeast *P. stipitis*. SSF of pretreated water hyacinth biomass (WHB) with *P. stipitis* instead of traditional yeast is found to be an effective biofuel production process.

Keywords: Water hyacinth; Hemicelluloses; Xylose; *T. reesei; F. oxysporum; P. stipitis*; SSF

Background

The global depletion of fossil fuels that are the dominant sources for supplying cheap energy for the world's economy has prompted recent significant research efforts in finding viable and sustainable alternatives [1]. Among various options, conversion of abundant lignocellulosic biomass to biofuels has received significant attention. Currently, bioethanol production from corn and sugarcane has posed a threat to the food supply [2], and the cost of these raw materials accounts for up to 40% to 70% of the production cost [3]. Lignocellulosic biomass serves as a cheap and abundant feedstock [4], which has the potential to produce low-cost bioethanol at a large scale. In recent days, screening of such substrates for biofuel has gained new speed and still there are many factors to be taken into consideration for the large scale production.

The performance of enzymatic saccharification is one of the foremost limiting factors which may strongly be dependent on the diverse species, complex chemical compositions, and structural characteristics of the feedstock

* Correspondence: pothi2005@yahoo.com
[1]PG and Research Department of Botany, Alagappa Government Arts College (Alagappa University), Karaikudi, Tamilnadu 630 003, India
Full list of author information is available at the end of the article

materials. The sugar yields from enzymatic hydrolysis vary from plant to plant as a result of the differences mainly in cellulose content [5]. Like cellulose, hemicellulose is also a viable source of fermentable sugars such as xylose for biorefining applications. It was suggested that the production of fuel-grade ethanol from xylose requires a microorganism capable of producing 50 to 60 g/L ethanol within 36 h with a yield of at least 0.4 g ethanol per gram of sugar [6]. But only few xylose-fermenting microorganisms have been reported earlier [7], and it is generally known that *Pichia stipitis* is superior to all other yeast species for ethanol production from xylose.

Water hyacinth (*Eichhornia crassipes*) is a fast growing perennial aquatic weed invasively distributed throughout the world. This tropical plant can cause infestations over large areas of water resources and consequently leads to series of problems like reduction of biodiversity, blockage of rivers and drainage system, depletion of dissolved oxygen, and alteration on water chemistry that leads to severe environmental pollution. In the past, attempts have been geared towards the use of biological, chemical, and mechanical approaches for preventing the spread of, or eradication of, water hyacinth. On the other hand, much attention has been focused on the potentials and constrains of using water hyacinth for a variety of applications since it has a lignocellulosic composition of 48% hemicelluloses, 18% cellulose, and 3.5% lignin [8,9]. Since the biomass productivity of this plant is very high, it can be a suitable feedstock for ethanol production.

The technologies for the possible conversion of water hyacinth to biogas or fuel ethanol using fungal extracellular enzymes are well documented in a number of developing countries [10-13]. *Saccharomyces cerevisiae* and *Zymomonas mobilis* are being used as candidate organisms in the large-scale production of ethanol from cellulosic biomass. These organisms are capable of utilizing hexose sugars efficiently but not the pentoses, which are the second dominant sugar source in lignocellulosic biomass [14]. From earlier research, *P. stipitis* has been identified as an efficient strain for the conversion of pentose sugars into alcohol [15]. Fermentation technologies utilizing strains of *P. stipitis* instead of traditional yeast have been proposed by a number of authors [14,15], as they have been shown to ferment under fully anaerobic conditions with faster specific rates of pentose sugar uptake and ethanol production as well as an ethanol yield close to theoretical yield. The present study, therefore, was carried out to screen the feasibility of using hexose- and pentose-utilizing fungal strains (*Trichoderma reesei*, *Fusarium oxysporum*, and *Pichia stipitis*) for the effective conversion of water hyacinth biomass into ethanol.

Methods

Biomass and culture organisms

Fresh water hyacinth biomass (WHB) was collected from a local pond at Karaikudi, Tamilnadu, India (10.07°N, 78.78°E). The collected samples were washed to remove adhering dirt, cut into small piece (2 or 3 mm) thicknesses, and dried in sunlight. The proximate analysis for biomass was done using standard methods for moisture content, ash, crude protein, crude fibre, cellulose, hemicelluloses, and lignin [16,17]. The fungal strains of *T. reesei* and *F. oxysporum* were isolated by primary selection from a naturally contaminated water hyacinth, and the isolates were confirmed by their morphology and colony characteristics [18]. The isolated organisms were maintained on modified potato dextrose agar (PDA) slants at 4°C. Fresh colonies were used for saccharification and fermentation studies. The pure culture of *P. stipitis* (NCIM 3497) was procured from the National Collection of Industrial Microorganisms, Pune, India.

Alkaline pretreatment

The dried WHB (10% *w/v*) was pretreated with calcium hydroxide solution (0.5% *w/v*) with a soaking time of 3 h at 100°C. The pretreated WHB washed to neutrality with distilled water, oven dried to a constant weight, and then milled to powder was used for enzymatic hydrolysis and fermentation [19].

Experimental design

Two modes of bioconversion methodologies for ethanol production were trialed in the present study. Mode I comprised of a separate hydrolysis and fermentation (SHF) process using crude fungal enzymes with yeast. Mode II was designed to conduct a simultaneous saccharification and fermentation (SSF) process using mono and co-cultures of selected fungal strains.

Separate hydrolysis and fermentation (SHF)

The cellulolytic enzymes (cellulases and xylanases) were produced by growing the isolated fungal strains of *T. reesei* and *F. oxysporum* separately at 35°C in a simple liquid medium (4.2 g L^{-1} (NH$_4$)$_2$SO$_4$, 2 g L^{-1} KH$_2$PO$_4$, 0.05 g L^{-1} yeast extract, 2 mL L^{-1} Tween-80, 2% (*w/v*) poultry manure with 1.6% total N, pH 4.8) containing 100 g L^{-1} water hyacinth biomass as the chief C source for 5 days as optimized earlier [20]. The culture supernatants were separated at the end of the incubation period from each organism and used as crude enzymes source for hydrolysis. Cellulase and xylanase activities were measured in the culture supernatant as per standard methods. Cellulase was measured according to the IUPAC methods [21] using Whatman filter paper no. 1 as the substrate and glucose as the standard. Xylanase was assayed by the optimized method described by Bailey et al. [22], using 1% birchwood xylan as the substrate and xylose as the standard. One unit (IU) of enzyme activity is defined as the amount of enzyme releasing 1 μmol glucose or xylose/mL per minute

Enzymatic hydrolysis was carried out by incubating the pretreated WHB (10% *w/v*) with the crude fungal enzymes (10% (*v/v*)) of *T. reesei* and *F. oxysporum* separately at 35°C for 48 h with agitation at 200 rpm [23]. The pH of the reaction mixture (6.0) was maintained at constant. Samples were aliquoted from hydrolysates at a regular interval (24 and 48 h) to estimate the released sugar content using standard methods [24,25]. The hydrolysates obtained after 48 h from both the fungal cultures were centrifuged at 10,000 rpm for 10 min. The supernatants were collected separately and supplemented with basal medium (1 g L^{-1} yeast extract; 2 g L^{-1} $(NH_4)SO_4$; 1 g L^{-1} $MgSO_4 \cdot 7H_2O$) (pH 6.0) [23]. The culture suspension of *P. stipitis* (10% *v/v*) was added to initiate the fermentation by incubating the mixture at 35°C for 48 h with agitation at 200 rpm.

Simultaneous saccharification and fermentation (SSF)

SSF represents a single step process in which fermentable sugars get released by enzymatic hydrolysis and are simultaneously exploited by yeasts for fermentation in the same medium. The microbial fermentation was carried out using mono and co-cultures as previously described [9]. The influences of various parameters such as microbial biomass (5% to 25%), temperature (25°C to 45°C), and incubation time (24, 36, 48, 60, 72 h) on SSF were also optimized by step-wise experiments where the specified parameters were changed by keeping all other parameters constant. The pH of the reaction mixture in all the optimization experiments was kept constant at 6.0

Mono and co-culture fermentations

For monoculture experiments (F1 and F2), previously sterilized (121°C for 60 min) pretreated WHB supplemented with a basal medium (without C source) was inoculated with late log-phase cultures of *T. reesei* (F1) and *F. oxysporum* (F2), separately. For co-culture fermentation (F3 and F4), separate sets of reaction mixtures consisting of pretreated WHB supplemented with basal medium were treated with *P. stipitis* simultaneously with *T. reesei* (F3) and *F. oxysporum* (F4). The fermentation process was carried out at optimized conditions.

Estimations

Samples were withdrawn from the fermenting media at regular intervals of time for the determination of ethanol, residual sugar concentration, and microbial biomass. Estimation of xylose was done by the Trinder method [24] and glucose by the DNS method [25]. Ethanol estimation was done spectrophotometrically by potassium

dichromate method [26]. The microbial biomass was determined by harvesting cells by centrifugation, drying them at 70°C under vacuum to a constant weight, and expressed as gram dry cell weight (DCW) per liter [27]. The kinetic parameters of ethanol fermentation were determined followed by Abate et al. [28] as follows:

$$\text{Ethanol concentration } (Ec) = [\text{ethanol produced (g)}/\text{volume of reaction mixture (L)}]$$

$$\text{Ethanol productivity } (E_p) = [\text{ethanol produced (g)}/\text{volume of reaction mixture (L)}/\text{time (h)}]$$

$$\text{Ethanol Yield } (E_y) = [\text{ethanol produced (g)}/\text{weight of substrate (g)}]$$

$$\text{Specific ethanol yield } (E_{sy}) = [\text{ethanol produced (g)}/\text{sugar consumed (g)}]$$

The results obtained were analyzed by using analysis of variance (ANOVA), and the group means were compared with Duncan's Multiple Range Test (DMRT) [29].

Fourier transform infrared (FTIR) analysis

Fourier transform infrared spectra were studied on treated and untreated WHB using a Shimadzu spectrometer (Shimadzu, Kyoto, Japan). For this, 3.0 mg of the sample was dispersed in 300 mg of spectroscopic grade KBr and subsequently pressed into disks at 10 MPa for 3 min. The spectra were obtained with an average of 25 scans and a resolution of 4 cm^{-1} in the range of 4,000 to 400 cm^{-1}.

X-ray diffraction (XRD) analysis

The crystallinity of cellulose in the pretreated and treated water hyacinth was analyzed by X-ray diffraction method in a PANalytical X'pert[3] PRO Diffractometer (PANalytical B.V., Almelo, Netherlands) set at 40 KV, 30 mA; radiation was Cu Kα($\lambda = 1.54$Å) and the grade range between 10 to 30^0 with a step size of 0.03^0. The crystallinity index (CrI) was determined based on the equation shown below [30]:

$$\text{CrI} = \frac{I_{002} - I_{am}}{I_{am}} \times 100$$

where I_{002} is the intensity of the diffraction from the 002 plane at $2\theta = 22.6°$ and I_{am} is the intensity of the background scatter measured at $2\theta = 18.7°$. It is known that the I_{002} peak corresponds to the crystalline fraction and the I_{am} peak corresponds to the amorphous fraction [31].

Results and discussion

The lignocellulosic biomass composition of WH includes cellulose (20.2 g 100 g^{-1} dry matter (DM)), hemicellulose (34.3 g 100 g^{-1} DM), lignin (4.4 g 100 g^{-1} DM), crude protein (13.3 g 100 g^{-1} DM), crude fibre (18.2 g 100 g^{-1}

Table 1 Proximate composition of water hyacinth biomass (WHB) comparable with earlier literatures

Content	Present study [g 100 g^{-1} DM]	Previous literature		
		[g 100 g^{-1} DM] [8]	[g 100 g^{-1} DM] [12]	[g 100 g^{-1} DM] [32]
Moisture (%)	92.8	-	-	-
Ash	15.4	20.0	-	20.0
Crude protein	13.3	18.0	10.2	18.0
Crude fibre	18.2	-	-	-
Cellulose	20.2	25.0	19.02	25.0
Hemicellulose	34.3	35.0	32.69	35.0
Lignin	4.4	10.0	4.37	10.0

DM, dry matter.

DM), and ash (15.4 g 100 g^{-1} DM) (Table 1). The results obtained in the present study on the proximate composition of WHB are basically consistent with previous literatures [8,12,32]. The digestibility of lignocelluloses is hindered by many physicochemical, structural, and compositional factors which required a suitable pretreatment in order to enhance the susceptibility of biomass for hydrolysis. It is highly essential for the economical production of ethanol that both the cellulose and hemicellulosic sugars present in the biomass should be utilized efficiently. The FTIR and XRD data in the present study clearly suggested that the pretreatment with lime could increase the availability of polysaccharide for enzymatic hydrolysis. Among different pretreatment methods used in earlier researches for water hyacinth, maximum reducing sugar was observed in diluted H$_2$SO$_4$ (0.342 g g^{-1} biomass) [33], HCl (0.277 g g^{-1} biomass), acetic acid (0.097 g g^{-1} biomass), and formic acid (0.088 g g^{-1} biomass) [31,34]. In comparison with the above reports, it is evident that the lime pretreatment used in the present study is a promising method for higher sugar yield. The pretreatment with Ca(OH)$_2$ is preferable because it is less expensive, more safe as compared to NaOH, and it can be easily recovered from the hydrolysate by reaction with CO$_2$. Lime has been used to pretreat many lignocellulosic materials such as wheat straw [35], poplar wood [36], and corn stover [37].

XRD - cellulose crystallinity

Cellulose crystallinity, usually measured as CrI, is considered an important parameter determining the enzymatic hydrolysis susceptibility of cellulose. The CrI of a cellulose sample is an indication of the degree of formed crystallinity in the sample when the cellulose aggregates. The crystallinity has been found to have a greater impact on enzymatic hydrolysis than other structural characteristics such as the degree of polymerization (DP) of cellulose or the specific surface area (SSA) [38]. The XRD profile of WHB indicated that the CrI of untreated WHB is 28.6% and alkali-treated WHB is 66.7% (Figure 1). The X-ray diffractogram clearly revealed that the lime pretreatment

increased the crystallinity of cellulose in water hyacinth. Similar results were reported earlier by Kim and Holtzapple [39] who found that the degree of crystallinity of corn stover slightly increased from 43% to 60% through delignification with calcium hydroxide and by Li et al. [40] who have reported high cellulose CrI of 70.6% in *Metasequoia* chips by nitric acid-ethanol method. The increase in CrI of alkali-treated WHB might be due to the removal of amorphous components including lignin during the pretreatment process [41,42]. According to Satyanagalakshmi et al. [33], the amorphous cellulose portions in aquatic plants are more prone to recrystallization to form crystalline cellulose, resulting in greater increases in CrI.

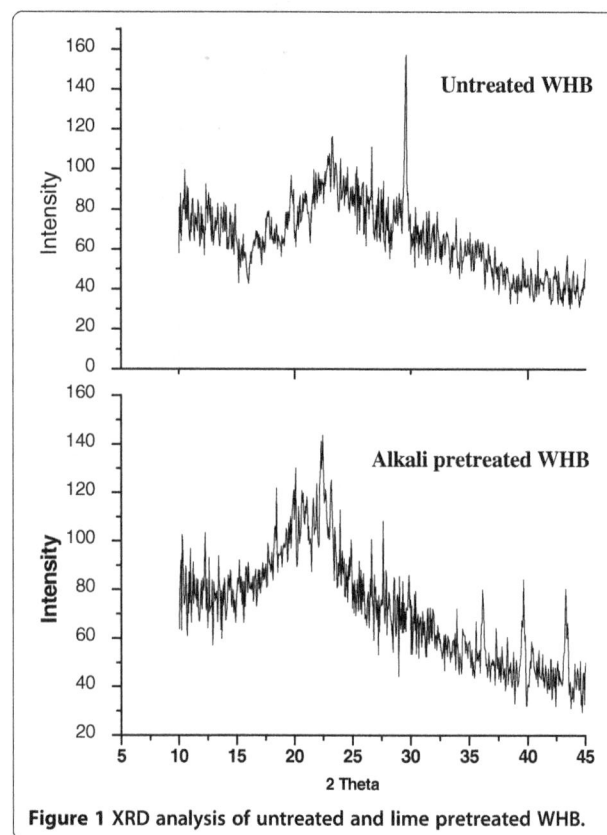

Figure 1 XRD analysis of untreated and lime pretreated WHB.

FTIR analysis

FTIR spectra of the untreated and treated samples indicated structural changes in the biomass upon pretreatment (Figure 2). The increased absorption bands at 1,000 to 1,200 cm^{-1} were related to structural features of cellulose and hemicelluloses [43]. The spectra of alkali-treated WHB sample (Figure 3) showed increase in absorbance in the above-mentioned range. The peak at 1,635 cm^{-1} was observed due to either the acetyl and uronic ester linkage of carboxylic group of the ferulic and p-coumeric acids of lignin and/or hemicelluloses [44]. A sharp band at 896 cm^{-1}, corresponding to the C1 group frequency or ring frequency, was attributed to the glycosidic linkages between xylose units in hemicelluloses [45]. The peaks in the pretreated sample had the highest absorbance suggesting increase in cellulose and hemicellulose content. In the FTIR spectrum, the peaks observed at 1,092 and 842 cm^{-1} were attributed to C-O stretching and C-H rocking vibration of the cellulose structure.

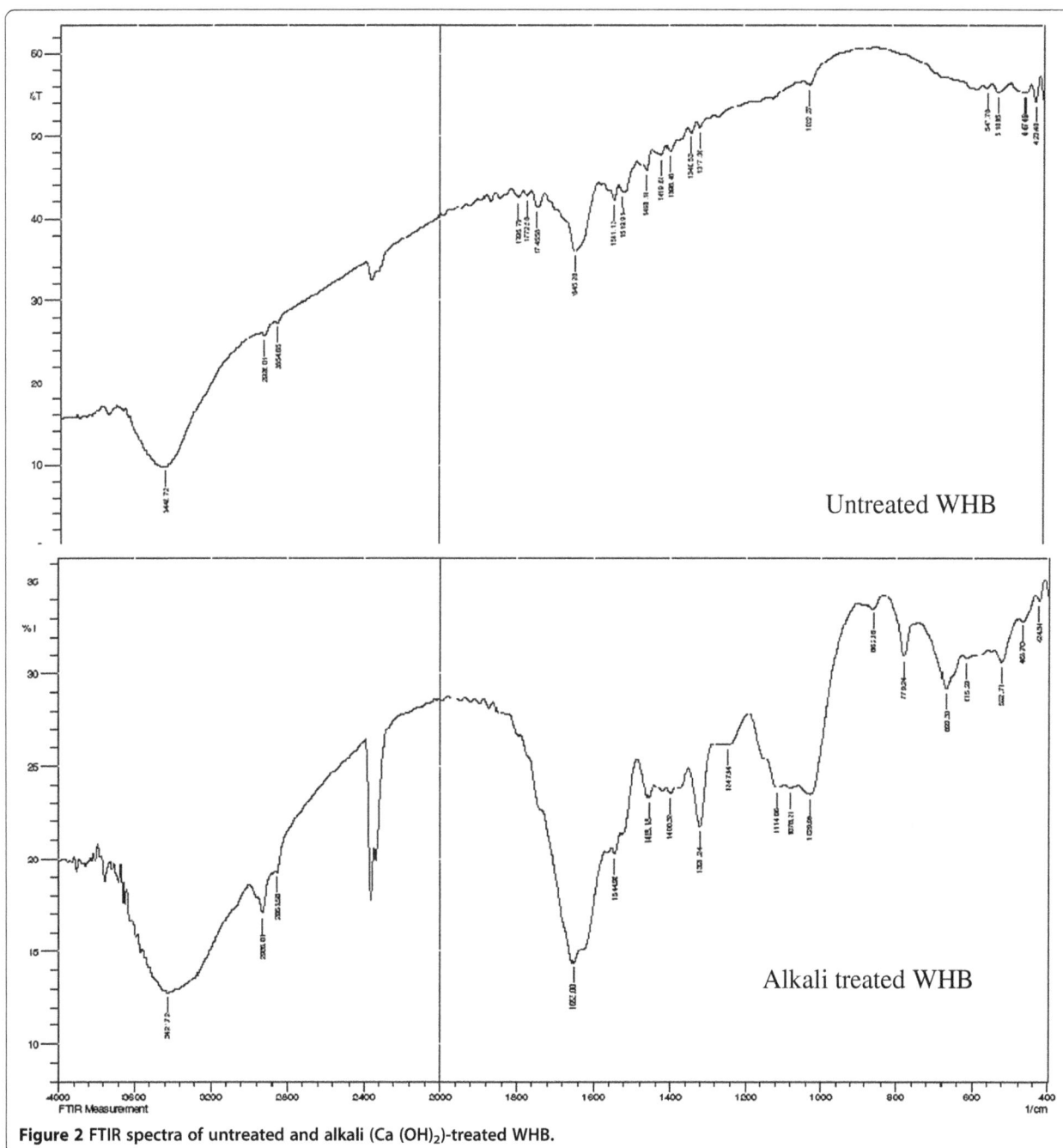

Figure 2 FTIR spectra of untreated and alkali (Ca (OH)$_2$)-treated WHB.

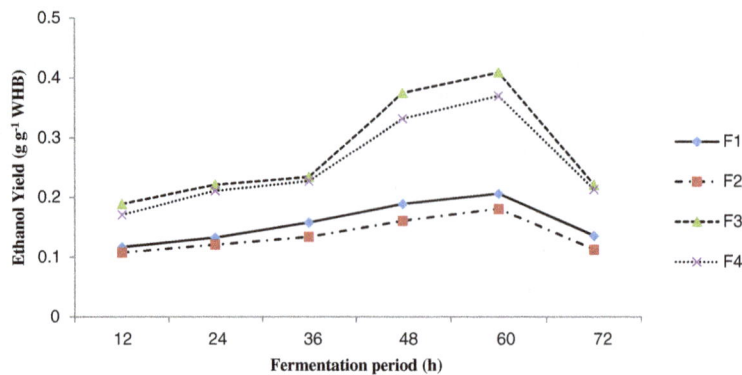

Figure 3 Effect of fermentation period on ethanol yield from pretreated WHB using mono and co-cultures. F1, *T. reesei*; F2, *F. oxysporum*; F3, *T. reesei* + *P. stipitis*; F4, *F. oxysporum* + *P. stipitis*.

Enzyme activity

In recent decades, the use of fungi in bioprocesses has grown in importance because of the production of numerous enzymes with different biochemical properties and excellent potential for biotechnological application. The cellulase and xylanase activities reached their maximum values on the 6th day of incubation for both the fungal isolates. Cellulase production on WHB with nutrient supplements indicated higher cellulase production by *T. reesei* (0.923 IU/mL) compared to *F. oxysporum* (0.432 IU/mL). However, it is less than the value of 1.35 IU/mL reported by Deshpande et al. [46] on the substrate water hyacinth with Toyama-Ogawa medium [47]. The xylanase production was slight, but significantly higher in *F. oxysporum* (0.764 IU/mL), compared to *T. reesei* (0.611 IU/mL). According to Kang et al. [48], high xylanase production in some fungi has been shown to be linked strictly to the ratio of cellulose to xylan of the growth substrate and substrate degradation due to time course or incubation period.

According to Polizeli et al. [49], filamentous fungi are widely utilized as enzyme producers and are generally considered more potent xylanase producers than bacteria or yeast. Several mesophilic fungal species have been evaluated in relation to xylanase production, including members of *Aspergillus*, *Trichoderma*, and *Penicillium*. Currently, most commercial xylanolytic preparations are produced by genetically modified *Trichoderma* or *Aspergillus* strains [50].

Sugar yield

The yield of sugars from enzymatic hydrolysis of WHB using crude enzymes produced by fungal isolates was summarized in Table 2. The saccharification was significantly higher (40.8%) while using crude enzyme from *T. reesei* when compared to *F. oxysporum* (38.2%). The release of total sugars by the crude enzymes of both monocultures increased slowly to reach a peak value at 48 h of incubation. The maximum yield of total sugar (0.531 g g^{-1} WHB) including glucose (0.444 g g^{-1} WHB) and xylose (0.057 g g^{-1} WHB) was observed after 48 h of hydrolysis using crude enzymes of *T. reesei*. The crude enzymes obtained from *F. oxysporum* produced comparably lower reducing sugar (0.428 g g^{-1} WHB) and xylose (0.038 g g^{-1} WHB). Thus, it substantiates that the amount of sugar released increases with time which may be due to the increased action of cellulolytic and xylanolytic enzymes *of T. reesei* and *F. oxysporum* [51]. The cellulolytic fungus *T. reesei* looks promising for on-site cellulase production due to its superior features, i.e., capability to produce all components of cellulase complex, endocellulase, exocellulase, and β-glucosidase in good proportions as well as production of other enzymes such as xylanases or laccases in comparison to other enzyme producers [52].

Ethanol

The optimization studies in SSF showed that the yield of ethanol is found to be proportional to fermentation time where the yield increases with the increase in time up to 60 h and then declines (Figure 3). Maximum yield of

Table 2 Sugar composition (g g^{-1} WHB) of enzymatic hydrolysates of pretreated WHB at 48 h

Enzyme source	Glucose	Xylose	Total sugar	Saccharification %
T. reesei	0.444^a ± 0.12	0.057^a ± 0.09	0.531^a ± 0.12	40.2
F. oxysporum	0.428^b ± 0.31	0.038^b ± 0.11	0.488^b ± 0.17	38.2

Values are the mean of three replicates ± SE. Means followed by the same letter within treatment do not differ significantly ($p = 0.05$).

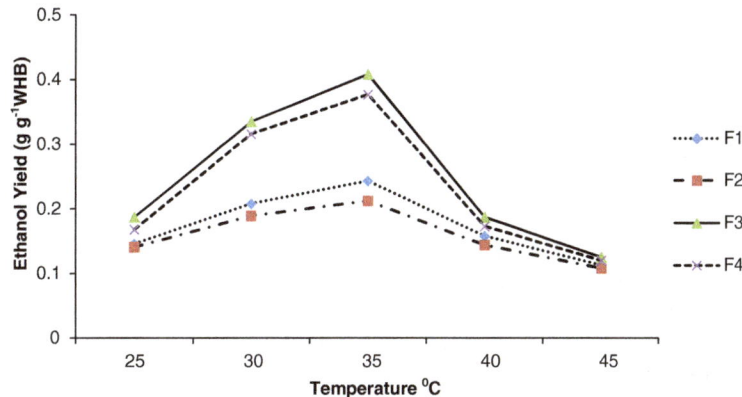

Figure 4 Effect of temperature on ethanol yield from pretreated WHB using mono and co-cultures. F1, *T. reesei*; F2 *F. oxysporum*; F3 *T. reesei + P. stipitis*; F4, *F. oxysporum + P. stipitis*.

ethanol is 0.413, 0.378, 0.194, and 0.187 g g^{-1} of WHB at 60 h of fermentation for F3, F4, F1, and F2, respectively. After 60 h of time, the yield of ethanol decreases in all treatments, and therefore, fermentation time of 60 h is taken as the optimum time for ethanol fermentation. With the increase in temperature, the yield of ethanol increased up to 35°C and then it decreased (Figure 4). At high temperature (>35°C), death rate exceeds the growth rate, which causes a net decrease in concentration of viable fungal populations with lower generation of ethanol. With the increase in loading of biomass, the yield of ethanol increased up to 10% and then decreased in all the samples (Figure 5). The decrease in ethanol yield with the increase in biomass loading can be attributed to the inhibitory effect of either the product or the biomass. Inhibitory compounds limit efficient utilization of hydrolysates by the fermenting organism resulting in less ethanol production [33].

In the SHF process, a maximum of 14.3 g L^{-1} ethanol was produced at the end of the process (96 h) which is equivalent to 0.143 g g^{-1} WHB (Table 3). The minimum production of ethanol observed in the submerged

fermentation of pretreated water hyacinth biomass using monocultures of *T. reesei* and *F. oxysporum* was due to the inability of these organisms to convert pentose sugars into ethanol. A similar finding was reported earlier where 0.11 g ethanol was obtained from alkalipretreated water hyacinth through SHF [23]. According to Preez et al. [53] *P. stipitis* is known to produce ethanol up to 33 to 57 g/L; however, 30 g/L is known as a critical concentration above which cells cannot grow at 30°C.

In SSF, monocultures of *T. reesei* (F1) and *F. oxysporum* (F2) produced 19.3 and 17.8 g L^{-1} ethanol, respectively, after 60-h fermentation (Table 4). Simultaneous co-culturing of *T. reesei* (F3) and *F. oxysporum* (F4) with *P. stipitis* resulted in a higher ethanol production (40.8 and 36.8 g L^{-1}, respectively) at the same time. The maximal ethanol yield was 0.411 g g^{-1} WHB when *P. stipitis* was used along with *T. reesei* which is positively correlated to the theoretical yield 0.429 g on the basis of biomass. Since xylose was present as a predominant sugar in the WHB hydrolysate, *P. stipitis* was used to make the biomass-to-ethanol process more economical. Mishima et al. [34], on

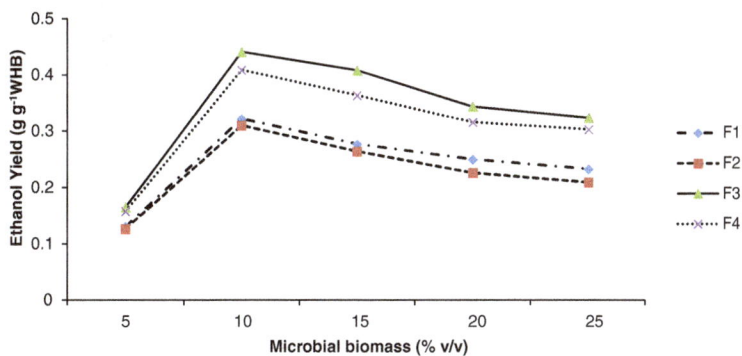

Figure 5 Effect of microbial biomass on ethanol yield from pretreated WHB using mono and co-cultures. F1, *T. reesei*; F2, *F. oxysporum*; F3, *T. reesei + P. stipitis*; F4, *F. oxysporum + P. stipitis*.

Table 3 Ethanol production in SHF process using crude fungal enzymes and _P. stipitis_ with pretreated WHB at 96 h

Sample (s)	E_c	E_p	E_y	E_{sy}	E_{TE}
S_1	$14.3^c \pm 0.12$	$0.24^a \pm 0.09$	$0.143^c \pm 0.11$	$0.322^c \pm 0.11$	$0.261^c \pm 0.11$
S_2	$12.8^d \pm 0.13$	$0.21^a \pm 0.17$	$0.128^d \pm 0.08$	$0.299^d \pm 0.08$	$0.243^d \pm 0.17$

E_c, ethanol concentration (g L^{-1}); E_p, ethanol productivity (g L^{-1} h^{-1}); E_y, ethanol yield (g g^{-1} WHB); E_{sy}, specific ethanol yield (g g^{-1} sugar); E_{TE}, theoretical yield (g g^{-1} WHB); S1, hydrolysate from _T. reesei_; S2, hydrolysate from _F. oxysporum_. Values are the mean of three replicates ± SE. Mean followed by the same letter within treatment do not differ significantly (p = 0.05).

the other hand, reported a lesser ethanol yield of 0.14 g g^{-1} dry substrate through SSF of pretreated water hyacinth using commercial cellulase and _S. cerevisiae_. The overall production could be enhanced by co-culture rather than monoculture of test organisms. Similarly, direct microbial conversion of cellulosic or lignocellulosic biomass into ethanol using co-cultures had been reported by several authors [32,34,54]. _S. cerevisiae_ or _Z. mobilis_ utilize glucose or sucrose efficiently but their inability to utilize pentose sugars make them inappropriate candidates for refineries, but the candidate organism _P. stipitis_ used in the present study showed efficient conversion of pentose sugars into alcohol. Among the pentose-fermenting organisms, _P. stipitis_ has been shown to have the most promise for industrial applications [55]. Earlier reports showed that the hemicellulosic hydrolysates of _Prosopis juliflora_ (18.24 g sugar/L broth) when fermented with _P. stipitis_ produced 7.13 g/L ethanol [56]. Kuhad et al. [57] observed 0.33 g g^{-1} ethanol yield from detoxified xylose-rich hydrolysate of _Lantana camara_ fermented with _P. stipitis_ at pH 5 for 36 h. Similarly, the detoxified water hyacinth hemicellulose acid hydrolysate (rich in pentose sugars) fermented with _P. stipitis_ NCIM-3497 at pH 6.0 and 30°C resulted in 0.425 g ethanol/g lignocelluloses [15]. The yield of ethanol per unit biomass of water hyacinth obtained through the bioprocess in the present study was comparable to or even better than those reported earlier. The current results clearly demonstrated the saccharification potential of _T. reesei and F. oxysporum_, where the performance of both strains in co-cultures with _P. stipitis_ was significantly higher than their respective single culture.

Microbial biomass

All the co-culture processes reached a higher value of microbial biomass than the single fermentation process. A maximum of 3.12 g DCW L^{-1} biomass content was obtained in the co-culture of _T. reesei_ and _P. stipitis_ at 60-h fermentation (Table 4). Inoculation of _P. stipitis_ with _F. oxysporum_ resulted in a biomass content of 2.64 g DCW L^{-1} over the monocultures. Statistically, a less significant difference was observed with monoculture's fermentation when compared with co-culture [58].

Conclusions

The fermentation of bioethanol from pretreated water hyacinth biomass with mono and co-cultures of fungal strains along with _P. stipitis_ is found to be an effective biofuel production process. The yield of ethanol recovered from WHB through enzymatic hydrolysis and fermentation from simultaneous inoculation of co-cultures of fungal isolates with _P. stipitis_ was significantly higher than that recovered through monocultures. The optimum parameters for bioethanol fermentation are as follows: time 60 h, temperature 35°C, and WHB loading 100 g L^{-1}. The maximum yield of ethanol in the fermentation process was found to be 0.411 g g^{-1} of WHB which is equivalent to a specific yield of 0.456 g g^{-1} total sugar consumed. The use of crude fungal enzymes produced on-site would be a cost-effective approach towards enzymatic hydrolysis of alkali-pretreated WHB biomass instead of using commercial cellulases. The aquatic menace water hyacinth, which is currently being used in waste water treatment for its unique ability to absorb heavy metal pollutants, could also be utilized as abundant cheap feedstock for the production of fuel ethanol. This study proved that water hyacinth has a potential renewable and low-cost biomass for alcohol production on the commercial scale. Present cost effectiveness of respective process at a commercial scale needs to be standardized, and the water hyacinth biomass could be a better substrate source for alcohol production.

Table 4 Ethanol production in mono and co-culture fermentation process (SSF) using pretreated WHB at 60 h

Culture (s)	E_c	E_p	E_y	E_{sy}	Microbial biomass (g DCW L^{-1})
T. reesei	$19.3^c \pm 0.12$	$0.32^c \pm 0.09$	$0.196^c \pm 0.11$	$0.377^c \pm 0.11$	$2.14^c \pm 0.01$
F. oxysporum	$17.8^d \pm 0.13$	$0.29^d \pm 0.17$	$0.176^d \pm 0.08$	$0.348^d \pm 0.08$	$2.06^c \pm 0.08$
T. reesei + P. stipitis	$40.8^a \pm 0.09$	$0.68^a \pm 0.14$	$0.411^a \pm 0.03$	$0.798^a \pm 0.11$	$3.12^a \pm 0.12$
F. oxysporum + P. stipitis	$36.8^b \pm 0.06$	$0.61^b \pm 0.12$	$0.371^b \pm 0.07$	$0.720^b \pm 0.08$	$2.64^b \pm 0.20$

E_c, ethanol concentration (g L^{-1}); E_p, ethanol productivity (g L^{-1} h^{-1}); E_y, ethanol yield (g g^{-1} WHB); E_{sy}, specific ethanol yield (g g^{-1} sugar). Values are the mean of three replicates ± SE. Mean followed by the same letter within treatment do not differ significantly ($p = 0.05$).

Sustaining ethanol production from lime pretreated water hyacinth biomass using mono and co-cultures...

137

Competing interests

The authors declare that they have no competing interests.

Authors' contributions

CP carried out the submerged fermentation processes and helped design the whole study. RA carried out proximate analysis of biomass and helped in manuscript preparation. RMG helped in characterization of biomass using FTIR and XRD. All authors read and approved the final manuscript.

Acknowledgements

The authors are thankful to all faculty members of the PG and Research Department of Botany, Alagappa Government Arts College (Alagappa University), Karaikudi. We also acknowledge the Head, Department of physics, Alagappa University, Karaikudi and the Department of Chemistry, VHNSN College, Virudhunagar, Tamilnadu for helping us with XRD and FTIR analysis, respectively.

Author details

[1]PG and Research Department of Botany, Alagappa Government Arts College (Alagappa University), Karaikudi, Tamilnadu 630 003, India. [2]Department of Microbiology, VHNSN College, Virudhunagar, Tamilnadu 626 001, India.

References

1. Chang KL, Thitikorn-amorn J, Hsieh JF, Ou BM, Chen SH, Ratanakhanokchai K, Huang PJ, Chen ST (2011) Enhanced enzymatic conversion with freeze pretreatment of rice straw. Biomass Bioenergy 35(1):90–95
2. Guragain YN, De Coninck J, Husson F, Durand A, Rakshit SK (2011) Comparison of some new pretreatment methods for second generation bioethanol production from wheat straw and water hyacinth. Bioresour Technol 102(6):4416–4424
3. Quintero JA, Montoya MI, Sánchez OJ, Giraldo OH, Cardona CA (2008) Fuel ethanol production from sugarcane and corn: comparative analysis for a Colombian case. Energy 33(3):385–399
4. Balat M (2011) Production of bioethanol from lignocellulosic materials via the biochemical pathway: a review. Energy Convers Manage 52(2):858–875
5. Sukumaran RK, Singhania RR, Mathew GM, Pandey A (2009) Cellulase production using biomass feed stock and its application in lignocelluloses saccharification for bio-ethanol production. Renew Energy 34(2):421–424
6. Lee Y, Kim M, Kim K, Park K, Ryu Y, Seo J (2000) A parametric study on ethanol production from xylose Pichia stipitis. Biotechnol Bioprocess Eng 5:27–31
7. Ganguly A, Das S, Bhattaacharya A, Dey A, Chatterjee PK (2013) Enzymatic hydrolysis of water hyacinth biomass for the production of ethanol: optimization of driving parameters. Ind J Exp Biol 51:556–566
8. Gunnarsson CC, Petersen CM (2007) Water hyacinth as a resource in agriculture and energy production: a literature review. Waste Man 27:117–129
9. Nigam JN (2002) Bioconversion of water hyacinth (Eichhornia crassipes) hemicellulose acid hydrolysate to motor fuel ethanol by xylose-fermenting yeast. J Biotech 97:107–116
10. Zabala I, Ferrer A, Ledesma A, Aiello C (1994) Microbial protein production by submerged fermentation of mixed cellulolytic cultures In: Galindo E, Ramirez OT (eds) Advances in Bioprocess Engineering. Kluwer Academic Publishers, The Netherlands, pp 455-460
11. Singhal V, Rai JP (2003) Biogas production from water hyacinth and channel grass used for phytoremediation of industrial effluents. Bioresour Technol 86:221–225
12. Sornvoraweat B, Kongkiattikajorn J (2010) Separated hydrolysis and fermentation of water hyacinth leaves for ethanol production. KKU Res J 15(9):794–802
13. Idrees M, Adman A, Sheikh S, Qureshi FA (2013) Optimization of dilute acid pretreatment of water hyacinth biomass for enzymatic hydrolysis and ethanol production. EXCLI J 12:30–40
14. Bhattacharya A, Ganguly A, Das S, Chatterjee PK, Dey A (2013) Fungal isolates from local environment: isolation, screening and application for the production of ethanol from water hyacinth. Int J Emerg Tech Adv Engi 3(3):58–65

15. Kumari N, Bhattacharya A, Dey A, Ganguly A, Chatterjee PK (2014) Bioethanol production from water hyacinth biomass using isolated fungal strain from local environment. Biolife 2(2):516–522
16. Association of Official Analytical Chemists (AOAC) (1975) Methods of analysis of the Association of Official Analytical Chemists. Association of Official Analytical Chemists, Washington DC
17. Robertson JB, van Soest PJ (1981) The Detergent System of Analysis and its application to human foods. In: James WPT, Thiander O (eds) The analysis of dietary fibers in food. Marcel Dekker, New York, pp 123–158
18. Alexopoulos CJ, Beneke ES (1962) Laboratory manual for introductory mycology. 1, 2 Burgess publishing Co, Minneapolis
19. Chang VS, Burr B, Holtzapple MT (1997) Lime pretreatment of switchgrass. Appl Biochem Biotechnol 63(65):3–19
20. Mukhopadhyay S, Nandi B (2001) Cellulase production by Trichoderma reesei on pretreated water hyacinth: effect of nutrients. J Mycopathol Res 39(1):57–60
21. Ghose TK (1987) Measurement of cellulose activities. Pure Applied Chem 59(257):268
22. Bailey M, Buchert J, Viikari L (1993) Effect of pH on production of xylanase of Trichoderma reesei on xylan- and cellulose-based media. Appl Microbiol Biotechnol 40:224–229
23. Mukhopadhyay S, Mukherjee PS, Chatterjee NC (2008) Optimization of enzymatic hydrolysis of water hyacinth by Trichoderma reesei vis-à-vis production of fermentable sugars. Acta Aliment 37(3):367–377
24. Trinder P (1975) Micro-determination of xylose in plasma. Analyst 100:12–15
25. Miller GL (1959) Use of DNS reagent for the determination of reducing sugars. Anal Chem 31:426–428
26. Caputi A, Veda M, Brown T (1968) Spectrophotometric determination of ethanol in wine. American J Enol Viticul 19:160–165
27. Doelle HW, Greenfield PF (1985) The production of ethanol from sucrose using Zymomonas mobilis. Appl Microbial Biotechnol 22:405–441
28. Abate C, Callieri D, Rodriguez E, Garro O (1996) Ethanol production by mixed culture of flocculent strains of Zymomonas mobilis and Saccharomyces sp. Appl Microbial Biotechnol 45:580–583
29. Duncan BD (1957) Multiple range tests for correlated and heteroscedastic means. Biometrics 13(2):359–364
30. Segal L, Creely JJ, Martin AE, Conrad CM (1959) An empirical method for estimating the degree of crystallinity of native cellulose using X –Ray diffractometer. Text Res J 29:786–794
31. Wang LS, Zhang YZ, Gao PJ, Shi DX, Liu HW, Gao HJ (2006) Changes in the structural properties and rate of hydrolysis of cotton fibers during extended enzymatic hydrolysis. Biotechnol Bioeng 93(3):443–456
32. Aswathy US, Sukumaran RK, Lalitha Devi G, Rajasree KP, Singhania RR, Pandey A (2010) Bio-ethanol from water hyacinth biomass: an evaluation of enzymatic saccharification strategy. Biores Technol 101:925–930
33. Satyanagalakshmi K, Sindhu R, Binod P, Janu KU, Sukumaran RK, Pandey A (2011) Bioethanol production from acid pretreated water hyacinth by separate hydrolysis and fermentation. J Sci Ind Res 70:156–161
34. Mishima D, Kuniki M, Sei K, Soda S, Ike M, Fujita M (2008) Ethanol production from candidate energy crops: water hyacinth (Eichhornia crassipes) and water lettuce (Pistia stratiotes L.). Biores Technol 99:2495–2500
35. Chang VS, Nagwani M, Holtzapple MT (1998) Lime pretreatment of crop residues bagasse and wheat straw. Appl Biochem Biotechnol 74:135–159
36. Chang VS, Nagwani M, Kim CH, Holtzapple MT (2001) Oxidative lime pretreatment of high-lignin biomass. Appl Biochem Biotechnol 94:1–28
37. Karr WE, Holtzapple MT (1998) The multiple benefits of adding non-ionic surfactant during the enzymatic hydrolysis of corn stover. Biotechnol Bioeng 59:419–427
38. Peng HD, Li HQ, Luo H, Xu J (2013) A novel combined pretreatment of ball milling and microwave irradiation for enhancing enzymatic hydrolysis of microcrystalline cellulose. Bioresour Technol 130:81–87
39. Kim S, Holtzapple MT (2006) Effect of structural features on enzyme digestibility of corn stover. Bioresource Technol 97:583–591
40. Li L, Wenbing Z, Hongwei W, Yun Y, Fen L, Duanwei Z (2014) Relationship between crystallinity index and enzymatic hydrolysis performance of cellulose separated from aquatic and terrestrial plant materials. Bioresources 9(3):3993–4005
41. Converse AO, Kwartneg IK, Grethlein HE, Ooshima H (1989) Kinetics of thermochemical pretreatment of lignocellulosic materials. Appl Biochem Biotechnol 20(21):63–78

42. Maeda RB, Serpa VI, Rocha RAL, Mesquita RAA, Anna LMMS, De Carlo AM, Driemeier CE, Pereira N, Polikarpov I (2011) Enzymatic hydrolysis of pretreated sugar cane baggase using *Penicillium funiculosum* and *Trichoderma harzianum* cellulases. J Process Biochem 30:5–8

43. Hu J, Arantes J, Saddler JN (2011) The enhancement of enzymatic hydrolysis of lignocellulosic substrates by the addition of accessory enzymes such as xylanase: is it an additive or synergistic effect. Biotechnol Biofuels 4:36

44. Langkilde FW, Svantesson A (1995) Identification of celluloses with Fourier-transform (FT) mid-infrared, FT-Raman and near-infrared spectrometry. J Pharm Biomed Anal 13:409

45. Marimuthu TS, Atmakuru R (2012) Isolation and characterization of cellulose nanofibers from the aquatic weed water hyacinth-*Eichhornia crassipes*. Carbohydr Polym 87:1701

46. Deshpande SK, Bhotmange MG, Chakrabarti T, Shastri PN (2008) Production of cellulase and xylanase by *Trichoderma reesei* (QM 9414 mutant), *Aspergillus niger* and mixed culture by solid state fermentation (SSF) of water hyacinth (*Eichhornia crassipes*). Ind J Chem Tech 15:449–456

47. Toyama M, Ogawa K (1977) Cellulase production of *Trichoderma viride* in solid and submerged culture methods. In: Ghosh TK, Ghosh TK (eds) Proc. symp. On bioconversion of cellulosic substrates into energy, chemical and microbial protein. IIT, New Delhi, India, pp 305–312

48. Kang SW, Park YS, Lee JS, Hong SI, Kim SW (2004) Production of cellulose and hemicellulases by *Aspergillus niger* KK2 from lignocellulosic biomass. Bioresour Technol 91:153–156

49. Polizeli MLT, Rizzati ACS, Monti R, Terenzi HF, Jorge JA, Amorin DS (2005) Xylanases from fungi: properties and industrial applications. Appl Microbiol Biotechnol 67(5):577–591

50. Mussatto SI, Teixeira JA (2010) Lignocellulose as raw material in fermentation processes. In: Méndez-Vilas A (ed) Current research, technology and education topics in applied microbiology and microbial biotechnology. Formatex Research Center, Badajoz, pp 897–907

51. Sun YC, Weu JL, Xu F, Sun RC (2011) Structure and thermal characterization of hemicelluloses isolated by organic solvents and alkaline solutions from *Tamarix austromongolica*. Biores Technol 102:5947

52. Arantes V, Saddler JN (2010) Access to cellulose limits the efficiency of enzymatic hydrolysis: the role of amorphogenesis. Biotechnol Biofuels 3(4):1–11

53. Preez JC, Bosch M, Prior BA (1987) Temperature profiles of growth and ethanol tolerance of xylose fermenting yeasts *Candida shehatae* and *Pichia stipitis*. Appl Microbiol Biotechnol 25:521–525

54. Mukhopadhyay S, Chatterjee NC (2010) Bioconversion of water hyacinth hydrolysate into ethanol. BioResourses 5(2):1301–1310

55. Agbogbo FK, Coward-Kelly G, Torry-Smith M, Wenger KS (2006) Fermentation of glucose/xylose mixtures using *Pichia stipitis*. Process Biochem 41:2333–2336

56. Gupta R, Sharma KK, Kuhad RC (2009) Separate hydrolysis and fermentation (SHF) of *Prosopis juliflora*, woody substrate for the production of cellulosic ethanol by *Saccharomyces cerevisiae* and *Pichia stipitis*-NCIM 3498. Bioresour Technol 100:1214–1220

57. Kuhad RC, Gupta R, Khasa YP, Singh A (2010) Bioethanol production from *Lantana camara* (red sage): pretreatment, saccharification and fermentation. Bioresour Technol 101:8348–8354

58. Manilal VB, Narayanan CS, Balagopalan C (1991) Cassava starch effluent treatment with concomitant SCP production. World J Microbiol Biotechnol 7:185–190

16

The influence of hydroxypropyl-β-cyclodextrin on the enantioselective hydrolysis of 2-amino phenylpropionitrile catalyzed by recombinant nitrilase

Ming-Yang Li and Xue-Dong Wang[*]

Abstract

Background: Hydrolysis of 2-amino phenylpropionitrile by nitrilase is a fundamental biochemical reaction that produces chiral phenylalanine. For practical application of this biochemical reaction, researchers have attempted to improve enzyme enantioselectivity and the reaction rate.

Results: The substrate concentration was increased from 100 to 200 mM without substrate inhibition because of the formation of a substrate-hydroxypropyl-β-cyclodextrin (HP-β-CD) complex. Meanwhile, the activity of recombinant nitrilase increased 2.5 times because the addition of HP-β-CD solubilized hydrophobic substrates in the aqueous system. Furthermore, the formation of the substrate-HP-β-CD inclusion improved the enantioselectivity of the enzymatic reaction toward producing L-phenylalanine (L-Phe). The enantiomeric excess (*e.e.*) value of L-Phe increased from 65% to 83% when the conversion rate reached 50%.

Conclusions: The recombinant nitrilase enantioselectively hydrolyzed 2-amino phenylpropionitrile to produce L-Phe. The addition of HP-β-CD to the reaction system enhanced the solubility and bioavailability of hydrophobic substrates as well as the enantioselectivity. The results showed that this additive has potential advantages in biochemical reactions of hydrophobic substrates, particularly for enantioselective biosynthesis.

Keywords: HP-β-CD; Recombinant nitrilase; 2-Amino phenylpropionitrile; Enantioselectivity; *e.e.* value

Introduction

L-Phenylalanine (L-Phe) is an essential amino acid generally used in food industries, human nutrients, and pharmaceuticals. For example, it is a precursor of some anticancer drugs and the dipeptide sweetener aspartame [1]. In early industrial processes, L-Phe was mainly produced by chemical synthesis. Because of the specific demand for the stereospecific form and the consideration of eco-friendly chemical synthesis, this approach was gradually replaced with bioprocesses, such as microbial fermentation and enzymatic transformation [2].

Nitrilase (EC 3.5.5.1) is an enzyme that converts nitrile to its carboxylic acid or amide [3]. Some important

chiral pharmaceutical intermediates and bulk products, such as acrylic acid [4], (*R*)-(−)-mandelic acid [5], 3-hydroxyvaleric acid [6], and nicotinic acid [7], are produced by nitrile hydrolysis. For example, (*R*)-(−)-mandelic acid, widely used for the production of semisynthetic cephalosporins and anti-obesity agents, is produced by hydrolyzing mandelonitrile [5,8]. In addition, other studies focused on strategic optimizations for increasing the reaction rate, reducing substrate inhibition, and improving enzyme enantioselectivity [9-11].

Cyclodextrin (CD) is the generic term for cyclic oligosaccharides that are used frequently as host molecules in supramolecular chemistry. Despite the outside of the CD molecule being hydrophilic, CD contains a hydrophobic cavity that entraps most hydrophobic molecules to form inclusion complexes [12]. Thus, it improves the solubility and bioavailability of hydrophobic compounds. Its

---
* Correspondence: xdwang@ecust.edu.cn
State Key Laboratory of Bioreactor Engineering, East China University of Science and Technology, 130 Meilong Road, Shanghai 200237, People's Republic of China

use is of interest for reactions in which hydrophobic compounds are to be delivered. In previous reports, CD has been proven to increase the availability of insoluble substrates, reduce substrate inhibition, and enhance the efficiency of catalysis by increasing the reaction rate in other catalytic reactions [13]. It has also been used in an enzymatic enantioselective reaction to increase the enantiomeric excess (*e.e.*) value of the product [14].

In our previous study, a novel nitrilase from *Rhodobacter* sp. LHS-305 was cloned and expressed in *Escherichia coli* [15]. This nitrilase displayed high activity toward both aliphatic and aromatic nitriles, similar to the nitrilase from *Rhodococcus rhodochrous* ATCC 33278 [16]. It also showed regioselectivity toward dinitriles to produce cyanocarboxylic acids. Because this nitrilase shows properties different from those of typical nitrilases, it potentially has industrial applications in the future.

In this study, the stereoselective hydrolysis of 2-amino phenylpropionitrile by this novel nitrilase was investigated. The influences of CD on this hydrolysis in terms of substrate inhibition and the *e.e.* value of the product were found to improve the catalytic reaction. The investigations comprehensively increased our knowledge of this unique nitrilase for its application.

Materials and methods
Chemicals
Hydroxypropyl-β-cyclodextrin (HP-β-CD) (≥98.5%) was purchased from Shanghai Lingfeng Chemical Reagent Company (Shanghai, China), biochemical-grade L-Phe from Sigma-Aldrich (St. Louis, MO, USA), and glucose and other chemicals from Sinopharm Chemical Reagent Co., Ltd. (Shanghai, China).

Microorganism and culture media
Recombinant *E. coli* expressing a novel nitrilase gene from *Rhodobacter sphaeroides* LHS-305 were used in this study.

Seed medium (g/L) contained peptone 10, yeast extract 5, and NaCl 10, pH 7.0 to 7.5.

Fermentation medium (g/L) contained yeast extract 10, peptone 5, NaCl 5, glucose 3, and MgSO4·7H₂O 3, pH 7.0 to 7.5. Media were sterilized for 20 min at 115°C.

Growth conditions
Precultures cultured in the seed medium for 3 to 4 h at 37°C were inoculated into 100 mL of fermentation medium in a 500-mL Erlenmeyer flask, with the addition of α-lactose (1 g/L), and cultured on a shaker at 200 rpm and 20°C for 12 h.

Preparation of recombinant nitrilase
The recombinant nitrilase was purified by affinity chromatography as described in our previous report [15].

Preparation of the 2-amino phenylpropionitrile/HP-β-CD complex
2-Amino phenylpropionitrile (73 mg) was added to 100 μL methanol to obtain a 5 M substrate stock solution. The stock solution was added to 50 mM PB solution (pH 7.0) with HP-β-CD and stirred for 20 min at 40°C. This yielded the 2-amino phenylpropionitrile/HP-β-CD inclusion complex.

Assay of enzyme activity toward bioconversion of 2-amino phenylpropionitrile
The catalytic reaction was performed in 1 mL of sodium phosphate buffer (50 mM, pH 7.0) containing the substrate 2-amino phenylpropionitrile (20 mM, final concentration) and nitrilase (1 mg/mL, final concentration). Reaction mixtures were incubated at 40°C for 10 min, and reactions were quenched by the addition of 10% (*v/v*) 1 mol/L HCl. Enzyme activity (U) was defined as the amount of enzyme required for the hydrolysis of 1 μmol of the 2-amino phenylpropionitrile substrate to the corresponding acid within 1 min. All experiments were performed in triplicate.

Analysis of L-Phe
The enantiomeric purity of Phe was determined by reversed-phase HPLC (Agilent, Santa Clara, CA, USA) equipped with a Chirobiotic T column (Sigma-Aldrich Co.) at a flow rate of 0.5 mL/min with a solvent system (75:25, *v/v*) of phosphate buffer (25 mM, adjusted to pH 3.5 with H_3PO_4) and methanol. Peaks were detected using an ultraviolet detector at 210 nm.

Results and discussion
Effects of HP-β-CD addition on the catalytic reaction
SDS-PAGE for the purified nitrilase indicated a single band at 40 kDa, naturally corresponding to 14 subunits of identical size [15]. The optimal reaction pH and temperature of the purified nitrilase were 7.0 and 40°C, respectively [15]. In the hydrolysis of 2-amino phenylpropionitrile to L-Phe by the purified nitrilase, the reaction rate was limited by the low solubility of 2-amino phenylpropionitrile. HP-β-CD can form a complex with hydrophobic substrates. It may enhance the solubility and bioavailability of such compounds. The diminished transmittances of the cyano group at 2,240 to 2,222 cm^{-1} by infrared spectrometry indicated that the substrate molecule had been included by HP-β-CD, as shown in Figure 1. The effects of HP-β-CD/substrate molar ratios on enzyme activity were examined, as shown in Figure 2. The reaction rate was improved significantly

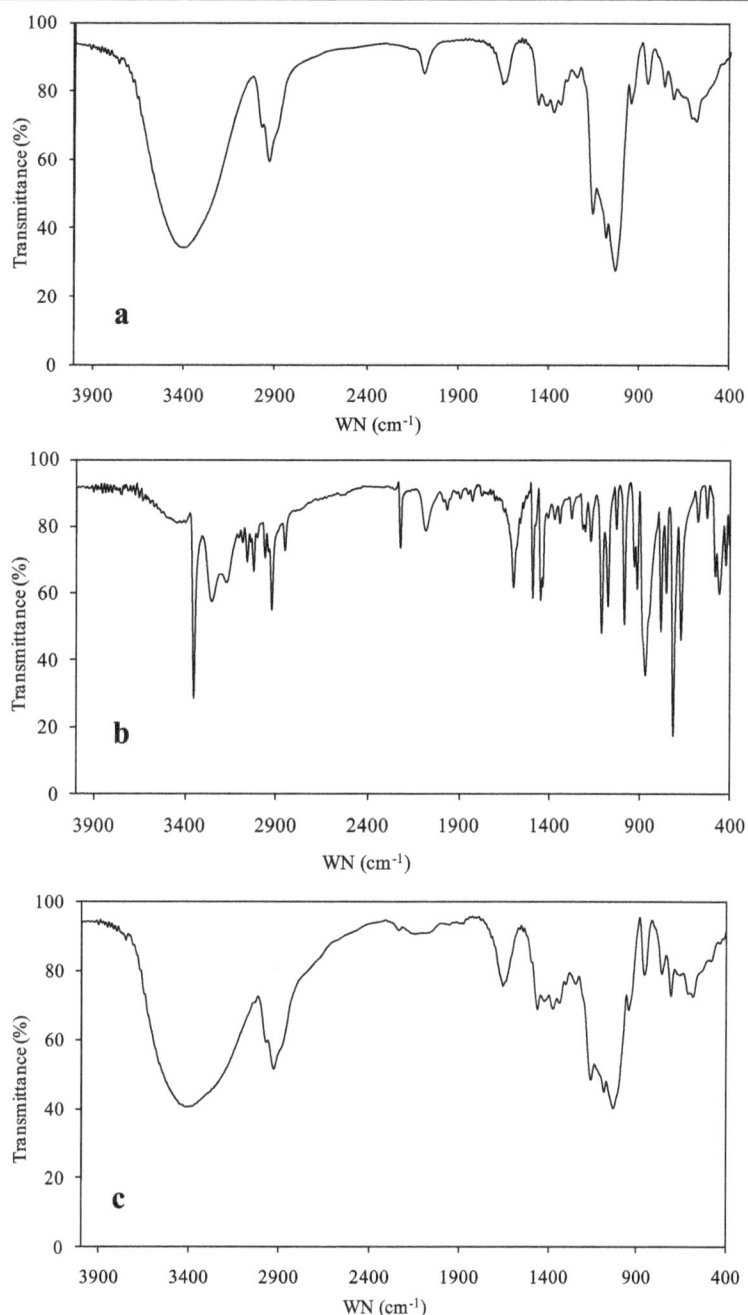

Figure 1 Infrared spectroscopy of the HP-β-CD, substrate, and HP-β-CD-substrate inclusion complex. (a) Spectroscopy of HP-β-CD. **(b)** Spectroscopy of substrate. **(c)** Spectroscopy of HP-β-CD-substrate inclusions.

by the addition of HP-β-CD, consistent with a previous study in which HP-β-CD was used as a solubilizer [17]. The highest relative activity (245%) was obtained at a ratio of 0.2. However, the host-guest complex was present in a dynamic equilibrium with free substrates and the inclusion complex. The concentration of free substrates decreased with the increase in CD available for enzymatic attack. Meanwhile, HP-β-CD

addition also affected the *e.e.* value of the product. As shown in Figure 2, the *e.e.* value of the product decreased when the ratio was lower than 0.2. These results showed that the HP-β-CD/substrate ratio was a sensitive parameter affecting the reaction rate and the enantiopurity of the product. The optimum HP-β-CD/substrate ratio of 0.2 was selected in subsequent experiments.

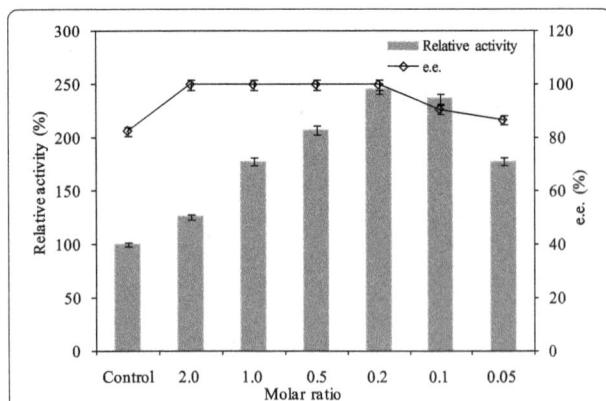

Figure 2 Effect of the HP-β-CD/substrate molar ratio on the catalytic reaction. Reactions were carried out in 50 mM sodium phosphate buffer (pH 7.0) at 40°C with a substrate concentration of 20 mM. HP-β-CD was added according to the molar HP-β-CD/substrate ratio (0.05 to 2.0). The relative activity was tested when the reaction had proceeded for 10 min, and the *e.e.* value was determined when the conversion rate reached 40%. The actual activity at 100% was 33.8 U/mL.

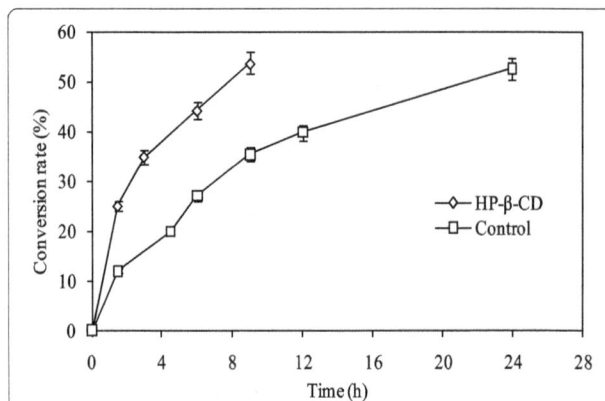

Figure 4 Time course of the biotransformation process. Reactions were carried out in 50 mM sodium phosphate buffer (pH 7.0) at 40°C with a substrate concentration of 200 mM. A constant molar HP-β-CD/substrate ratio of 0.2 was applied.

Effect of HP-β-CD inclusions on substrate inhibition

The inhibitory effects of substrates with or without HP-β-CD were investigated, as shown in Figure 3. Significant substrate inhibition was observed, and the initial reaction rate decreased correspondingly as substrate concentration exceeded 100 mM without the addition of HP-β-CD. It was necessary to eliminate the inhibitory effect of substrate on recombinant nitrilase activity in this biotransformation. Substrate inhibition was reduced by the toluene-water biphasic system or *in situ* production removal by the addition of resin [10,18]. Here, the substrate inhibition was effectively reduced by addition

of HP-β-CD in the reaction system. HP-β-CD could form host-guest complexes with hydrophobic substrates. The CD-substrate inclusion complex can function as a reservoir to slowly and continuously release free substrate for the enzymatic reaction. After the substrate was allowed to interact with HP-β-CD, substrate concentration for the maximum initial rate increased from 100 to 200 mM. The maximum initial rate increased from 32 μmol/(min·mg) at a substrate concentration of 100 mM to 62 μmol/(min·mg) at a substrate concentration of 200 mM. The reaction time at which the conversion rate reached 50% was shortened to 9 h at a substrate concentration of 200 mM in the presence of HP-β-CD, as shown in Figure 4. The HP-β-CD-substrate reaction system greatly increased the bioprocess efficiency.

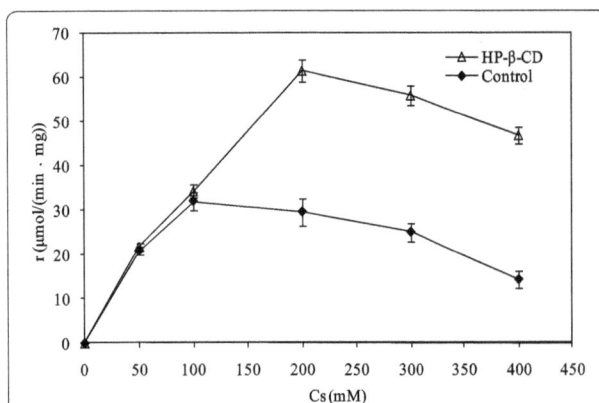

Figure 3 Substrate concentration against initial rate of reaction with and without HP-β-CD which was catalyzed by nitrilase. Reactions were carried out for 10 min in 50 mM sodium phosphate buffer (pH 7.0) at 40°C. A constant molar HP-β-CD/substrate ratio of 0.2 was applied.

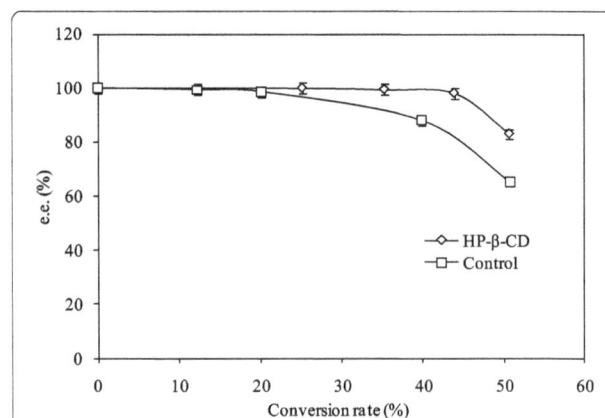

Figure 5 The *e.e.* value of the product catalyzed by nitrilase with or without HP-β-CD. Reactions were carried out in 50 mM sodium phosphate buffer (pH 7.0) at 40°C with a substrate concentration of 200 mM. A constant molar HP-β-CD/substrate ratio of 0.2 was applied.

Effect of HP-β-CD inclusion on the *e.e.* value of the product

Enantiomerically pure α-hydroxy carboxylic acid or amino acid can be produced by the enzymatic hydrolysis of the cyano group and transesterification or the hydrolysis of an ester group [19]. The recombinant nitrilase enantioselectively hydrolyzed racemic 2-amino phenylpropionitrile to L-Phe. However, the *e.e.* value of the product was only 65% when the conversion rate reached 50% with a substrate concentration of 200 mM. In previous approaches, CDs were used for chiral separation of enantiomer and the enhancement of enantioselectivity in an enzymatic reaction by the formation of CD-substrate inclusion complexes [14,20]. When HP-β-CD was added to the system to create an inclusion complex, the *e.e.* value of the product increased to 83% when the conversion rate reached 50%, as shown in Figure 5. Although the *e.e.* value of the product decreased with the increasing conversion rate, an *e.e.* value of 99% for the product was obtained when the biotransformation rate reached 40% after 5 h of hydrolysis, and the *E* value was 254 under these optimal reaction conditions. The results showed that the catalytic asymmetric synthesis of L-Phe by the recombinant nitrilase is a typical dynamic kinetic reaction favoring the D or L enantiomer. We inferred that the D type more readily formed inclusion complexes with HP-β-CD because of the difference in molecular recognition of enantiomer [21]. The D type was contained within the hydrophobic cavity, and the L type was free outside the hydrophobic cavity and was exposed to enzymatic attack. Consequently, the hydrophobic cavity protected and suppressed the hydrolysis of the D enantiomer and led to the enantioselective hydrolysis of an enantiomeric molecule.

Conclusions

The recombinant nitrilase enantioselectively hydrolyzed 2-amino phenylpropionitrile to produce L-Phe. Using the HP-β-CD-substrate reaction system, the reaction rate was greatly improved by enhancing the solubility and bioavailability of the hydrophobic substrate. Meanwhile, the *e.e.* value of the product was also improved significantly because of the formation of inclusion complexes. In this manner, HP-β-CD enhanced both the efficiency of the catalytic reaction and the optical purity of the product. The properties of CDs to form inclusion complexes with hydrophobic molecules led to their practical application in a biochemical reaction with a hydrophobic substrate. In particular, they were used to promote enantioselective biosynthesis.

Competing interests

We hold a Chinese patent relating to the content of the manuscript. The ownership of the patent belongs to East China University of Science and Technology.

Authors' contributions

MYL carried out the biotransformation process design, product analysis, and data processing. XDW conceived the study, participated in its design and coordination, and helped to draft the manuscript. Both authors read and approved the final manuscript.

Acknowledgements

We thank Guinan Li and Hualei Wang who provided the fundamental works, such as clone and expression of the novel nitrilase in *E. coli*.

References

1. Zhou H, Liao X, Wang T, Du G, Chen J (2010) Enhanced L-phenylalanine biosynthesis by co-expression of pheAfbr and aroFwt. Bioresour Technol 101:4151–4156
2. Liu DX, Fan CS, Tao JH, Liang GX, Gao SE, Wang HJ, Li X, Song DX (2004) Integration of *E. coli* aroG-pheA tandem genes into *Corynebacterium glutamicum* tyrA locus and its effect on L-phenylalanine biosynthesis. World J Gastroenterol 10:3683–3687
3. Fernandes BCM, Mateo C, Kiziak C, Chmura A, Wacker J, van Rantwijk F, Stolz A, Sheldon RA (2006) Nitrile hydratase activity of a recombinant nitrilase. Adv Synth Catal 348:2597–2603
4. Kamal A, Kumar MS, Kumar CG, Shaik TB (2011) Bioconversion of acrylonitrile to acrylic acid by *Rhodococcus ruber* strain AKSH-84. J Microbiol Biotechnol 21(1):37–42
5. Banerjee A, Dubey S, Kaul P, Barse B, Piotrowski M, Banerjee UC (2009) Enantioselective nitrilase from *Pseudomonas putida*: cloning, heterologous expression, and bioreactor studies. Mol Biotechnol 41(1):35–41
6. Wu SJ, Fogiel AJ, Petrillo KL, Hann EC, Mersinger LJ, DiCosimo R, O'Keefe DP, Ben-Bassat A, Payne MS (2007) Protein engineering of *Acidovorax facilis* 72W nitrilase for bioprocess development. Biotechnol Bioeng 97:689–693
7. Yang CS, Wang XD, Wei DZ (2011) A new nitrilase-producing strain named *Rhodobacter sphaeroides* LHS-305: biocatalytic characterization and substrate specificity. Appl Biochem Biotechnol 165:1556–1567
8. Zhang ZJ, Xu JH, He YC, Ouyang LM, Liu YY (2013) Cloning and biochemical properties of a highly thermostable and enantioselective nitrilase from *Alcaligenes* sp. ECU0401 and its potential for (R)-(–)-mandelic acid production. Bioproc Biosyst Eng 34(3):315–322
9. Sosedov O, Matzer K, Bürger S, Kiziak C, Baum S, Altenbuchner J, Chmura A, van Rantwijk F, Stolz A (2009) Construction of recombinant *Escherichia coli* catalysts which simultaneously express an (S)-oxynitrilase and different nitrilase variants for the synthesis of (S)-mandelic acid and (S)-mandelic amide from benzaldehyde and cyanide. Adv Synth Catal 351:1531–1538
10. Zhang ZJ, Pan J, Liu JF, Xu JH, He YC, Liu YY (2011) Significant enhancement of (R)-mandelic acid production by relieving substrate inhibition of recombinant nitrilase in toluene-water biphasic system. J Biotechnol 152:24–29
11. Qiu J, Su EZ, Wang W, Wei DZ (2014) High yield synthesis of D-phenylglycine and its derivatives by nitrilase mediated dynamic kinetic resolution in aqueous-1-octanol biphasic system. Tetrahedron Lett 55(8):1448–1451
12. Liu L, Guo QX (2004) Use of quantum chemical methods to study cyclodextrin chemistry. J Incl Phenom Macro Chemistry 50(1–2):95–103
13. Yue HY, Yuan QP, Wang WH (2007) Enhancement of L-phenylalanine production by β-cyclodextrin. J Food Eng 79:878–884
14. Mine Y, Zhang L, Fukunaga K, Sugimura Y (2005) Enhancement of enzyme activity and enantioselectivity by cyclopentyl methyl ether in the transesterification catalyzed by *Pseudomonas cepacia* lipase co-lyophilized with cyclodextrins. Biotechnol Lett 27(6):383–388
15. Wang HL, Li GN, Li MY, Wei DZ, Wang XD (2014) A novel nitrilase from *Rhodobacter sphaeroides* LHS-305: cloning, heterologous expression and biochemical characterization. World J Microbiol Biotechnol 30:245–252
16. Yeom SJ, Kim HJ, Lee JK, Kim DE, Oh DK (2008) An amino acid at position 142 in nitrilase from *Rhodococcus rhodochrous* ATCC 33278 determines the substrate specificity for aliphatic and aromatic nitriles. Biochem J 415:401–407
17. Shen Y, Wang M, Zhang L, Ma Y, Ma B, Zheng Y, Liu H, Luo JM (2011) Effects of hydroxypropyl-β-cyclodextrin on cell growth, activity, and integrity of steroid-transforming *Arthrobacter simplex* and *Mycobacterium sp.* Appl Microbiol Biotechnol 90(6):1995–2003

18. Xue YP, Liu ZQ, Xu M, Wang YJ, Zheng YG, Shen YC (2010) Enhanced biotransformation of (R, S)-mandelonitrile to (R)-(−)-mandelic acid with in situ production removal by addition of resin. Biochem Eng J 53:143–149

19. Gröger H (2001) Enzymatic routes to enantiomerically pure aromatic α-hydroxy carboxylic acids: a further example for the diversity of biocatalysis. Adv Synth Catal 343:547–558

20. Tang K, Miao JB, Zhou T, Liu YB, Song LT (2011) Reaction kinetics in reactive extraction for chiral separation of α-cyclohexyl-mandelic acid enantiomers with hydroxypropyl-β-cyclodextrin. Chem Eng Sci 66(3):397–404

21. Aree T, Arunchai R, Koonrugsa N (2012) Fluorometric and theoretical studies on inclusion complexes of β-cyclodextrin and D-, L-phenylalanine. Spectrochim Acta A 96:736–743

17

Leaf extract mediated green synthesis of silver nanoparticles from widely available Indian plants: synthesis, characterization, antimicrobial property and toxicity analysis

Priya Banerjee[1], Mantosh Satapathy[2], Aniruddha Mukhopahayay[1] and Papita Das[3*]

Abstract

Background: In recent years, green synthesis of silver nanoparticles (AgNPs) has gained much interest from chemists and researchers. In this concern, Indian flora has yet to divulge innumerable sources of cost-effective non-hazardous reducing and stabilizing compounds utilized in preparing AgNPs. This study investigates an efficient and sustainable route of AgNP preparation from 1 mM aqueous AgNO$_3$ using leaf extracts of three plants, *Musa balbisiana* (banana), *Azadirachta indica* (neem) and *Ocimum tenuiflorum* (black tulsi), well adorned for their wide availability and medicinal property.

Methods: AgNPs were prepared by the reaction of 1 mM silver nitrate and 5% leaf extract of each type of plant separately. the AgNPs were duely characterized and tested for their antibacterial activity and toxicity.

Results: The AgNPs were characterized by UV-visible (vis) spectrophotometer, particle size analyzer (DLS), scanning electron microscopy (SEM), transmission electron microscopy (TEM) and energy-dispersive spectroscopy (EDS). Fourier transform infrared spectrometer (FTIR) analysis was carried out to determine the nature of the capping agents in each of these leaf extracts. AgNPs obtained showed significantly higher antimicrobial activities against *Escherichia coli* (*E. coli*) and *Bacillus* sp. in comparison to both AgNO$_3$ and raw plant extracts. Additionally, a toxicity evaluation of these AgNP containing solutions was carried out on seeds of Moong Bean (*Vigna radiata*) and Chickpea (*Cicer arietinum*). Results showed that seeds treated with AgNP solutions exhibited better rates of germination and oxidative stress enzyme activity nearing control levels, though detailed mechanism of uptake and translocation are yet to be analyzed.

Conclusion: In totality, the AgNPs prepared are safe to be discharged in the environment and possibly utilized in processes of pollution remediation. AgNPs may also be efficiently utilized in agricultural research to obtain better health of crop plants as shown by our study.

Keywords: Green synthesis; Silver nanoparticles; SEM; TEM; Antimicrobial property; Toxicity

Background

Indian greeneries are the chief and cheap source of medicinal plants and plant products. From centuries till date, these medicinal plants have been extensively utilized in Ayurveda. Recently, many such plants have been gaining importance due to their unique constituents and their versatile applicability in various developing fields of research and development. Nanobiotechnology is presently one of the most dynamic disciplines of research in contemporary material science whereby plants and different plant products are finding an imperative use in the synthesis of nanoparticles (NPs). In general, particles with a size less than 100 nm are referred to as NPs. Entirely novel and enhanced characteristics such as size, distribution and morphology have been revealed by these particles in comparison to the larger particles of the mass material that they have been prepared from [1]. NPs of noble metals like gold, silver and platinum are well recognized to have significant applications in electronics, magnetic, optoelectronics and information storage [2-5]. One such

* Correspondence: papitasaha@gmail.com
[3]Department of Chemical Engineering, Jadavpur University, Kolkata 700 032, India
Full list of author information is available at the end of the article

important member of the noble metal NPs are silver NPs (Ag NPs). They are also broadly applied in shampoos, soaps, detergents, cosmetics, toothpastes and medical and pharmaceutical products and are hence directly encountered by human systems [6,7]. Earlier, the antifungal properties of silver and silver nitrate were well incorporated in the field of medical science. Also, the medicinal importance of innumerable plants and plant parts were known. But the plant-mediated silver nanoproduct is a relatively newer concept. Nanobiotechnology and their derived products are unique not only in their treatment methodology but also due to their uniqueness in particle size, physical, chemical, biochemical properties and broad range of application as well. This current emerging field of nanobiotechnology is at the primary stage of development due to lack of implementation of innovative techniques in large industrial scale and yet has to be improved with the modern technologies. Hence, there is a need to design an economic, commercially feasible as well environmentally sustainable route of synthesis of Ag NPs in order to meet its growing demand in diverse sectors.

Various approaches available for the synthesis of silver NPs include chemical [8], electrochemical [9], radiation [10], photochemical methods [11] and Langmuir-Blodgett [12,13] and biological techniques [14]. In this race of Ag NP preparation, plant-mediated green biomimetic synthesis of silver nanoparticle is considered a widely acceptable technology for rapid production of silver nanoparticles for successfully meeting the excessive need and current market demand and resulting in a reduction in the employment or generation of hazardous substances to human health and the environment. Studies have shown that *Alfalfa* roots can absorb Ag (0) from agar medium and are able to transport it to the plant shoot in the same state of oxidation [15]. Existing literature also reports successful synthesis of silver nanoparticles through a green route where the reducing and capping agent selected was the latex obtained from *Jatropha curcas* [16]. Ag NPs were also obtained using *Aloe vera* [17], *Acalypha indica* [18], *Garcinia mangostana* [19] leaf extracts. *Crataegus douglasii* fruit extract [20] as well as various other plant extracts [21] as reducing agent. Here we have developed a rapid, eco-friendly and convenient green method for the synthesis of silver nanoparticles from silver nitrate using leaf extracts of three Indian medicinal, namely, *Musa balbisiana* (banana), *A. indica* (neem) and *O. tenuiflorum* (black tulsi), by microwave irradiation method. In this research, the plant mediated synthesized Ag NPs were characterized and studied in details with all of their properties significant to current science and prevailing technologies.

The antimicrobial effects of leaf extract of these three medicinal plants and their respective biologically synthesized Ag NPs was evaluated by disc diffusion method. Comparative studies were also performed to analyze the toxicity of these biologically synthesized Ag NPs on two legume plants of the family *Fabaceae*, namely Moong Bean (*Vigna radiata*) and Chickpea (*Cicer arietinum*) as they are native to the Indian Subcontinent and widely consumed as pulses. Seeds were treated with four different concentrations of AgNP suspensions, to study the effect of the same on germination parameters and oxidative stress in the respective seeds. Discovering these new biological sources for synthesis of silver nanoparticles are more advantageous than contemporary physical or chemical procedures as these sources are abundantly available, cost-effective and conveniently utilizable. Results obtained from analysis of antimicrobial property and toxicity of these AgNPs ensure that they are safe to be discharged in the environment and hence fit to be applied for pollution remediation.

Methods

Preparation of the leaf extract

Three Indian medicinal plants, *M. balbisiana* (banana), *A. indica* (neem) and *O. tenuiflorum* (black tulsi), were selected from West Bengal, India, on the basis of cost-effectiveness, ease of availability and medicinal property. Fresh and healthy leaves were collected locally and rinsed thoroughly first with tap water followed by distilled water to remove all the dust and unwanted visible particles, cut into small pieces and dried at room temperature. About 10 g of these finely incised leaves of each plant type were weighed separately and transferred into 250 mL beakers containing 100 mL distilled water and boiled for about 20 min. The extracts were then filtered thrice through Whatman No. 1 filter paper to remove particulate matter and to get clear solutions which were then refrigerated (4°C) in 250 mL Erlenmeyer flasks for further experiments. In each and every steps of the experiment, sterility conditions were maintained for the effectiveness and accuracy in results without contamination.

Silver nanoparticle (Ag NP) synthesis

Aqueous solution (1 mM) of silver nitrate (AgNO$_3$) was prepared in 250 mL Erlenmeyer flasks and leaf extract was added for reduction into Ag$^+$ ions for each type of leaf extract. The composite mixture was then kept on turntable of the microwave oven for complete bioreduction at a power of 300 W for 4 min discontinuously to prevent an increase of pressure. In the mean time, the colour change of the mixture from faint light to yellowish brown to reddish brown to colloidal brown was monitored periodically (time and colour change were recorded along with periodic sampling and scanning by UV-visible spectrophotometry) for maximum 30 min. This was separately performed with each type of plant extract. The reactions were carried out in darkness (to avoid photoactivation of $AgNO_3$) at room temperature. Suitable controls were maintained all through the conduction of experiments. Complete reduction of $AgNO_3$ to Ag^+ ions was confirmed by the change in colour from colourless to colloidal brown. After irradiation, the dilute colloidal solution was cooled to room temperature and kept aside for 24 h for complete bioreduction and saturation denoted by UV-visible spectrophotometric scanning. Then, the colloidal mixture was sealed and stored properly for future use. The formation of Ag NPs was furthermore confirmed by spectrophotometric analysis.

UV-vis spectra analysis

Samples (1 mL) of the suspension were collected periodically to monitor the completion of bioreduction of Ag^+ in aqueous solution, followed by dilution of the samples with 2 ml of deionized water and subsequent scan in UV-visible (vis) spectra, between wave lengths of 200 to 700 nm in a spectrophotometer (Beckman - Model No. DU - 50, Fullerton, CA, USA), having a resolution of 1 nm. UV-vis spectra were recorded at intervals of 0 min, 15 min, 30 min, 45 min, 60 min and 24 h.

Characterization of Ag NPs

FTIR analysis

FTIR analysis of the dried Ag NPs was carried out through the potassium bromide (KBr) pellet (FTIR grade) method in 1:100 ratio and spectrum was recorded using Jasco FT/IR-6300 Fourier transform infrared spectrometer equipped with JASCO IRT-7000 Intron Infrared Microscope using transmittance mode operating at a resolution of 4 cm^{-1} (JASCO, Tokyo, Japan).

SEM analysis

Each of the colloidal solution containing Ag NPs were centrifuged at 4,000 rpm for 15 min, and the pellets was discarded and the supernatants were again centrifuged at 25,900 rpm for 30 min. This time, the supernatants were discarded and the final pellets were dissolved in 0.1 mL of deionized water. The pellet was mixed properly and carefully placed on a glass cover slip followed by air-drying. The cover slip itself was used during scanning electron microscopy (SEM) analysis. The samples were then gold coated using a coater (JEOL, Akishima-shi, Japan, and Model No. JFC-1600). The images of NPs were obtained in a scanning electron microscope (ZEISS EVO-MA 10, Oberkochen, Germany). The details regarding applied voltage, magnification used and size of the contents of the images were implanted on the images itself.

TEM analysis

Transmission electron microscopy (TEM) technique was used to visualize the morphology of the Ag NPs. The 200 kV ultra-high-resolution transmission electron microscope (JEOL, Model No. JEM 2100 HR with EELS). TEM grids were prepared by placing a 5 μL of the AgNP solutions on carbon-coated copper grids and drying under lamp. Additionally, addition presence of metals in the sample was analyzed by energy-dispersive spectroscopy (EDS) using INCA Energy TEM 200 with analysis software (JEOL) was used.

Estimation of antibacterial activity

The comparative antibacterial activities of the plant leaf extracts and of the Ag NPs synthesized from the respective extracts were effectively accessed against one Gram (+) ve (*Bacillus*) bacteria and one Gram (-) ve (*Escherichia coli* (*E. coli*)) bacteria as test microorganisms procured from IMTEch Chandigarh. Disc diffusion method [22] was followed for testing each type of plant leaf extract and their respective Ag NPs containing solution. The discs were soaked with double distilled water, plant leaf extracts, silver nitrate solution and solution containing silver nanoparticles of each type separately. Then the discs were air dried in sterile condition. The plates containing nutrient agar media were prepared by swabbing them with the microbial cultures. Plates containing media as well as culture were divided in to four equal parts and previously prepared discs were placed on each part of the plate. The discs were placed in the following order: disc soaked with double distilled water as negative control, disc soaked with plant leaves extract, disc soaked with 1 mM silver nitrate solution and disc soaked with solution containing plant leaves mediated synthesized silver nanoparticles. The plates were incubated at 37°C for 24 to 48 h. Then, the maximum zone of inhibition were observed and measured for analysis against each type of test microorganism.

Toxicity assay

Four different concentrations of (25%, 50%, 75% and 100% (v/v)) of three different Ag NP dispersions were prepared in distilled water. The germination test was carried out in sterile Petri dishes of 12 cm diameter by placing a Whatman® no. 3 filter paper on them. Fifty seeds of each receptor crop, Moong Bean (*V. radiata*) and Chickpea (*C. arietinum*), were placed in the respective Petri dishes. The seeds were surface sterilized with 0.1% $HgCl_2$ solution and rinsed three times with distilled water. The solution of each concentration was added to each Petri dish of respective treatment daily in such an amount just enough to wet the seeds. The Petri dish were then placed in seed germinator in dark (Yorco Y58765) and maintained at 25°C. Seeds with root tip 1 mm and higher were considered as germinated. Percent germination and length of root and shoot (in mm) obtained following each 24 h up to 72 h after the germination of seeds were observed were calculated thereafter.

Oxidative stress enzyme activities in root and shoot samples (l00 mg) from control and treated seedlings were precisely rinsed with distilled water and homogenized with cold phosphate buffer (0.1 M, pH 7.0) and centrifuged at 12,000 rpm for 15 min at 4°C. Pellets were discarded and the supernatants were further used for enzyme assay. The protein content of sample extracts was assessed using bovine serum albumin (BSA; Sigma Chemical, St. Louis, MO, USA) as standard protein [23].

Catalase activity (CAT) in samples was recorded as a decline in absorbance at 240 nm for 1 min and expressed as μmol H_2O_2 consumed min^{-1} mg^{-1} protein [24]. The substrate mixture consisting of phosphate buffer (50 mM) and H_2O_2 (15 mM) was added to the enzyme extract.

Superoxide dismutase (SOD) activity was then assessed using nitroblue tetrazolium (NBT) in the presence of riboflavin [25]. Results were expressed as SOD units mg^{-1} protein. The reaction mixture was prepared by taking 50 μL enzyme extract and adding 1 mL NBT (50 μM), 500 μL methionine (13 mM), 1 mL riboflavin (1.3 μM), 950 μL (50 mM) phosphate buffer and 500 μL EDTA

(75 mM). This reaction was started by keeping reaction solution under 30 W fluorescent lamp illuminations and turning the fluorescent lamp on. The reaction stopped when the lamp turned off 5 min later. The NBT photo reduction produced blue formazane which was used to measure the increase in absorbance at 560 nm. The same reaction mixtures without enzyme extract in dark were used as blank.

Soluble peroxidase (POD) activity of the tissues was estimated by adding Guaiacol solution (0.25% guaiacol (*v/v*) in 10 mmol/L sodium phosphate buffer (pH 6.0) and 0.125% H_2O_2 (*v/v*)) to the enzyme extract. POD activity was recorded as the resultant increase in absorbance at 470 nm and expressed as $\Delta OD/min/mg$ protein [26]. All spectrophotometric measurements were done with UV-vis spectophotometer (Shimadzu UV -1601, Kyoto, Japan).

Statistical analysis

Analytical determinations were made in triplicate. All experimental data are expressed as mean ± standard deviation. All the statistical analyses were carried out

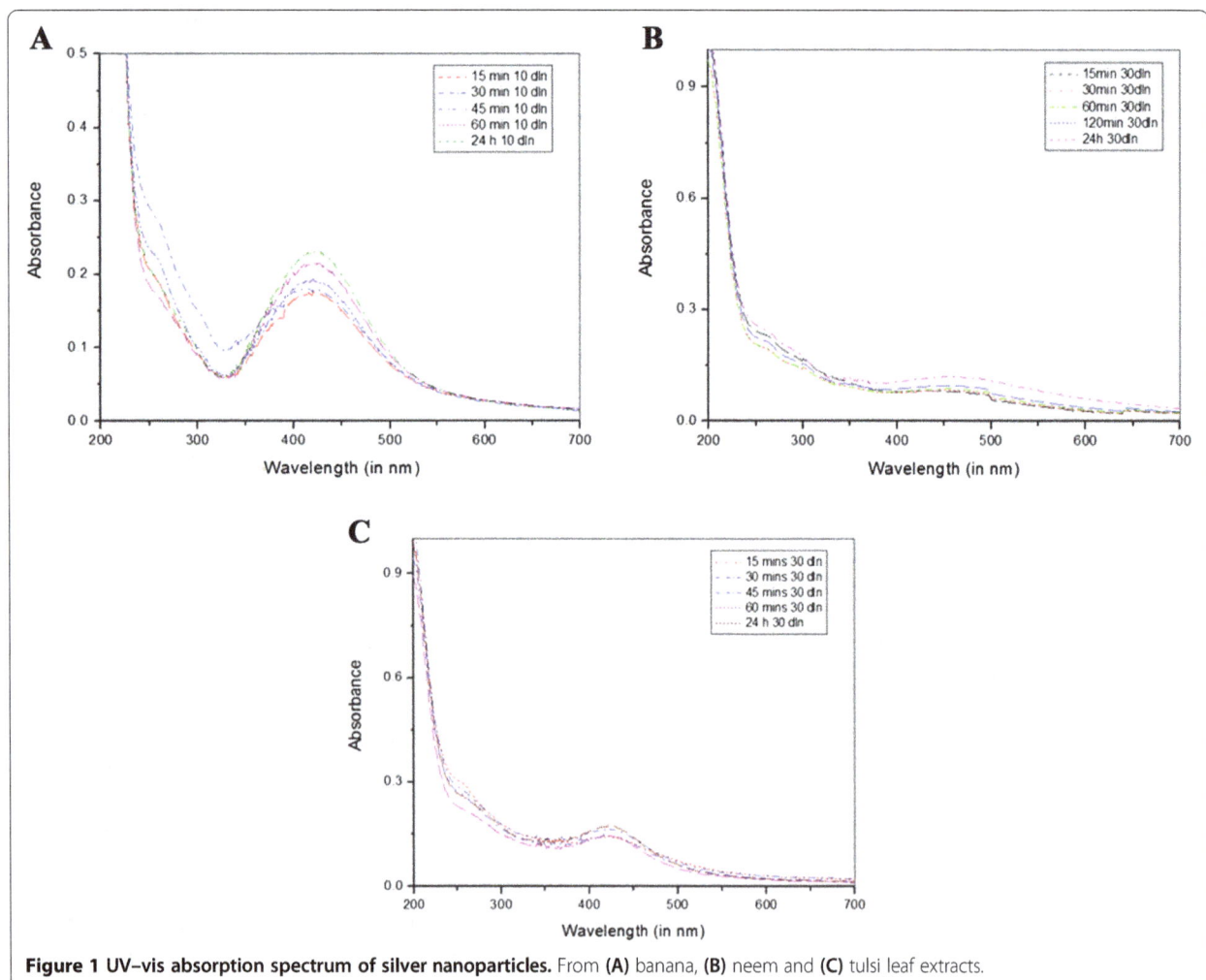

Figure 1 UV–vis absorption spectrum of silver nanoparticles. From **(A)** banana, **(B)** neem and **(C)** tulsi leaf extracts.

by using Origin software (version 7.0383; OriginLab Corporation, Northampton, MA 01060, USA). Statistical significance of the data were analyzed by two sample independent t tests and one-way ANOVA by using GraphPad InStat 3 (San Diego, CA, USA) with significance limit set at 5% probability.

Results and discussion
AgNP characterization
UV-vis analysis

Silver nanoparticles (AgNPs) appear yellowish brown in colour in aqueous medium as a result of surface plasmon vibrations [18]. As the different leaf extracts were added to aqueous silver nitrate solution, the colour of the solution changed from faint light to yellowish brown to reddish brown and finally to colloidal brown indicating AgNP formation. Similar changes in colour have also been observed in previous studies [27-31] and hence confirmed the completion of reaction between leaf extract and AgNO$_3$. The UV-vis spectra recorded after time intervals of 15 min, 30 min, 45 min, 60 min and 24 h from the initiation of reaction are shown in Figure 1. Absorption spectra of AgNPs formed in the reaction media has absorption maxima in the range of 425 to 475 nm due to surface plasmon resonance of AgNPs. The UV-vis spectra recorded, implied that most rapid bioreduction was achieved using banana leaf extract as reducing agent followed by tulsi and neem leaf extracts. This was denoted by broadening of the peak which indicated the formation of polydispersed large nanoparticles due to slow reduction rates [30,31]. The UV-vis spectra also revealed that formation of AgNPs occurred rapidly within the first 15mins only and the AgNPs in solution remained stable even after 24 h of completion of reaction.

FTIR analysis

FTIR analysis carried out to characterize the AgNPs obtained from each type of plant extract (curve A, banana; curve B, neem and curve C, tulsi) is shown in Figure 2. In all three AgNP solutions, prominent bands of absorbance were observed at around 1,025, 1,074, 1,320, 1,381, 1,610 and 2,263 cm^{-1}. The observed peaks denote -C-O-C-, ether linkages, -C-O-, germinal methyls, -C=C- groups or from aromatic rings and alkyne bonds, respectively. These bands denote stretching vibrational bands responsible for compounds like flavonoids and terpenoids [32,33] and so may be held responsible for efficient capping and stabilization of obtained AgNPs.

SEM analysis

The SEM images of the AgNPs are shown in Figure 3. It is seen that AgNPs of different shapes were obtained in case of different leaf extracts being used as reducing and capping agents. Banana, neem and tulsi extracts formed approximately spherical, triangular and cuboidal AgNPs, respectively. This may be due to availability of different quantity and nature of capping agents present in the different leaf extracts. This is also supported by the shifts and difference in areas of the peaks obtained in the FTIR analysis.

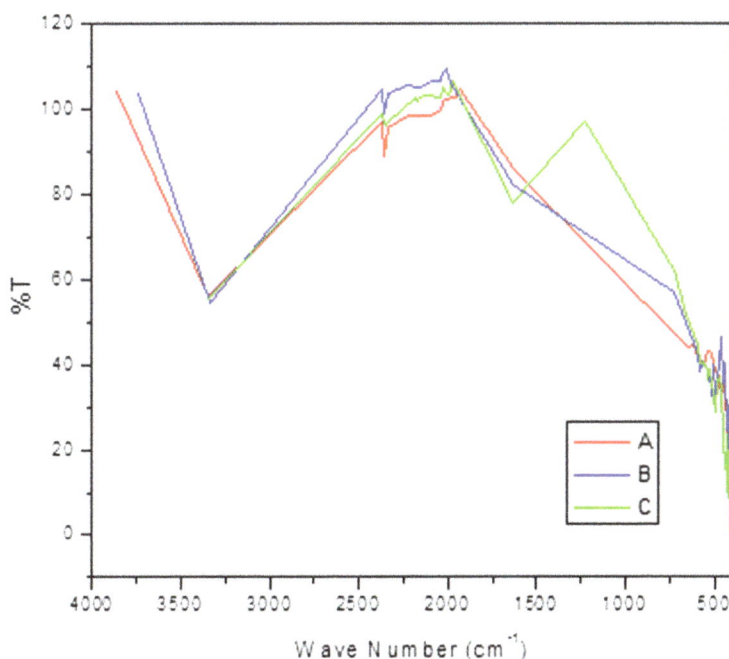

Figure 2 Graphs obtained from FTIR analysis of AgNPs obtained. From (curve A) banana. (curve B) neem and (curve C) Tulsi leaf extracts respectively.

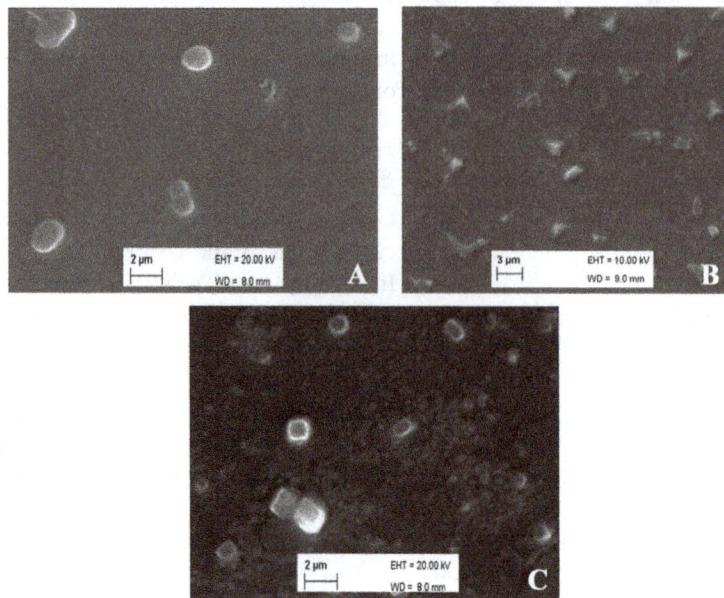

Figure 3 SEM images of silver nanoparticles. Formed by the reaction of 1 mM silver nitrate and 5% leaf extract of **(A)** banana. **(B)** neem and **(C)** tulsi leaves respectively.

TEM and EDS analysis

Figure 4 shows the TEM images obtained by the reaction of 5% of each type of leaf extract and 1 mM silver nitrate solution separately. A mixture of plates (triangles, pentagons and hexagons) and spheres was obtained though mainly spherical shapes were predominant. A similar trend is also observed in the SEM images (Figure 5). It is clear that the triangles, pentagons and hexagons are plate structures with sizes of up to 200 nm.

The EDS spectra recorded from the silver nanoparticles are shown inset Figure 4 (A, B and C).

The EDS profile shows a strong silver signal along with weak oxygen and carbon peaks, which may have originated from the biomolecules bound to the surface of the

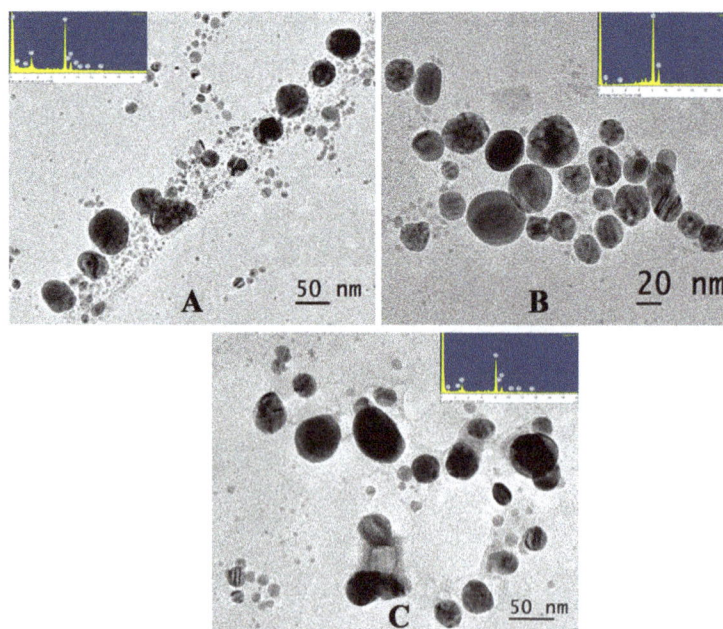

Figure 4 TEM images of silver nanoparticles. Formed by the reaction of 1 mM silver nitrate and 5% leaf extract of **(A)** banana. **(B)** neem and **(C)** tulsi leaves respectively.

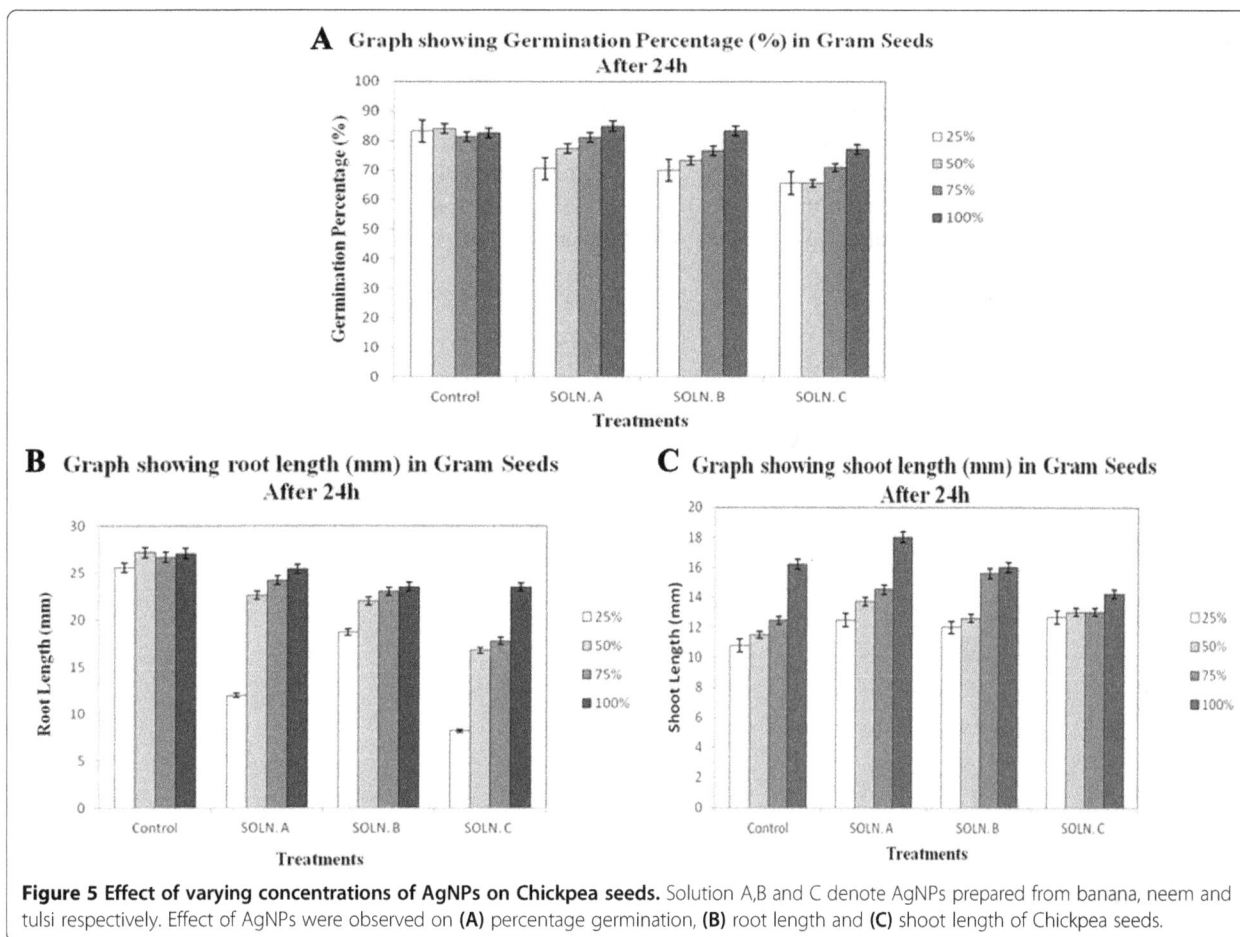

Figure 5 Effect of varying concentrations of AgNPs on Chickpea seeds. Solution A,B and C denote AgNPs prepared from banana, neem and tulsi respectively. Effect of AgNPs were observed on **(A)** percentage germination, **(B)** root length and **(C)** shoot length of Chickpea seeds.

silver nanoparticles. Carbon and copper peaks may be due to the same being present in the grids. It has been reported that nanoparticles synthesized using plant extracts are surrounded by a thin layer of some capping organic material from the plant leaf broth and are, thus, stable in solution up to 4 weeks after synthesis [34,35]. This is another advantage of nanoparticles synthesized using plant extracts over those synthesized using chemical methods.

Antibacterial property analysis

In this study, the antimicrobial property of AgNPs were investigated by growing *Bacillus* and *E. coli* colonies on nutrient agar plates supplemented with AgNPs. A control plate was separately maintained for both organisms in water. Results obtained are shown in Table 1. The inhibition zones obtained indicates maximum antibacterial activity of the prepared test sample. Results obtained in previous studies [28,36] also support the antibacterial potential of AgNPs. The zone of bacterial inhibition by AgNPs prepared from banana leaf extract show maximum inhibition for gm + ve *Bacillus,* which may be concluded

from the fact that these particles had the smallest diameter than those prepared from neem and tulsi leaf extracts, which in turn exhibited equal antimicrobial property. Also, in comparison to $AgNO_3$ and AgNPs, there is no such prominent antimicrobial activity in case of the plant extracts when used in crude form. No zone of inhibition was obtained in case of control.

Table 1 Showing zone of inhibitions found in *Bacillus* and *E. coli* cultures

Extracts	Bacteria	Zone of inhibition (in mm)		
		By leaf extract	By $AgNO_3$	By AgNPs
Banana (leaf)	+ (*Bacillus*)	6 ± 0.02	9 ± 0.015	16 ± 0.016
	− (*E. coli*)	6 ± 0.005	9 ± 0.004	14 ± 0.02
Neem (leaf)	+	8 ± 0.013	9 ± 0.02	14 ± 0.008
	−	8 ± 0.015	9 ± 0.013	12 ± 0.007
Tulsi (leaf)	+	8 ± 0.02	9 ± 0.007	14 ± 0.021
	−	8 ± 0.02	9 ± 0.02	14 ± 0.017

Cultures were treated with crude leaf extracts, $AgNO_3$ and AgNPs. Results were expressed as Mean ± SD.

AgNP toxicity analysis
Effect on seed germination

Toxicity analysis of the AgNPs prepared was carried out on Moong bean (*V. radiata*) and Chickpea (*C. arietinum*) seeds and their resultant root and shoot lengths were recorded. Seeds were considered to have germinated by observing the emergence of radicles. Maximum percentage germination of seeds was obtained after 24 h. Results

obtained varied significantly ($p < 0.05$) with each treatment. Chickpea seeds treated with AgNP solution prepared from banana extract (solution A) were significantly higher ($p < 0.05$) than those treated with AgNP solution prepared from tulsi extract (solution C), which were in turn significantly lower ($p < 0.05$) than those treated with AgNP solution prepared from neem extract (solution B). Root length and shoot length of seedlings were significantly

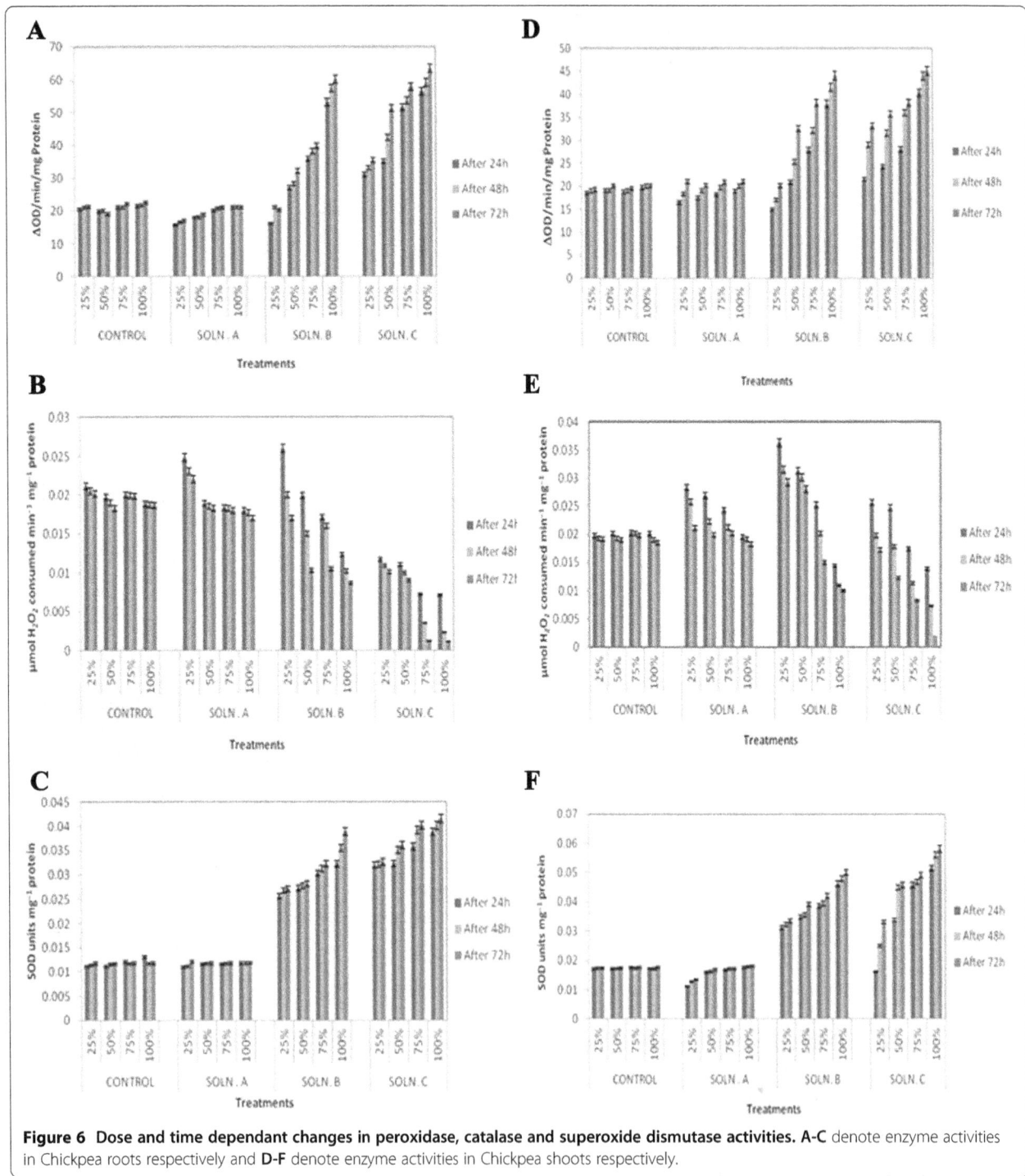

Figure 6 Dose and time dependant changes in peroxidase, catalase and superoxide dismutase activities. A-C denote enzyme activities in Chickpea roots respectively and **D-F** denote enzyme activities in Chickpea shoots respectively.

higher ($p < 0.05$) than distilled water-treated controls in seeds treated with solution A, were lower than controls in seeds treated with solution B and were further reduced in seeds treated with solution C. Results obtained in case of Chickpea seeds are shown in Figure 5. Moong bean seeds also displayed a similar trend (figure not shown). Similar results were found [37] in case of spinach seeds, where TiO_2 NPs had induced germination and plant growth by altering the vigor of aged seeds. This was also accompanied by an increase in the root and shoot length and lower oxidative stress enzyme levels (discussed below).

Analysis of oxidative stress

Peroxidases are those enzymes that catalyses dehydrogenation of large number of organic compounds like phenols, amines etc [38]. Spectrophometric analysis revealed that POD activity was significantly higher ($p < 0.01$) in AgNP-treated chickpea samples in comparison to the control treated ones. However, peroxidase activity was seen to augment with an increase in AgNP size and concentration. Increased peroxidase activity obtained in shoot and root tissues of chickpea seedlings exposed to solutions B and C may indicate an elevated level of ROS production in those organisms in comparison to that of the control and those treated with solution A. POD activity increased both in a time-dependent and dose-dependent manner. Two other enzymes, namely SOD and CAT are considered as first-defence system against oxidative stress whose activity may be altered by toxic stress. CAT activities significantly decreased ($p < 0.01$) while SOD activity in chickpeas was significantly elevated ($p < 0.01$) with respect to an increase in size and concentration of AgNPs. A succeeding increase in SOD activity supports the fact that suppression of CAT activity may be associated with a flux of superoxide radicals or inhibition of enzyme synthesis [39]. Changes in enzyme activities are shown in Figure 6. Similar results were obtained from Moong seeds as well (figure not shown).

Hence, a general pattern of increased health and growth was observed in seeds treated with solution A consisting of the smallest particles (80.2 nm) and was further affected with an increase in particle size. Oxidative stress was low and nearing control levels in both samples treated with solution A but appeared to undergo a significant increase with increase in concentrations of solutions B and C. These positive effects may possibly have occurred due to the antimicrobial properties of AgNPs, which can make plants strong and resistant to stress. At the same time, AgNPs, especially those with high specific surface area may be also be responsible for sequestering nutrients on their surfaces thereby serving as a nutrient stock to the germinating seeds and aid in their growth [37]. These results are also supported by other studies which have reported the antioxidant property [29] and non-toxic nature [40] of green synthesized AgNPs.

Conclusions

Silver nanoparticles (AgNPs) were successfully obtained from bioreduction of silver nitrate solutions using banana, neem and tulsi leaf extracts. Owing to varying properties of these three plant species, AgNPs obtained from them also varied in size, the smallest being yield using banana leaf extracts. AgNPs have been appropriately characterized using UV-vis spectroscopy, SEM, TEM and EDS analysis. Results denoted banana leaf extract to be a better reducing agent in comparison to neem and tulsi leaf extracts. FTIR analysis revealed the efficient capping and stabilization properties of these AgNPs. Besides, they also aided in plant germination and growth by sequestering nutrients for them and could hence be implemented for agricultural purposes. Hence, due to their benign and stable nature and antimicrobial property, these AgNPs may be well utilized in industrial and remedial purposes. However, plant uptake and utilization of AgNPs require more detailed research on many issues like uptake potential of various species, process of uptake and translocation and the activities of the AgNPs at the cellular and molecular levels.

Competing interests
The authors declare that they have no competing interests.

Authors' contributions
PB carried out synthesis of AgNPs and helped in manuscript preparation. MS carried out characterization of AgNPs. AM helped in designing toxicity analysis. PD helped design the whole study and manuscript preparation. All authors read and approved the final manuscript.

Acknowledgements
The authors are thankful to all members of the Department of Environmental Science, University of Calcutta. We also acknowledge Mr. Tridip Sen, Miss Urmila Goswami, Center for Research in Nanoscience and Nanotechnology and Mr. Tanmoy kumar Dey, Laboratory of Food Science and Technology, Food and Nutrition Division, University of Calcutta, for helping us with SEM, TEM and FTIR analysis, respectively.

Author details
[1]Department of Environmental Science, University of Calcutta, 51/2 Hazra Road, Kolkata 700 026, India. [2]Biotechnology Department, National Institute of Technology, Durgapur 713209, India. [3]Department of Chemical Engineering, Jadavpur University, Kolkata 700 032, India.

References
1. van den Wildenberg W (2005) Roadmap report on nanoparticles. W&W Espana sl, Barcelona, Spain
2. Gratzel M (2001) Photoelectrochemical cells. Nature 414:338–344
3. Okuda M, Kobayashi Y, Suzuki K, Sonoda K, Kondoh T, Wagawa A, Kondo A, Yoshimura H (2005) Self-organized inorganic nanoparticle arrays on protein lattices. Nano Lett 5:991–993
4. Dai J, Bruening ML (2002) Catalytic nanoparticles formed by reduction of metal ions in multilayered polyelectrolyte films. Nano Lett 2:497–501
5. Murray CB, Sun S, Doyle H, Betley T (2001) Monodisperse 3d transition-metal (Co, Ni, Fe) nanoparticles. MRS Bull 26:985–991

6. Bhattacharya R, Murkherjee P (2008) Biological properties of "naked" metal nanoparticles. Adv Drug Deliv Rev 60:1289–1306

7. Bhumkar DR, Joshi HM, Sastry M, Pokharkar VB (2007) Chitosan reduced gold nanoparticles as novel carriers for transmucosal delivery of insulin. Pharm Res 24:1415–1426

8. Sun Y, Yin Y, Mayers BT, Herricks T, Xia Y (2002) Uniform form silver nanowires synthesis by reducing AgNO₃ with ethylene glycol in presence of seeds and and poly(vinyl pyrrolidone). Chem Mater 14:4736–4745

9. Yin B, Ma H, Wang S, Chen S (2003) Electrochemical synthesis of silver nanoparticles under protection of poly (N-vinylpyrrolidone). J Phys Chem B 107:8898–8904

10. Dimitrijevic NM, Bartels DM, Jonah CD, Takahashi K, Rajh T (2001) Radiolytically induced formation and optical absorption spectra of colloidal silver nanoparticles in supercritical ethane. J Phys Chem B 105:954–959

11. Callegari A, Tonti D, Chergui M (2003) Photochemically grown silver nanoparticles with wavelength-controlled size and shape. Nano Lett 3:1565–1568

12. Zhang L, Shen YH, Xie AJ, Li SK, Jin BK, Zhang QF (2006) One-step synthesis of monodisperse silver nanoparticles beneath vitamin E Langmuir monolayers. J Phys Chem B 110:6615–6620

13. Swami A, Selvakannan PR, Pasricha R, Sastry M (2004) One-step synthesis of ordered two dimensional assemblies of silver nanoparticles by the spontaneous reduction of silver ions by pentadecylphenol Langmuir monolayers. J Phys Chem B 108:19269

14. Naik RR, Stringer SJ, Agarwal G, Jones S, Stone MO (2002) Biomimetic synthesis and patterning of silver nanoparticles. Nat Mater 1:169–172

15. Gardea-Torresdey JL, Gomez E, Peralta-Videa JR, Parsons JG, Troiani H, Jose-Yacaman M (2003) Alfalfa sprouts: a natural source for the synthesis of silver nanoparticles. Langmuir 19:1357–1361

16. Bar H, Bhui DK, Sahoo GP, Sarkar P, De SP, Misra A (2009) Green synthesis of silver nanoparticles using latex of Jatropha curcas. Colloids Surf A Physicochem Eng Asp 339:134–139

17. Chandran SP, Chaudhary M, Pasricha R, Ahmad A, Sastry M (2006) Synthesis of gold nanotriangles and silver nanoparticles using Aloe vera plant extract. Biotechnol Prog 22:577–583

18. Krishnaraj C, Jagan EG, Rajasekar S, Selvakumar P, Kalaichelvan PT, Mohan N (2010) Synthesis of silver nanoparticles using Acalypha indica leaf extracts and its antibacterial activity against water borne pathogens. Colloids Surf B: Biointerfaces 76:50–56

19. Veerasamy R, Xin TZ, Gunasagaran S, Xiang TFW, Yang EFC, Jeyakumar N, Dhanaraj SA (2010) Biosynthesis of silver nanoparticles using mangosteen leaf extract and evaluation of their antimicrobial activities. J Saudi Chem Soc 15:113–120

20. Ghaffari-Moghaddam M, Hadi-Dabanlou R (2014) Plant mediated green synthesis and antibacterial activity of silver nanoparticles using Crataegus douglasii fruit extract. J Indus Eng Chem 20:739–744

21. Ghaffari-Moghaddam M, Hadi-Dabanlou R, Khajeh M, Rakhshanipour M, Shameli K (2014) Green synthesis of silver nanoparticles using plant extracts. Korean J Chem Eng 31:548–557

22. Bao Q, Zhang D, Qi P (2011) Synthesis and characterization of silver nanoparticle and graphene oxide nanosheet composites as a bactericidal agent for water disinfection. J Colloid Interface Sci 360:463–470

23. Lowry OH, Rosebrough NJ, Farr AL, Randall RJ (1951) Protein measurement with the Folin phenol reagent. J Biol Chem 193:265–275

24. Chandlee JM, Scandalios JG (1984) Analysis of variance affecting the catalase development programme in maize scutellum. Theor Appl Genet 69:71–77

25. Beauchamp C, Fridovich I (1971) Superoxide dismutase: improved assays and an assay applicable to acrylamide gels. Anal Biochem 44:276–287

26. Cipollini DF (1998) The induction of soluble peroxidase activity in bean leaves by wind-induced mechanical perturbation. Amer J Bot 85:1586–1591

27. Shukla VK, Pandey S, Pandey AC (2010) Green synthesis of silver nanoparticles using neem leaf (Azadirachta indica) extract. In: Proceedings of International Conference On Advanced Nanomaterials And Nanotechnology. ICANN-2009, Guwahati, Assam (India). 9–11 December 2009

28. Namratha N, Monica PV (2013) Synthesis of silver nanoparticles using Azadirachta indica (Neem) extract and usage in water purification. Asian J Pharm Tech 3:170–174

29. Lalitha A, Subbaiya R, Ponmurugan P (2013) Green synthesis of silver nanoparticles from leaf extract Azhadirachta indica and to study its anti-bacterial and antioxidant property. Int J Curr Microbiol App Sci 2:228–235

30. Singhal G, Bhavesh R, Kasariya K, Sharma AR, Singh RP (2011) Biosynthesis of silver nanoparticles using Ocimum sanctum (Tulsi) leaf extract and screening its antimicrobial activity. J Nanoparticle Res 13:2981–2988

31. Philip D, Unni C (2011) Extra cellular biosynthesis of gold and silver nanoparticles using Krishna tulsi (Ocimum sanctum) leaf. Phys E 43:1318–1322

32. Siddiqui BS, Afshan F, Faizi GS, Naqui SNH, Tariq RM (2000) Two insecticidal tetranortriterpenoids from Azadirachta indica. Phytochemistry 53:371–376

33. Huang Q, Li D, Sun Y, Lu Y, Su X, Yang H, Wang Y, Wang W, Shao N, Hong J, Chen C (2007) Biosynthesis of silver and gold nanoparticles by novel sundried Cinnamomum camphora leaf. Nanotechnol 18:105104

34. Shankar SS, Rai A, Ahmad A, Sastry M (2004) Rapid synthesis of Au, Ag, and bimetallic Au core Ag shell nanoparticles using Neem (Azadirachta indica) leaf broth. J Colloid Interface Sci 275:496–502

35. Song JY, Kim BS (2009) Rapid biological synthesis of silver nanoparticles using plant leaf extract. Bioprocess Biosyst Eng 32:79–84

36. Rout Y, Behera S, Ojha AK, Nayak PL (2012) Green synthesis of silver nanoparticles using Ocimum sanctum (Tulashi) and study of their antibacterial and antifungal activities. J Microbiol Antimicro 4:103–109

37. Zheng L, Hong FS, Lu SP, Liu C (2005) Effect of nano-TiO₂ on strength of naturally and growth aged seeds of spinach. Biol Trace Elem Res 104:83–91

38. Bashir F, Mahmooduzzafar STO, Iqbal M (2007) The antioxidative response system in Glycine max (L.) Merr: exposed to Deltamethrin, a synthetic pyrethroid insecticide. Environ Poll 147:94–100

39. Malik CP, Singh MB (1980) Plant enzymology and histoenzymology. Kalyani Publishers, New Delhi, India

40. Daniel SCGK, Kumar R, Sathish V, Sivakumar M, Sunitha S, Sironmani TA (2011) Green synthesis (Ocimum tenuiflorum) of silver nanoparticles and toxicity studies in Zebra Fish (Danio rerio) model. Int J NanoSci Nanotechnol 2:103–117

18

Enhanced limonene production by optimizing the expression of limonene biosynthesis and MEP pathway genes in *E. coli*

Fu-Liang Du, Hui-Lei Yu, Jian-He Xu and Chun-Xiu Li[*]

Abstract

Background: Limonene is an important monoterpene used as a chemical commodity and precursor for producing biofuels, flavor and medicinal compounds.

Results: In this paper, we engineered *Escherichia coli* by embedding two exogenous genes encoding a limonene synthase (LS) and a geranyl diphosphate synthase (GPPS) for production of limonene. Out of 12 *E. coli* strains transformed with various plasmids, the best one with p15T7-*ls-gpps* produced limonene with a titer of 4.87 mg/L. In order to enhance the limonene production, two rate-limiting enzymes in the endogenous MEP pathway of *E. coli*, 1-deoxy-xylulose-5-phosphate synthase (DXS) and isopentenyl diphosphate isomerase (IDI), were overexpressed consecutively on vector pET21a+, resulting in a production of 17.4 $mg_{limonene}$/L at 48 h.

Conclusions: After the preliminary optimization of the medium in a two-phase culture system composed of *n*-hexadecane (1/50, V_{org}/V_{aq}), the final production of limonene was raised up to 35.8 mg/L, representing approximately a 7-fold improvement compared to the initial titer.

Keywords: Limonene; Biosynthesis; *Escherichia coli*; MEP pathway; Geranyl pyrophosphate synthase; Limonene synthase; Two-phase culture system

Background

Terpenoid, which has more than forty thousand kinds of chemicals, is the largest family of natural products [1]. Limonene, one of the simplest monocyclic *p*-menthane (1-methyl-4-isopropylcyclohexane) type monoterpenes, has been used as a flavor or fragrance with aroma value [2,3]. Limonene is also an important plant monoterpene precursor of several fine chemicals, flavorings, fragrances, and pharmaceuticals such as carveol, carvone, perillyl alcohol and menthol [2-6]. In addition, the hydrogenated form of limonene can be used as fuel [7]. However, the major supply of limonene is limited to plant sources currently. Although chemical processes for monoterpene synthesis have been well documented, very few pathway-engineered processes [3,4] have been reported so far for the production of limonene. Therefore, it is promising and

attractive to develop a route of microbial synthesis for limonene production.

Isopentenyl diphosphate (IPP) and its isomer dimethylallyl diphosphate (DMAPP) are the two essential building blocks to synthesize all isoprenoids (terpenoids). There are two biosynthetic pathways to produce IPP and DMAPP. The first one is mevalonate pathway (MVA pathway) present in all higher eukaryotes such as *Saccharomyces cerevisiae* [8-10] The second one is MEP pathway starting with the reaction between pyruvate and glyceraldehyde-3-phosphate to synthesize IPP and DMAPP [9,11]. Plants produce their terpenoids using MEP pathway, which takes place in their plastids. In addition, MEP pathway occurs in most bacteria including *Escherichia coli*.

Based on the IPP and DMAPP produced from the MEP pathway, limonene can be synthesized by the catalysis of two key enzymes [4,12], geranyl diphosphate synthase (GPPS) and limonene synthase (LS), which were reported to occur in *Abies grandis* [13] and *Mentha spicata* [2], respectively. GPPS catalyzes the condensation between IPP and DMAPP, forming a linear diphosphate intermediate,

---

* Correspondence: chunxiuli@ecust.edu.cn
Laboratory of Biocatalysis and Synthetic Biotechnology, State Key Laboratory of Bioreactor Engineering, East China University of Science and Technology, Shanghai 200237, P. R. China

geranyl diphosphate (GPP), the precursor of all the mono-terpenes. On the other hand, LS catalyzes the intramolecular cyclization of GPP to give limonene (Figure 1).

Although overexpression of rate-limiting enzymes in the MEP pathway has been used to produce many terpenoids including isoprene, taxol precursor and levopimaradiene [14-19], this method has not been used in the most common engineered strain *Escherichia coli* to produce limonene. In a previous study, an engineered *Escherichia coli*, harboring a heterologous GPPS gene from *Abies grandis* and a LS gene from *Mentha spicata*, was reported to produce limonene in a titer of merely 5 mg/L (24 h) [4]. A possible reason for the low limonene production might be the insufficient supply of IPP and DMAPP. In another report, an engineered *Escherichia coli* in which a heterologous MVA pathway was installed, can produce limonene with titers around 335 mg/L at 48 h [3].

In this paper, we engineered *E. coli* to produce (−)-limonene by expressing a GPPS, a LS, a 1-deoxy-d-xylulose-5-phosphate synthase (DXS) and an isopentenyl diphosphate isomerase (IDI). A series of modifications to optimize the expression of those enzymes and the medium have been done. The final production of limonene was raised up to 35.8 mg/L, representing approximately a 7-fold improvement compared to the initial titer.

Methods
Bacterial strains, genes and vectors
Escherichia coli BL21 (DE3) (Novagen, Germany) was used for gene cloning and expression. *Escherichia coli* K12 MG1655 was used for amplification of *dxs*, *idi*, and *ispA* genes. The gene sequences of GPPS and LS were obtained from *Abies grandis* and *Mentha spicata* (Genbank accession numbers: AF513112 and L13459). These genes were condon optimized for *E. coli* and

commercially synthesized in Generay (Shanghai, China). *N*-Terminal amino acids at positions 87 and 56 of GPPS and LS (plastid transit peptide) were removed [13,20]. Vectors pET-21a, pET-28a, pACYCDuet-1, pQE30 and pTrcHis2B (Novagen, Germany) were used for gene expression.

Genes cloning and plasmids construction
The PCR fragments of *gpps* and *ls* were cloned into the *Bam*HI–*Eco*RI and *Eco*RI–*Sal*I sites of the pTrcHis2B vector to create p40Trc-*ls-gpps* and p40Trc-*gpps-ls* plasmids. The generated *ls-gpps* and *gpps-ls* operons from p40Trc-*ls-gpps* and p40Trc-*gpps-ls* were then cloned into pACYCDueT-1 vector through *Bam*H I and *Sal* I sites to generate p15T7-*ls-gpps* and p15T7-*gpps-ls* plasmids. For constructing p40T7-*ls-gpps* and p40T7-*gpps-ls* plasmids, initially the operons *ls-gpps* and *gpps-ls* were inserted into pET-28a vector at *Bam*H I and *Sal* I sites. The ispA^{S80F}, which is designated as *Ec*GPPS here, is another geranyl pyrophosphate synthase. The method for constructing plasmids p40Trc-*ls-Ecgpps*, p40Trc-*Ecgpps-ls*, p15T7-*ls-Ecgpps*, p40T7-*ls-Ecgpps*, p40T7-*Ecgpps-ls* and p15T7-*Ecgpps-ls* was the same as the plasmids listed above. The *dxs-idi* operon was initially constructed by cloning each of the genes from the genome of *E. coli* K12 MG1655. The *dxs-idi* operon was cloned into pTrcHis2B vector by the *Bam*H I-*Kpn* I sites to create p40Trc-*dxs-idi* plasmid. The *dxs-idi* operon was sub-cloned into the pQE30 and pET-21a vectors through *Bam*H I and *Sal* I sites to construct p40T5-*dxs-idi* and p40T7-*dxs-idi* plasmids, respectively. The primers used were listed in Table 1. All plasmids in this study were shown in Table 2, and all engineered *E. coli* strains and the corresponding plasmids were shown in Table 3.

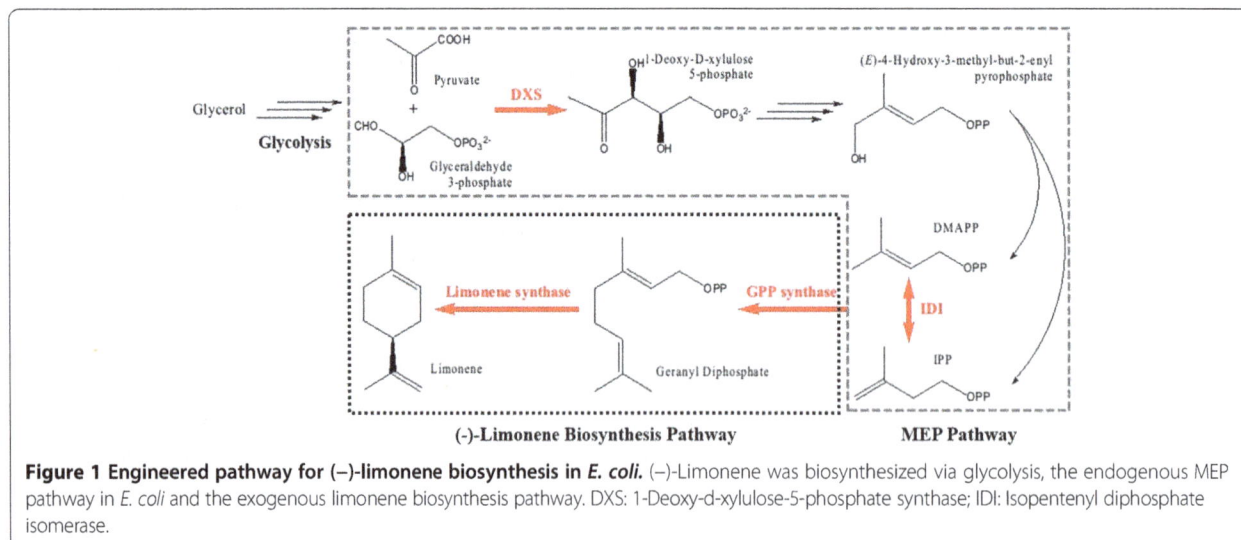

Figure 1 Engineered pathway for (−)-limonene biosynthesis in *E. coli*. (−)-Limonene was biosynthesized via glycolysis, the endogenous MEP pathway in *E. coli* and the exogenous limonene biosynthesis pathway. DXS: 1-Deoxy-d-xylulose-5-phosphate synthase; IDI: Isopentenyl diphosphate isomerase.

Table 1 The plasmids constructed in this study

Plasmids	Parent plasmids	Copies	Promoter	Antibiotic marker	Expression strength
p40Trc-ls-gpps	pTrcHis2B	40	Trc	ampicillin	40
p15T7-ls-gpps	pACYCDuet-1	15	T7	chloramphenicol	75
p40T7-ls-gpps	pET28a+	40	T7	kanamycin	200
p40Trc-gpps-ls	pTrcHis2B	40	Trc	ampicillin	40
p15T7-gpps-ls	pACYCDuet-1	15	T7	chloramphenicol	75
p40T7-gpps-ls	pET28a+	40	T7	kanamycin	200
p40Trc-ls-Ecgpps	pTrcHis2B	40	Trc	ampicillin	40
p15T7-ls-Ecgpps	pACYCDuet-1	15	T7	chloramphenicol	75
p40T7-ls-Ecgpps	pET28a+	40	T7	kanamycin	200
p40Trc-Ecgpps-ls	pTrcHis2B	40	Trc	ampicillin	40
p15T7-Ecgpps-ls	pACYCDuet-1	15	T7	chloramphenicol	75
p40T7-Ecgpps-ls	pET28a+	40	T7	kanamycin	200
p40T7-dxs-idi	pET21a+	40	T7	ampicillin	200
p40T5-dxs-idi	pQE30	40	T5	ampicillin	80
p40Trc-dxs-idi	pTrcHis2B	40	Trc	ampicillin	40

The expression strengths of vectors were estimated using published values of promoter strength and copy number. Promoter strengths were calculated as Trc = 1, T5 = 2, T7 = 5 (Brosius et al. [21], Brunner et al. [22]). Gene copy number was assigned by published copy numbers for origin of replication for the different plasmids used.

Culture media and conditions

Culture for limonene production was carried out in 2YT medium using a shaking incubator at 20°C and 180 rpm. Glycerol as the main carbon source, was added at 2% (w/v) concentration to all the media used in this study [23]. LB medium and TB medium were compared with

Table 2 List of engineered *E. coli* strains and the corresponding plasmids

Strain	MEP pathway *dxs-idi* operon		Limonene pathway *ls-gpps*, *gpps-ls*, *ls-Ecgpps*, or *Ecgpps-ls* operons	
	Plasmid	Expression strength	Plasmid	Expression strength
1	-	1	p40Trc-ls-gpps	40
2	-	1	p15T7-ls-gpps	75
3	-	1	p40T7-ls-gpps	200
4	-	1	p40Trc-gpps-ls	40
5	-	1	p15T7-gpps-ls	75
6	-	1	p40T7-gpps-ls	200
7	-	1	p40Trc-ls-Ecgpps	40
8	-	1	p15T7-ls-Ecgpps	75
9	-	1	p40T7-ls-Ecgpps	200
10	-	1	p40Trc-Ecgpps-ls	40
11	-	1	p15T7-Ecgpps-ls	75
12	-	1	p40T7-Ecgpps-ls	200
13	p40T7-dxs-idi	200	p15T7-ls-gpps	75
14	p40T5-dxs-idi	80	p15T7-ls-gpps	75
15	p40Trc-dxs-idi	40	p15T7-ls-gpps	75

For a full plasmids description, see Table 1.

2YT medium as basis media for limonene production. The addition of pyruvate as an auxiliary carbon source has been reported to increase terpenoid production [24]. Ampicillin (100 μg/mL), kanamycin (50 μg/mL) and chloramphenicol (50 μg/mL) were added to the culture as required. Cell culture was carried out in 100 mL of medium, and growth was determined by measuring the optical density at 600 nm (OD_{600}). For the two-phase culture for limonene production, IPTG was added when OD_{600} reached about 0.7, and 2 mL of *n*-hexadecane was layered over 100 mL of culture medium. The assay for glycerol was conducted by the Nash reagent [25].

Analysis of limonene

In the two-phase culture system with *n*-hexadecane overlay, limonene was extracted with the upper *n*-hexadecane phase. Then *n*-hexadecane containing the limonene was collected and centrifuged at 14,000 rpm for 10 min. The *n*-hexadecane extracts were analyzed with GC-MS (QP2010, Shimadzu, Kyoto, Japan) equipped with an InertCap DB-5 ms column. Limonene from splitless 1 μL injection was separated using a GC oven temperature program of 50°C for 3 min, followed by a 10°C/min ramp to 250°C. Injector and MS quadrupole detector temperatures were 250°C and 150°C, respectively. To increase the sensitivity and selectivity of detection, the MS was operated in selected ion-monitoring (SIM) mode using ions of 136, 68, and 93 m/z, which represent the molecular ion and two abundant fragmental ions of limonene.

Table 3 Primers used in this work

Primers	Sequences of oligonucleotides (5′–3′)
dxs (s[1])	GGGAATTCCATATGAGTTTTGATATTGCCAA
dxs (a[2])	GCGAATTCTTATGCCAGCCAGGCCTTGATTTTG
idi (s)	CGGAATTCGAAGGAGATATACATATGCAAACGGAACACGTCATTTTATTG
idi (a)	GCGCTCGAGGCTCACAACCCCGGCAAATGTCGG
dxsidiBamHI(s)	CGCGGATCCGATGAGTTTTGATATTGCCAA
dxsidiKpnI (a)	CGGGGTACCGCTCACAACCCCGGCAAATGTCGG
dxsidiBamH I 1(s)	CGCGGATCCATGAGTTTTGATATTGCCAA
LSBamH I(s)	CGCGGATCCGATGGAACGTCGTAGCGGTAA
LS(a)	CGTAGAATTCTTATGCAAATGGTTCAAACA
GPPS(S)	CCGGAATTCGAAGGAGATATACATATGTTCGACTTCAACAAATAC
GPPS(A)	ACGCGTCGACTCAGTTCTGACGGAATGCAAC
GPPSBamH I(s)	CGCGGATCCGATGTTCGACTTCAACAAATAC
GPPSBamH I 1(s)	CGCGGATCCATGTTCGACTTCAACAAATAC
GPPS1(A)	CCGGAATTCTCAGTTCTGACGGAATGCAAC
LS1(S)	CCGGAATTCGAAGGAGATATACATATGGAACGTCGTAGCGGTAA
LS1 (a)	CCGGAATTCTTATGCAAATGGTTCAAACA
S81F-rev	AATCATCATGAATTAAAAAGTAAGCGTGGATACAC
S81F-for	GTGTATCCACGCTTACTTTTTAATTCATGATGATT
EcFPPSBamHIs	CGCGGATCCGATGGACTTTCCGCAGCAACT
EcFPPSEcoRIa	CCGGAATTCTTATTTATTACGCTGGATG
LSEcoRISDs	CGTAGAATTCAGAAGGAGATATACATATGGAACGTCGTAGCGGTAA
EcFPPSEcoRIs	CGTAGAATTCAGAAGGAGATATACATATGGACTTTCCGCAGCAACT
EcFPPSalI	ACGCGTCGACTTATTTATTACGCTGGATG

[1]:Forward primer;
[2]:Reverse primer.

Results
Optimizing the expression of limonene biosynthesis genes in E. coli

(−)-Limonene can be produced by introducing the *gpps* and *ls* genes into *E. coli*. Previously, IspA^{S80F} has been known to make the ispA variant synthesize GPP only [26]. We cloned the ispA genes from *E. coli*, and got the IspA^{S80F} gene *Ecgpps*. Polycistronic operons consisting of *gpps* and *ls* or *Ecgpps* and *ls*, in which the genes can be assembled as one transcriptional unit, were constructed based on pTrcHis2B, pACYCDuet-1 and pET28a+. Then totally twelve limonene synthesis plasmids were constructed (Table 2) and introduced into the *E. coli*, respectively. Those twelve recombinant *E. coli* cell lines, each harboring one limonene biosynthesis plasmid, were cultured in 2YT medium for 48 h. The cells were cultured with 250 μM IPTG induction. As shown in Figure 2, only trace amounts of limonene were detected in the culture of *E. coli* harboring *Ecgpps-ls* operon or *gpps-ls* operon. While the four recombinant strains harboring *ls-Ecgpps* operon and *ls-gpps* operon (strain #2, #7, #8 and #9) could produce a quite significant amount of limonene, indicating that the limonene synthase gene should be put before the gene of geranyl

diphosphate synthase when they were coexpressed in one plasmid. Moreover, the strain #2 harboring the *ls-gpps* operon could produce more limonene than the other strains (strain #7, #8 and #9) harboring the *ls-Ecgpps* operon. This result showed that GPPS was more suitable for monoterpene biosynthesis compared to *Ec*GPPS. In addition, for the better expression of *ls-gpps* operon, a moderate expression strength (Table 2) was good for *E. coli* to produce limonene. We also found that the cell mass was greatly associated with the limonene production (Figure 2, C). This phenomenon indicated that the expression of plasmid p15T7-*ls-gpps*, which could affect the main metabolism, was conducive to cell growth. As a result, the limonene synthesis plasmid p15T7-*ls-gpps* was therefore selected for limonene production in further experiments. This optimized limonene pathway plasmid could lead the recombinant *E. coli* to accumulate 4.87 mg/L of limonene or 0.863 mg/g$_{DCW}$ after 48 h of cultivation.

Overexpression of rate-limiting enzymes in the endogenous MEP pathway

The limonene building blocks, IPP and DMAPP, can be synthesized in *E. coli* via the endogenous MEP pathway.

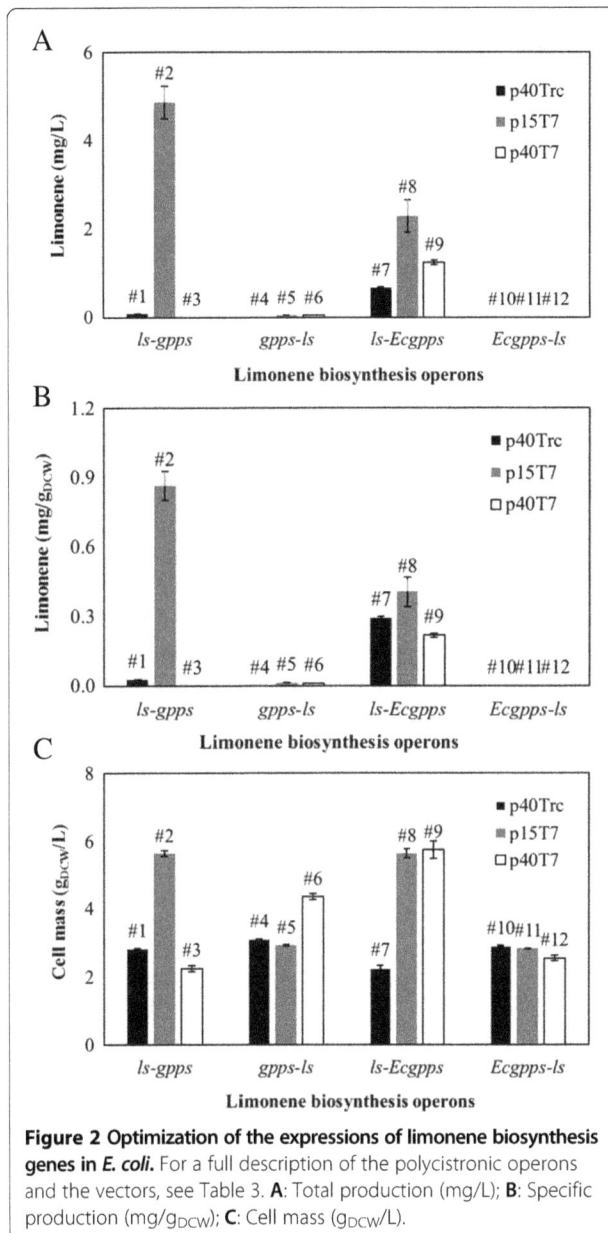

Figure 2 Optimization of the expressions of limonene biosynthesis genes in *E. coli*. For a full description of the polycistronic operons and the vectors, see Table 3. **A**: Total production (mg/L); **B**: Specific production (mg/g$_{DCW}$); **C**: Cell mass (g$_{DCW}$/L).

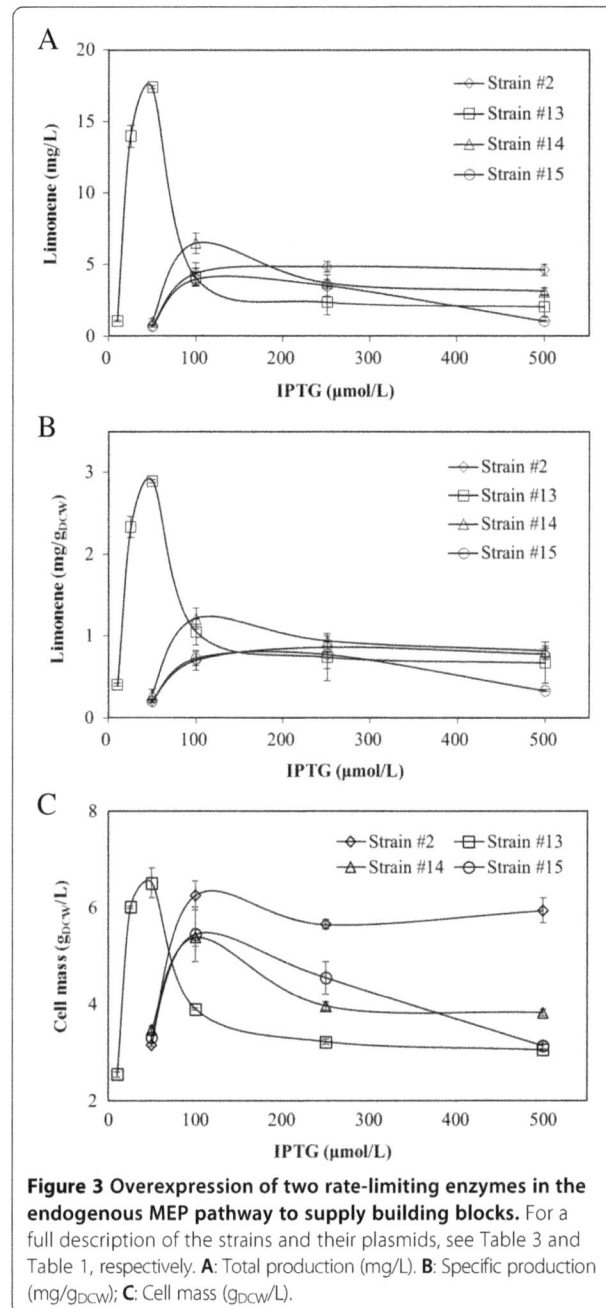

Figure 3 Overexpression of two rate-limiting enzymes in the endogenous MEP pathway to supply building blocks. For a full description of the strains and their plasmids, see Table 3 and Table 1, respectively. **A**: Total production (mg/L). **B**: Specific production (mg/g$_{DCW}$); **C**: Cell mass (g$_{DCW}$/L).

The reactions catalyzed by DXS and IDI have been reported as the most critical rate-limiting step in the endogenous MEP pathway of *E. coli* [27-30]. In this study, the polycistronic operon *dxs-idi* was constructed based on pTrcHis2B, pQE30 and pET21a + to form p40Trc-*dxs-idi*, p40T5-*dxs-idi* and p40T7-*dxs-idi*. These three plasmids were respectively introduced into the strain #2, resulting in another three strains designated as strains #13, #14 and #15. Then the strains #2, #13, #14 and #15 were cultured in 2YT medium for 48 h. Various inducer (IPTG) concentrations, from 10 μM to 500 μM, were also evaluated. As shown in Figure 3, the strain #13 containing both plasmids p40T7-*dxs-idi* and p15T7-*ls-gpps* could accumulate limonene up to 17.4 mg/L (or 2.89 mg/g$_{DCW}$) after 48 h

cultivation with IPTG induction at 50 μM, which represents approximately a 3.6-fold enhancement in total limonene production (from 4.87 to 17.4 mg/L) or a 3.3-fold improvement in specific production (from 0.863 to 2.89 mg/g$_{DCW}$). Meanwhile the cell mass was the largest. Additionally, limonene production significantly affected the cell mass and glycerol consumption (Figure 3C, D). Clearly, the higher productivity and more robust growth of strain #13 allowed higher limonene accumulation. Further improvements should be possible through medium optimization.

Effects of basis media and pyruvate on limonene production

Besides the genetic modulation described above, we also optimized the growth medium to further increase the limonene productivity. *E. coli* strain #13 was respectively cultured in 2YT, LB or TB media. The limonene production was measured after 24 h, 48 h and 72 h cultivation. Although the three media had no significant differences in the specific production of limonene, 2YT medium gave higher titers than the other two media (Figure 4). So 2YT medium was chosen as the basis medium for the cultivation of genetically engineered strain #13. In a previous work [24], pyruvate and dipotassium phosphate were found beneficial for isoprenoid production in *E. coli*. Therefore, the 2YT medium supplemented with different concentrations of pyruvate was designed and employed for the culture of strain #13. As shown in Figure 5, addition of 4 g/L pyruvate could stimulate the limonene production of up to 15.1 mg/L or 4.11 mg/g_{DCW} after 24 h cultivation, 32.5 mg/L or 5.59 mg/g_{DCW} after 48 h, and 35.8 mg/L or 5.76 mg/g_{DCW} after 72 h. Although the cell mass was basically the same (Figure 5, C), it is approximately a 2-fold increase in limonene production as compared with the un-optimized medium used in this work or a 7-fold increase in contrast to the initial titer (4.87 mg/L). Meanwhile, the consumption of glycerol was

also the largest (Figure 5, D). In addition, the effect of K_2HPO_4 was also examined (data not shown), although the improvement was not so significant as pyruvate.

Discussion

In this paper, we significantly increased the limonene yield in *E. coli* by optimizing the expression of the two limonene biosynthesis genes, overexpression of the two rate-limiting enzymes in the endogenous MEP pathway, and optimization of the culture medium (Figure 6). We established a limonene biosynthesis pathway in *E. coli* using four different polycistronic operons based on three vectors with varied expression strength. Previous researches on limonene biosynthesis focused merely on the expression of a certain polycistronic operon (e.g., *gpps-ls*) based on a specified vector [3,4]. In addition, only the GPPS from plant *Abies grandis* was tried for limonene production [3,4]. In this study, *Ecgpps*, a bacterial *gpps* gene that is native to *E. coli*, was tried for the first time for limonene production. Previous study indicated that limonene biosynthesis pathway flux was limited by the inherently low enzyme activity of the plant-originated GPPS [4]. However, our result showed that the protein-engineered *EcFPPS*$_{S80F}$ from *E. coli* also suffers a problem of low activity for limonene production. The two genes connected by IRES sequences can be expressed from a single promoter

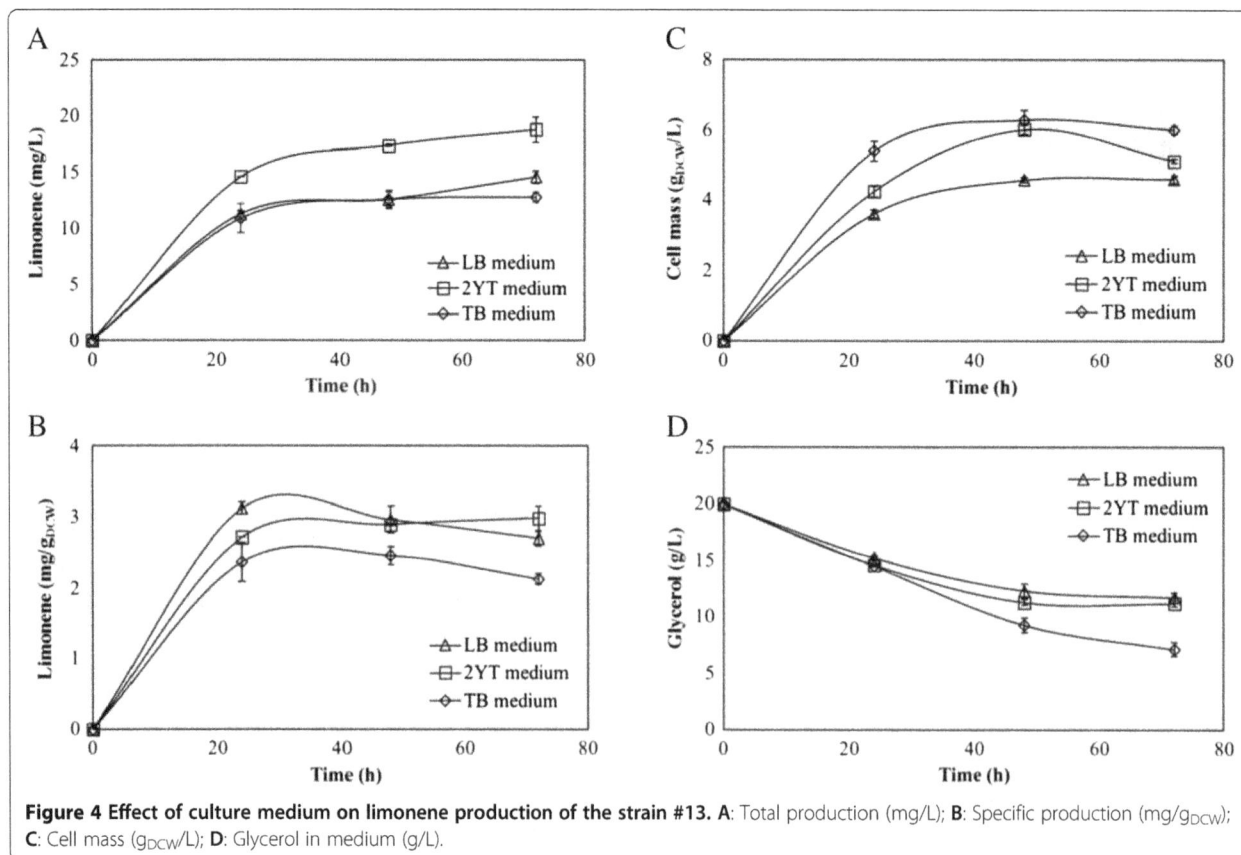

Figure 4 Effect of culture medium on limonene production of the strain #13. A: Total production (mg/L); **B**: Specific production (mg/g_{DCW}); **C**: Cell mass (g_{DCW}/L); **D**: Glycerol in medium (g/L).

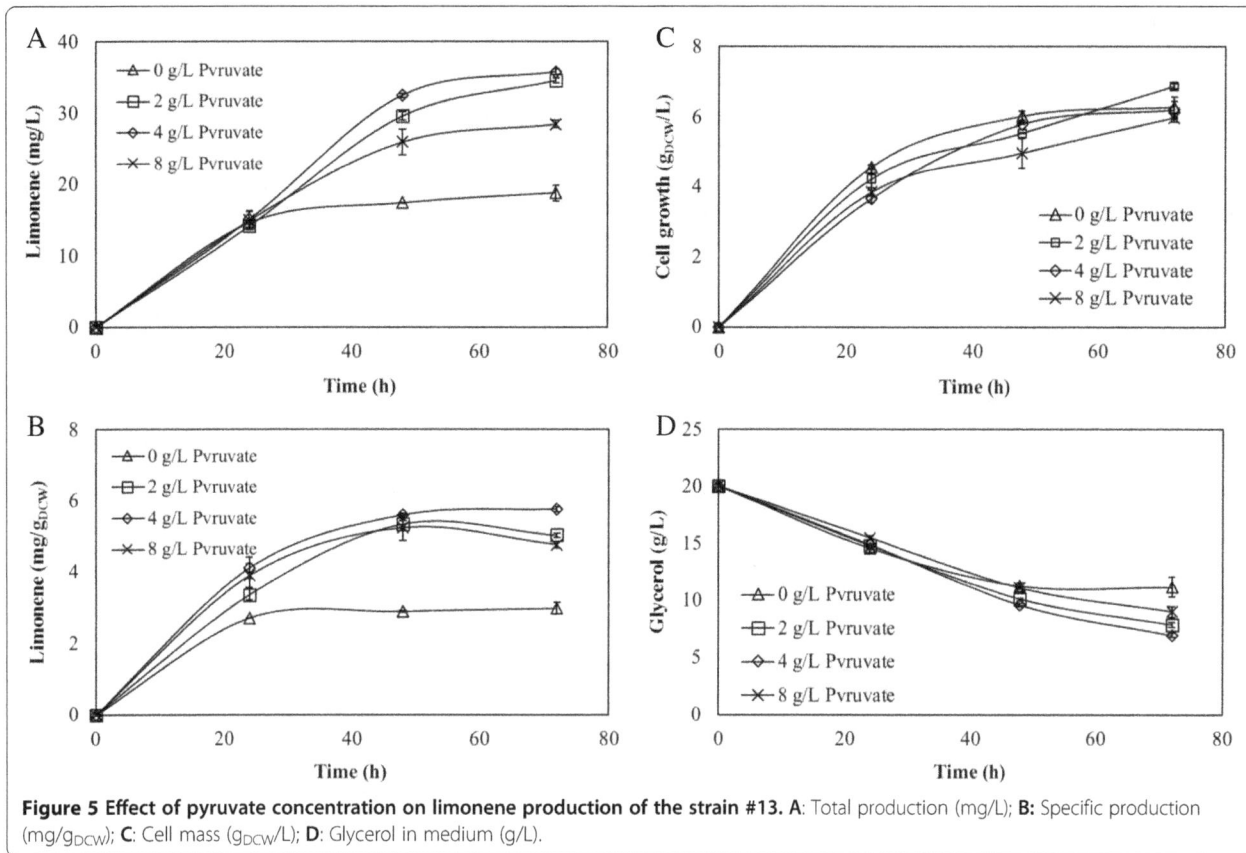

Figure 5 Effect of pyruvate concentration on limonene production of the strain #13. A: Total production (mg/L); **B:** Specific production (mg/g$_{DCW}$); **C**: Cell mass (g$_{DCW}$/L); **D**: Glycerol in medium (g/L).

in IRES method used in this study. In the prokaryotic expression system, the gene ranked in front will be transcripted and translated primarily, followed by the transcription and translation of the subsequent gene in the polycistronic operons [31]. So the order of the limiting genes in polycistronic operons has a lot to do with the expression of the genes and the quantity of final product [14,15,32]. Our result showed that those *E. coli* strains expressing *ls-gpps* or *ls-Ecgpps* produced larger amounts of limonene than the strains expressing *gpps-ls* or *Ecgpps-ls*. This result disputes the current notion that the order of

the limiting genes in polycistronic operons should correspond to that of terpenoid metabolic pathway in the natural and fine regulation mechanism of microorganisms [19]. In addition, the expression strength of vectors used for limonene biosynthesis was also indicated to be a very important determinant for the final limonene yield.

Our work represents the first report on limonene biosynthesis via overexpression of rate-limiting enzymes in MEP pathway. When the two limiting genes (*dxs* & *idi*) were coexpressed, both the expression strength and IPTG concentration were very important for achieving a high

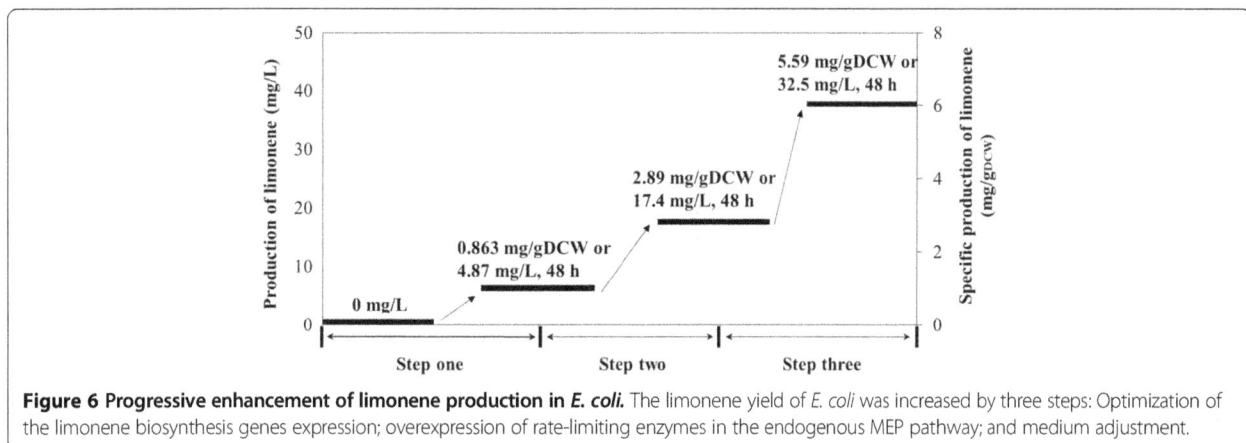

Figure 6 Progressive enhancement of limonene production in *E. coli*. The limonene yield of *E. coli* was increased by three steps: Optimization of the limonene biosynthesis genes expression; overexpression of rate-limiting enzymes in the endogenous MEP pathway; and medium adjustment.

limonene yield. The polycistronic operon *dxs-idi* was expressed on various vectors, which resulted in different limonene productions. This phenomenon was also reported in the researches on taxadiene and levopimaradiene biosynthesis [14,15]. Once again, an appropriate expression strength of genes made tremendous contribution to the overall yield of the final product. Although a direct comparison of limonene yield with previous studies on isoprene or terpenoid is difficult because of the differences in products and assay conditions used, our results agree well with the published values of increases in isoprene or terpenoid yields when overexpressing enzymes of MEP pathway (Table 4). Although it might have become a popular point that the MEP pathway is not effective for high-level production of terpenoid [9,33], the MEP pathway is a native pathway and the stoichiometry of MEP pathway is more efficient as compared to MVA pathway for IPP/DMAPP synthesis [34]. The limonene production by the MEP pathway in this work was much lower than the reported result by the MVA pathway [3]. It was partly because only two bottleneck enzymatic steps (*dxs* and *idi*) were targeted to increase the flux through the MEP pathway. Previous work [35] revealed that *ispB*, *ispD* and *ispF* also produce rate-limiting enzymes impacting the MEP flux. Second, previous results showed that the expression of *B. subtilis* bottleneck enzymes in *E. coli* resulted in an enhancement of total isoprene production [17]. The low enzyme activity and the native regulation of the native bottleneck enzymes might be the reasons of low limonene

production in this work. Third, different *E. coli* strains might result in the difference in limonene biosynthesis. *E. coli* BL21 (DE3) might not be as suitable as *E. coli* DH1 for terpene biosynthesis [3].

Besides genetic modulations, we also systematically optimized the culture medium to further increase limonene production. Pyruvate and dipotassium phosphate were chosen as supplements. The addition of pyruvate as the auxiliary carbon source has been reported to increase terpenoid production [24]. Since pyruvate was the immediate precursor of MEP pathway, the addition of pyruvate greatly increased the DMAPP synthesis so as to significantly enhance the limonene production. In addition, the addition of pyruvate affects the main metabolism. Namely, keto-acid such as pyruvate, OAA, and 2KG etc. affect Cya and in turn affect cAMP level. In the case of using glycerol as a carbon source, cAMP-Crp level is high in the wild type, while this may be decreased when pyruvate was added, and thus the glycerol consumption rate may be affected, since glpFKD is under control of cAMP-Crp. Dipotassium phosphate was also found to be beneficial for limonene production in *E. coli*. It supports the results of the previous study on terpenoid biosynthesis [24].

To prevent limonene volatilizing and to extract it from the culture medium, a two-phase culture system composed of *n*-hexadecane was adopted in this study. Hexadecane was chosen for its high hydrophobicity (log $P_{O/W}$, 8.8) for the extraction of limonene, and low volatility, which prevents loss due to evaporation. A ratio of 1/50 (V_{org}/V_{aq}, 2 mL of hexadecane was layered over 100 mL of culture

Table 4 The terpenoid production increased in this study and previous studies

Products	Production increased	Genes overexpressed	References
Isoprene	0.4-fold	*dxs* and *dxr*[a]	Xue et al. [16]
Phytoene	2.1-fold	*idi*	Kajiwara et al. [36]
Isoprene	2.3-fold	*dxs* and *dxr*	Zhao et al. [17]
β-Carotene	2.7-fold	*idi*	Kajiwara et al. [36]
Abietadiene	2.8-fold	*dxs*, *dxr* and *idi*	Morrone et al. [33]
Squalene	2.9-fold	*dxs* and *idi*	Ghimire et al. [37]
Limonene	**3.4-fold**	***dxs* and *idi***	**This work**
Carotenoids	3.5-fold	*dxs*, *dxr* and *idi*	Albrecht et al. [38]
Isoprene	4.0-fold	*dxs*, *ispG*[b], *ispH*[c], *idi*, *ispE*[d], *dxr*, *ispD*[e] and *ispF*[f]	Zurbriggen et al. [18]
Lycopene	4.5-fold	*idi*	Kajiwara et al. [36]
Isoprene	4.8-fold	*dxs*, *dxr* and *idi*	Lv et al. [19]
Taxadiene	15.6-fold	*dxs*, *idi ispD* and *ispF*	Ajikumar et al. [14]
Levopimaradiene	611.3-fold	*dxs*, *idi ispD* and *ispF*	Leonard et al. [15]

The rate-limiting enzymes in the MEP pathway were overexpressed in all the studies listed in this table.
[a]The gene of 1-deoxy-d-xylulose-5-phosphate reductoisomerase;
[b]The gene of (*E*)-4-hydroxy-3-methyl-but-2-enyl pyrophosphate synthase;
[c]The gene of (*E*)-4-hydroxy-3-methyl-but-2-enyl pyrophosphate reductase;
[d]The gene of 4-diphosphocytidyl-2-c-methyl-d-erythritol kinase;
[e]The gene of 2-c-methyl-d-erythrtol-4-phosphate cytidyltransferase;
[f]The gene of 2-c-methyl-d-erythritol 2, 4-cyclodiphosphate synthase.

broth) was adopted in this two-phase culture system. Limonene was extracted into the hexadecane phase, and negligible amounts of limonene were detected in the cell mass and culture broth (data not shown). As a result, limonene production was measured only from the hexadecane phase. The two-phase culture system using hydrophobic and low volatile organic solvent was a very useful and common method used in terpenoid biosynthesis [3,14,15].

Conclusions

Although the limonene production was enhanced significantly in this study, more efforts are still needed to achieve a higher production and to make it more economic. On one hand, the metabolic flux of the endogenous MEP pathway can be enhanced by overexpressing the rate-limiting enzymes from other bacteria such as *Bacillus subtilis* [17] and/or blocking the competing metabolic pathways. On the other hand, the problem of low GPPS/LS activities needs also to be addressed by directed evolution or rational protein engineering. Besides, the culture medium should be more efficient and economic, and the metabolically engineered strain should be cultivated in a bioreactor for commercial production of limonene in the future.

Competing interests
The authors declare that they have no competing interests.

Authors' contributions
F-LD has made substantial contributions to conception and design, and acquisition of data, and analysis and interpretation of data; H-LY has been involved in drafting the manuscript and revising it critically for important intellectual content; J-HX has given final approval of the version to be published; C-XL has made substantial contributions to design and has been involved in drafting the manuscript and revising it critically for important intellectual content. All authors read and approved the final manuscript.

Acknowledgements
This work was financially supported by the National Natural Science Foundation of China (No. 21276082), Ministry of Science and Technology, P. R. China (Nos. 2011CB710800), and Shanghai Commission of Science and Technology (No. 11431921600). The authors are grateful to Dr. Yunpeng Bai for proof-reading of the manuscript and to professor Hongwei Yu at Zhejiang University for his generous gift of vectors and constructive advices.

References
1. Misawa N (2011) Pathway engineering for functional isoprenoids. Curr Opin Biotechnol 22:627–633
2. Colby SM, Alonso WR, Katahira EJ, Mcgarvey DJ, Croteau R (1993) 4S-Limonene synthase from the oil glands of spearmint (*Mentha spicata*): cDNA isolation, characterization, and bacterial expression of the catalytically active monoterpene cyclase. J Biol Chem 268:23016–23024
3. Alonso-Gutierrez J, Chan R, Batth TS, Adams PD, Keasling JD, Petzold CJ, Lee TS (2013) Metabolic engineering of *Escherichia coli* for limonene and perillyl alcohol production. Metab Eng 19:33–41
4. Carter OA, Peters RJ, Croteau R (2003) Monoterpene biosynthesis pathway construction in *Escherichia coli*. Phytochemistry 64:425–433
5. Duetz WA, Bouwmeester H, Beilen JB, Witholt B (2003) Biotransformation of limonene by bacteria, fungi, yeasts, and plants. Appl Microbiol Biotechnol 61:269–277
6. Keasling JD (2010) Manufacturing molecules through metabolic engineering. Science 330:1355–1358
7. Tracy NI, ChenD CDW, Price GL (2009) Hydrogenated monoterpenes as diesel fuel additives. Fuel 88:2238–2240
8. Campos N, Rodriguez-Concepcion M, Sauret-Gueto S, Gallego F, Lois LM, Boronat A (2001) *Escherichia coli* engineered to synthesize isopentenyl diphosphate and dimethylallyl diphosphate from mevalonate: a novel system for the genetic analysis of the 2-c-methyl-d-erythritol-4-phosphate pathway for isoprenoid biosynthesis. Biochem J 353:59–67
9. Martin VJ, Pitera DJ, Withers ST, Newman JD, Keasling JD (2003) Engineering a mevalonate pathway in *Escherichia coli* for production of terpenoids. Nat Biotechnol 21:796–802
10. Pitera DJ, Paddon CJ, Newman JD, Keasling JD (2007) Balancing a heterologous mevalonate pathway for improved isoprenoid production in *Escherichia coli*. Metab Eng 9(2):193–207
11. Hunter WN (2007) The non-mevalonate pathway of isoprenoid precursor biosynthesis. J Biol Chem 282(30):21573–21577
12. Croteau RB, Davis EM, Ringer KL, Wildung MR (2005) (−)-Menthol biosynthesis and molecular genetics. Naturwissenschaften 92:562–577
13. Burke C, Croteau R (2002) Geranyl diphosphate synthase from *Abies grandis*: cDNA isolation, functional expression, and characterization. Arch Biochem Biophys 405:130–136
14. Ajikumar PK, Xiao WH, Tyo KEJ, Wang Y, Simeon F, Leonard E, Mucha O, Phon TH, Pfeifer B, Stephanopoulos G (2010) Isoprenoid pathway optimization for taxol precursor overproduction in *Escherichia coli*. Science 330:70–74
15. Leonarda E, Ajikumara PK, Thayer K, Xiao WH, Mo JD, Tidorb B, Stephanopoulosa G, Thayer K, Prather KLJ (2010) Combining metabolic and protein engineering of a terpenoid biosynthetic pathway for overproduction and selectivity control. Proc Natl Acad Sci 107:13654–13659
16. Xue J, Ahring BK (2011) Enhancing isoprene production by genetic modification of the 1-deoxy-d-xylulose-5-phosphate pathway in *Bacillus subtilis*. Appl Environ Microbiol 77:2399–2405
17. Zhao Y, Yang J, Qin B, Li Y, Sun Y, Su S, Xian M (2011) Biosynthesis of isoprene in *Escherichia coli* via methylerythritol phosphate (MEP) pathway. Appl Microbiol Biotechnol 90:1915–1922
18. Zurbriggen A, Kirst H, Melis A (2012) Isoprene production via the mevalonic acid pathway in *Escherichia coli* (bacteria). BioEnergy Research 5:814–828
19. Lv X, Xu H, Yu H (2013) Significantly enhanced production of isoprene by ordered coexpression of genes dxs, dxr, and idi in *Escherichia coli*. Appl Microbiol Biotechnol 97:2357–2365
20. Williams DC, McGarvey DJ, Katahira EJ, Croteau R (1998) Truncation of limonene synthase preprotein provides a fully active 'pseudomature' form of this monoterpene cyclase and reveals the function of the amino-terminal arginine pair. Biochemistry 37:12213–12220
21. Brosius J, Erfle M, Storella J (1985) Spacing of the −10 and −35 regions in the tac promoter: Effect on its in vivo activity. J Biol Chem 260:3539–3541
22. Brunner M, Bujard H (1987) Promoter recognition and promoter strength in the *Escherichia coli* system. EMBO J 6:3139–3144
23. Jang HJ, Yoon SH, Ryu HK, Kim JH, Wang CL, Kim JY, Oh DK, Kim SW (2011) Retinoid production using metabolically engineered *Escherichia coli* with a two-phase culture system. Microb Cell Fact 10:59
24. Zhou K, Zou R, Zhang C, Stephanopoulos G, Too HP (2013) Optimization of amorphadiene synthesis in *Bacillus subtilis* via transcriptional, translational, and media modulation. Biotechnol Bioeng 110:2556–2561
25. Song HB, Demain AL (1977) An improved colorimetric assay for polyols. Anal Biochem 81(1):18–20
26. Reiling KK, Yoshikuni Y, Martin VJ, Newman J, Bohlmann J, Keasling JD (2004) Mono and diterpene production in *Escherichia coli*. Biotechnol Bioeng 87:200–212
27. Lois LM, Rodriguez-Concepcion M, Gallego F, Campos N, Boronat A (2000) Carotenoid biosynthesis during tomato fruit development: regulatory role of 1-deoxy-d-xylulose-5-phosphate synthase. Plant J 22(6):503–513
28. Miller B, Heuser T, Zimmer W (2000) Functional involvement of a deoxy-xylulose-5-phosphate reductoisomerase gene harboring locus of *Synechococcus leopoliensisin* isoprenoid biosynthesis. FEBS Lett 481:221–226
29. Estevez JM, Cantero A, Reindl A, Reichler S, Leon P (2001) 1-Deoxy-d-xylulose-5-phosphate synthase, a limiting enzyme for plastidic isoprenoid biosynthesis in plants. J Biol Chem 276(25):22901–22909
30. Das A, Yoon SH, Lee SH, Kim JY, Oh DK, Kim SW (2007) An update on microbial carotenoid production: application of recent metabolic engineering tools. Appl Microbiol Biotechnol 77(3):505–512

31. Zaslaver A, Mayo AE, Rosenberg R, Bashkin P, Sberro H, Tsalyuk M, Surette MG, Alon U (2004) Just-in-time transcription program in metabolic pathways. Nat Genet 36:486–491

32. Nishizaki T, Tsuge K, Itaya M, Doi N, Yanagawa H (2007) Metabolic engineering of carotenoid biosynthesis in *Escherichia coli* by ordered gene assembly in *Bacillus subtilis*. Appl Environ Microbiol 73(4):1355–1361

33. Morrone D, Lowry L, Determan MK, Hershey DM, Xu M, Peters RJ (2010) Increasing diterpene yield with a modular metabolic engineering system in *E. coli*: comparison of MEV and MEP isoprenoid precursor pathway engineering. Appl Microbiol Biotechnol 85(6):1893–1906

34. Meng H, Wang Y, Hua Q, Zhang S, Wang X (2011) *In silico* analysis and experimental improvement of taxadiene heterologous biosynthesis in *Escherichia coli*. Biotechnol Bioprocess Eng 16(2):205–215

35. Yuan LZ, Rouvière PE, Larossa RA, Suh W (2006) Chromosomal promoter replacement of the isoprenoid pathway for enhancing carotenoid production in *E. coli*. Metab Eng 8(1):79–90

36. Kajiwara S, Fraser PD, Kondo K, Misawa N (1997) Expression of an exogenous isopentenyl diphosphate isomerase gene enhances isoprenoid biosynthesis in *Escherichia coli*. Biochem J 324:421–426

37. Ghimire GP, Lee HC, Sohng JK (2009) Improved Squalene Production via Modulation of the Methylerythritol 4-Phosphate Pathway and Heterologous Expression of Genes from Streptomyces peucetius ATCC 27952 *Escherichia coli*. Appl Environ Microbiol 75(22):7291–7293

38. Albrecht M, Misawa N, Sandmann G (1999) Metabolic engineering of the terpenoid biosynthetic pathway of *Escherichia coli* for production of the carotenoids β-carotene and zeaxanthin. Biotechnol Lett 21(9):791–795

19

Improving the mda-7/IL-24 refolding and purification process using optimized culture conditions based on the structure characteristics of inclusion bodies

Xiaojuan Wang, Chaogang Bai, Jian Zhang, Aiyou Sun[*], Xuedong Wang[*] and Dongzhi Wei

Abstract

Background: The melanoma differentiation-associated gene-7 (mda-7)/interleukin-24 (IL-24) can induce apoptosis in a wide variety of tumor cell types, whereas it has no toxicity in normal cells. However, recombinant human mda-7/IL-24 is difficult to obtain from *Escherichia coli* because of its insolubility.

Results: In this study, we improved the structure of inclusion bodies (IBs) by optimizing the induction temperature, pH, concentrations of inducer, and metal ion additives. Statistically designed experimental analyses of three metal ion factors were performed using the Box-Behnken design. Induction temperature of 30°C, pH 7.0, and 0.1 mM isopropyl-β-D-thiogalactopyranoside (IPTG) were selected, and the optimized levels for the factors predicted by the model comprised the following: Mg^{2+} (15.7 mM), Ca^{2+} (16.6 mM), and Mn^{2+} (3.0 mM). The optimized culture conditions improved the structure of the IBs, which was validated by scanning electron microscopy (SEM) and the increase of IBs solubility.

Conclusions: After optimization, IB solubility and renatured mda-7/IL-24 increased by 51% and 84%, respectively. This study also provided a simple purification method of specific IB washing steps. Manipulating the fermentation parameters to optimize the refolding and purification process is likely to be widely applicable to other proteins.

Keywords: *Escherichia coli*; Expression; Inclusion body; Interleukin-24; Purification; Refolding

Background

The World Cancer Report 2014 emphasized the urgent need for efficient prevention strategies to curb cancer. In 2012, the global cancer burden rose to an estimated 14 million new cases per year, which is expected to rise to 22 million per year within the next 20 years. Thus, studies are searching for the 'Holy grail' of cancer treatment, which can selectively destroy tumor cells without eliciting harmful effects on normal cells or tissues, and melanoma differentiation-associated gene-7 (mda-7)/ interleukin-24 (IL-24) appears to be a possible candidate [1]. The mda-7 is a novel tumor cell-specific apoptosis-inducing gene, and it was identified via a subtraction hybridization approach from human melanoma cells that were stimulated into growth arrest and terminal differentiation by treatment with fibroblasts of IFN and mezerein [2]. Subsequently, mda-7 has been classified as a member of the expanding IL-10 gene family and it was designated as IL-24 [3,4]. mda-7/IL-24 has apoptosis-inducing properties in a broad spectrum of human tumors [5-7]. It also has a potent 'bystander' activity [8], inhibits tumor angiogenesis [9], induces antitumor immunity, and synergizes with radiation and other chemotherapeutic agents [10]. Gene therapy of mda-7/IL-24 is currently undergoing phase II clinical trials [11,12], and it is a promising target in the field of antitumor drug research.

The *Escherichia coli* expression system is most widely used for producing large quantities of recombinant proteins [13]. However, mda-7/IL-24 protein is prone to form pyknotic inclusion bodies (IBs) when expressed in

* Correspondence: sunaiyou@ecust.edu.cn; xdwang@ecust.edu.cn
State Key Laboratory of Bioreactor Engineering, East China University of Science and Technology, 130 Meilong Road, Shanghai 200237, People's Republic of China

E. coli, which are very difficult to refold [14-16]. Solubilization and renaturation steps have an extremely significant effect on the overall process for preparation of biological active protein. Inefficient solubilization results in the reformation of aggregates during the subsequent renaturation steps [17,18]. Thus, the structural characteristics of IBs could affect their solubilization efficiency [19]. There were many studies about the refolding methods of IBs *in vitro* [20-22]. However, no previous studies have attempted to improve the structure of mda-7/IL-24 IBs by optimizing the culture conditions to facilitate the subsequent IB renaturation and purification process. In this study, we evaluated the correlation between the IB solubilization behavior and culture conditions. Furthermore, we assessed the effects of metal ions of Mg^{2+}, Ca^{2+}, and Mn^{2+} and their interactions using a statistically designed experiment (Box-Behnken design) and simplified the purification process based on the IB characteristics.

Materials and methods
Construction of pET28a-mda-7/IL-24

The entire cDNA of the human mda-7/IL-24 gene [GenBank:NM-006850] was stored in our laboratory. Two oligonucleotide primers were designed, forward (5′-GGGAAT TCCATATGGCCCAGGGCCAAGAATTCCACT-3′) and reverse (5′-CCCAAGCTTGGGTCAGAGCTTGTAGAAT TTCTG-3′), to introduce the *Nde* I and *Hin*d III restriction endonucleases (Takara Bio Inc., Otsu, Shiga, Japan), respectively. The mda-7/IL-24 gene was obtained by PCR with the primers. The resulting products were subcloned into the pET28a (+) expression vector after double digestion. The recombinant plasmid pET28a-mda-7/IL-24 was transformed into BL21 (DE3) pLysS competent cells. The sequence was confirmed by DNA sequencing (Shanghai Jieli Biotechnology Co., Ltd., Shanghai, China).

Expression of the recombinant human mda-7/IL-24

The recombinant *E. coli* BL21 (DE3) pLysS was inoculated into 250 mL Luria-Bertani (LB) medium, incubated at 37°C with shaking at 200 rpm. When the OD_{600} of the cultured medium reached 0.6, the culture was induced with isopropyl-β-D-thiogalactopyranoside (IPTG) at a final concentration of 1 mM for 4 h. Further, the cells were harvested by centrifugation at 8,000×*g* for 10 min at 4°C, and the pellets were resuspended in lysate buffer (20 mM Tris, 10 mM EDTA, pH 8.5) before disruption, where the sonication program comprised a series of 200 × 3 s bursts at 300 to 400 W power with a 6-s pause between each burst. The samples were then subjected to centrifugation at 10,000×*g* for 30 min. Protein expression was analyzed by sodium dodecyl sulfate-polyacrylamide gel electrophoresis (SDS-PAGE: 15% separating gel and 4% stacking gel).

Culture conditions

The effects of the culture conditions were studied, including the induction temperatures (18°C, 25°C, 30°C, and 37°C), pH (6.0, 7.0, 8.0, and 9.0), IPTG concentrations (0.1, 0.5, 1.0, and 1.5 mM), and the addition of different metal ions (Mg^{2+}, Fe^{3+}, Ca^{2+}, Na^+, NH_4^+, Mn^{2+}, Cu^{2+}, and K^+ at concentrations of 0.50, 1.00, 2.00, and 3.00 g/L). The corresponding compounds were $MgSO_4$, $Fe_2(SO_4)_3$, $CaCl_2$, $NaCl$, NH_4Cl, $MnSO_4·H_2O$, $CuSO_4·5H_2O$, and KCl. Each experiment was repeated twice with three replicates.

Furthermore, we optimized the concentrations of the metal ions using a Box-Behnken design [23,24] to minimize the IB solubility. The Design-Expert Software (version 8.0, Stat-Ease Inc., Minneapolis, MN, USA) was used for the experimental design and data analysis. Three variables were designed at three levels, which required 17 experiments. The minimum and maximum levels are described in Table 1. All experiments were performed in triplicate, where the response Y represented the average value. The optimal values were obtained by solving the regression equation and analyzing three-dimensional (3D) response surface plots and contour plots.

Solubility of the IBs

mda-7/IL-24 was produced in different culture conditions using 30 mL LB medium in a 250-mL shaker flask. The IBs were obtained by centrifugation after disruption. The same amount of enriched IBs (1 mg) was resuspended in 1 mL solubilization buffer (8 M urea) and vortexed for 2 h at room temperature. The IB solubility will be affected by their structures, which will result in difference in turbidity when IBs were dissolved in denaturation buffer. The turbidity of IB resuspension in the solubilization buffer was chosen as the index to reflect the solubility of IBs, and lower turbidities were associated with higher IB solubilities [25,26]. The turbidity of the suspension was measured at 480 nm.

The samples were centrifuged at 13,000×*g* for 30 min and filtrated through a 0.45-μm Millipore filter (Millipore Co., Billerica, MA, USA). The protein content was estimated by using micro-BCA kit (Beijing Solarbio Science & Technology Co., Ltd., Beijing, China).

Table 1 Coded values of variables used in Box-Behnken design

Independent variable name(g/L)	Level		
	−1	0	1
A: $MgSO_4$	1.50	2.00	2.50
B: $CaCl_2$	1.00	1.50	2.00
C: $MnSO_4$	0.50	0.75	1.00

Table 2 The components of '3 + 1' washing buffers

Washing buffer	Urea (M)	EDTA (mM)	Tris (mM)	NaCl (mM)	Triton X-100 (%)	pH of the buffer
Buffer I	2	10	20	100	1	8.5
Buffer II	4	10	20	0	2	8.5
Buffer III	2	10	20	0	1	9.5
Deionized water	0	0	0	0	0	6.5 to 7.0

Washing, denaturation, and refolding of IBs

After disruption, the enriched IBs were solubilized and refolded. mda-7/IL-24 IBs were washed using '3 + 1' washing buffers (about 1 g IBs were solubilized in 30 mL washing buffer): buffer I, buffer II, buffer III, and deionized water (Table 2). In each step, the IBs were resuspended in washing buffer and agitated for 2 h. Then, the IBs were harvested by centrifugation at 10,000×g for 30 min. After these washing steps, high-purity IBs of mda-7/IL-24 were obtained.

Further, the high-purity IBs were dissolved in a buffer solution (20 mM Tris, 10 mM EDTA, and 8 M urea, pH 9.0) and stirred overnight to facilitate denaturation of the mda-7/IL-24 protein. Renatured mda-7/IL-24 protein was achieved by dialysis (the concentration of mda-7/IL-24 was 0.10 mg/mL at the beginning of dialysis) with the decrease of urea concentrations (6, 3, 1.5, 0.75, and 0 M), where each dialysis buffer was applied for 12 h at 4°C with mild agitation. Finally, the bioactive mda-7/IL-24 protein was dialyzed in phosphate-buffered saline (PBS) (50 mM, pH 7.4).

Western blot assay

The SDS-PAGE analysis was performed using 15% SDS-polyacrylamide gels. The proteins retained in the gels were detected using Coomassie brilliant Blue R-250 or they were transferred by electrophoresis to nitrocellulose membranes for western blot analysis [27]. The membranes were probed with anti-mda-7/IL-24 polyclonal antibodies (R&D Systems, Minneapolis, MN, USA). A secondary antibody conjugated to horseradish peroxidase was used to enhance the chemiluminescence.

Scanning electron microscopy

Optimized and non-optimized IBs were also analyzed by scanning electron microscopy (SEM). The IBs after water dialysis were fixed and frozen [19] and dehydrated using a vacuum freeze dryer (Scientz-10N, Ningbo Scientz Biotechnology Co., Ltd., Ningbo City, China). Then, the samples were covered with gold, viewed, and photographed using a scanning electron microscope (S-3400N), Hitachi, Ltd., Chiyoda, Tokyo, Japan).

MTT assay

Proliferation inhibition was measured based on the 3-(4,5-dimethylazol-2-yl)-2,5-diphenyltetrazolium bromide (MTT) assay. Human breast cancer cell line MCF-7, human malignant melanoma cell line A-375, and normal human lung fibroblast (NHLF) cell line were plated onto 96-well microtiter plates (8×10^3 cells/well) in Dulbecco's modified Eagle's medium (DMEM) supplemented with 10% fetal bovine serum and incubated at 37°C with 5% CO_2. After 24 h, the cells were treated with mda-7/IL-24 at varying concentrations for 48 h and PBS as control. The surviving cells were detected by adding 20 µL MTT (5 mg/mL) and incubating for 4 h. The resulting absorbance measured at 490 nm was proportional to the number of viable cells.

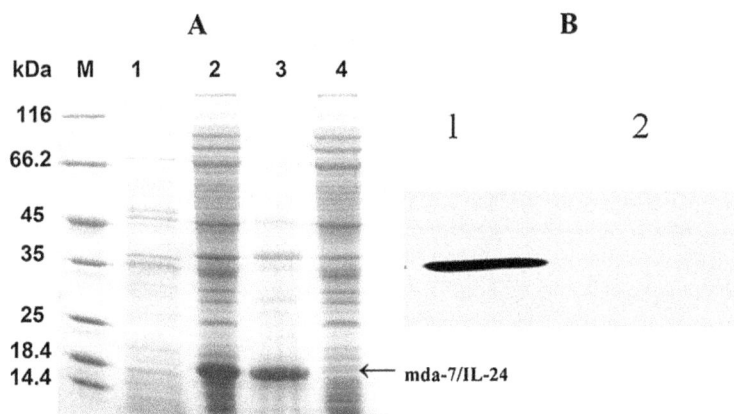

Figure 1 SDS-PAGE and western blot analysis of mda-7/IL-24. (A) SDS-PAGE analysis of mda-7/IL-24 expressed in *E. coli*. Lane M, protein molecular weight markers; lanes 1 to 2, whole cell lysate of pET28a-mda-7/IL-24 uninduced and induced; lane 3, insoluble fractions of cell lysate after induction by IPTG; lane 4, soluble fractions of cell lysate after induction by IPTG. **(B)** Western blot analysis of mda-7/IL-24 protein. Lane 1, western blot analysis of mda-7/IL-24; lane 2, blank control without anti-human mda-7/IL-24 antibody.

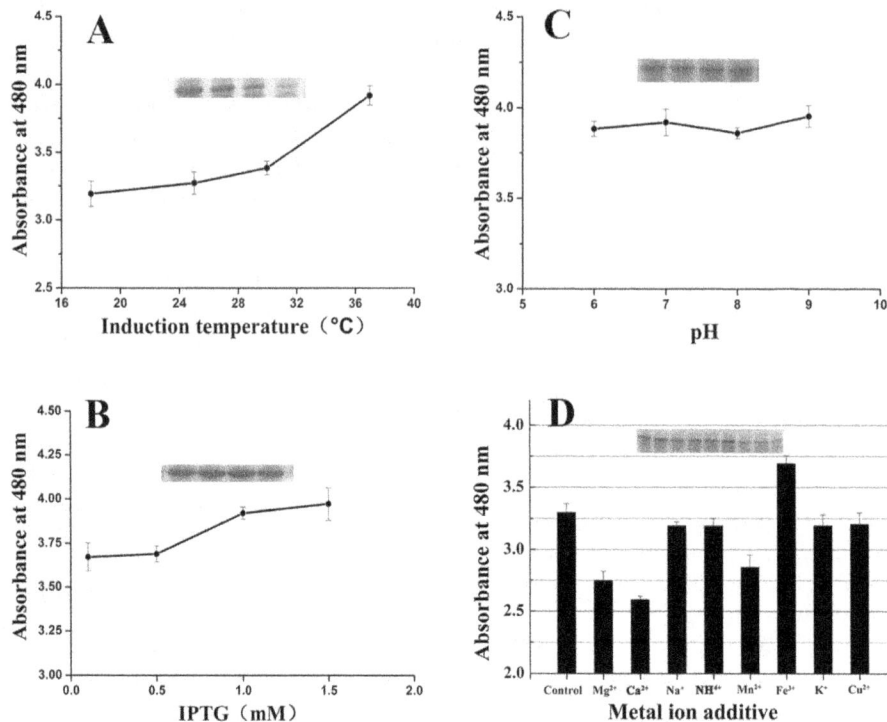

Figure 2 Influence of culture conditions on mean turbidity (±SD). (A) The effects of different induction temperatures on turbidity. **(B)** The effects of different IPTG concentrations on turbidity. **(C)** The effects of different pH values on turbidity. **(D)** The effects of different metal ions (2.00 g/L) on turbidity. The mda-7/IL-24 expressions analyzed by SDS-PAGE in corresponding conditions are placed above each graph.

Results

Expression of mda-7/IL-24

Recombinant plasmid pET28a-mda-7/IL-24 was obtained successfully, which contained a sequence of 474-bp coding sequence for 158 amino acid residues, and produced approximately 18-kDa protein (Figure 1A). The mda-7/IL-24

Table 3 Box-Behnken design of variables for process optimization

STD order	Factors coded			Response OD$_{480}$	
	A	B	C	Actual value	Predicted value
1	−1	−1	0	3.492	3.562
2	1	−1	0	2.738	2.769
3	−1	1	0	2.274	2.244
4	1	1	0	2.536	2.466
5	−1	0	−1	2.419	2.393
6	1	0	−1	2.203	2.216
7	−1	0	1	2.574	2.561
8	1	0	1	2.139	2.165
9	0	−1	−1	3.319	3.275
10	0	1	−1	1.945	2.002
11	0	−1	1	2.928	2.871
12	0	1	1	2.479	2.523
13	0	0	0	2.013	2.075
14	0	0	0	2.273	2.075
15	0	0	0	1.996	2.075
16	0	0	0	1.892	2.075
17	0	0	0	2.202	2.075

Table 4 Analysis of variance (ANOVA) for quadratic polynomial model

Source	Sum of squares	df	Mean square	F value	Prob > F	Remark
Model	3.3000	9	0.3700	20.8900	0.0003	Significant
A	0.1633	1	0.1633	9.3161	0.0185	Significant
B	1.3146	1	1.3146	74.9956	<0.0001	Significant
C	0.0068	1	0.0068	0.3905	0.5519	
AB	0.2581	1	0.2581	14.7218	0.0064	Significant
AC	0.0120	1	0.0120	0.6840	0.4355	
BC	0.2139	1	0.2139	12.2027	0.0101	Significant
A^2	0.1295	1	0.1295	7.3897	0.0298	Significant
B^2	1.0926	1	1.0926	62.3285	<0.0001	Significant
C^2	0.0291	1	0.0291	1.6607	0.2835	
Residual	0.1227	7	0.0175			
Lack of fit	0.0238	3	0.0079	0.3208	0.8113	Not significant
Pure error	0.0989	4	0.0247			
Cor total	3.4189	16				

protein comprised up to 30% of the total cellular protein, most of which was present as IBs. The western blot analysis confirmed that mda-7/IL-24 was recognized by the anti-IL-24 monoclonal antibody (Figure 1B).

Culture condition optimization by response surface methodology

The factors that are known to affect the protein solubility include plasmid copy number, culture media, bacterial

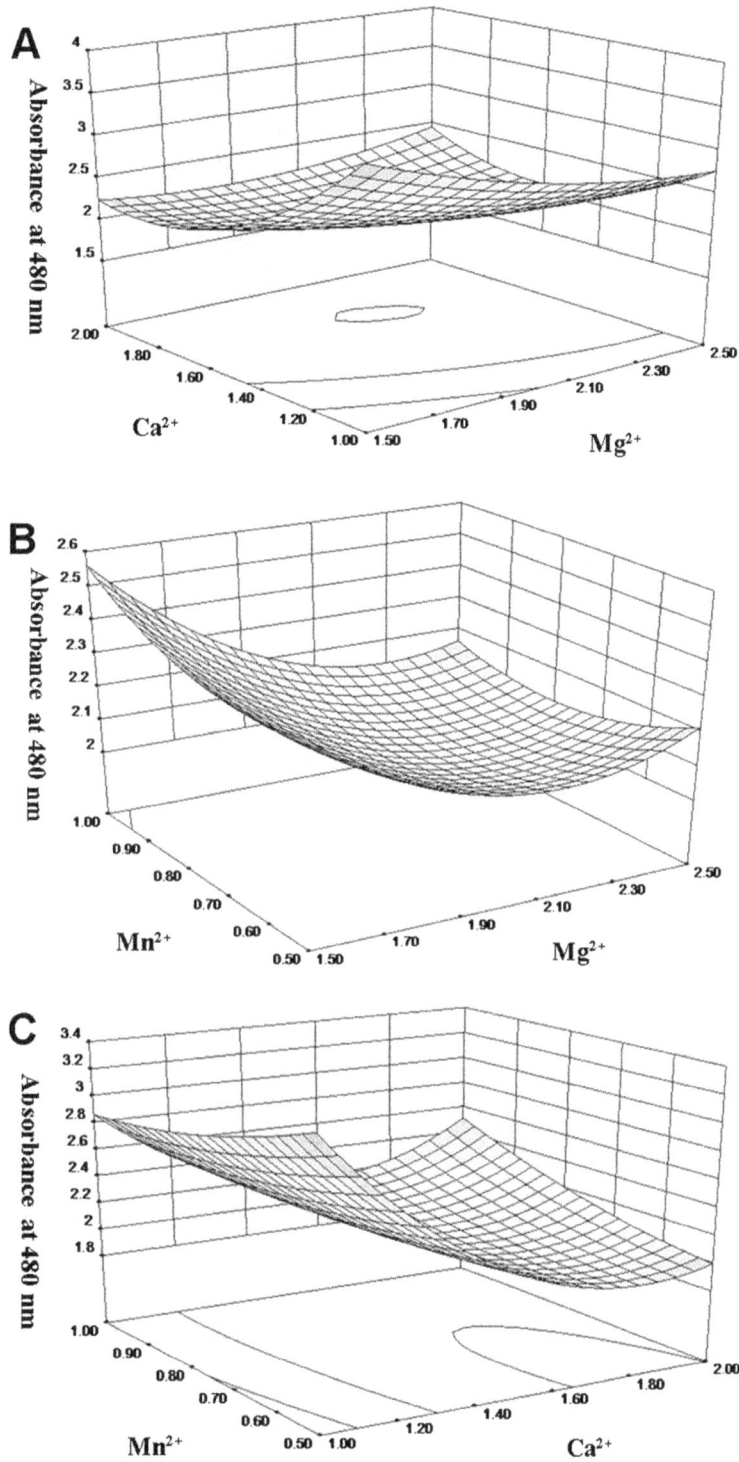

Figure 3 Three-dimensional response surface curve showing the effects of metal ion additives. (A) Mg^{2+} and Ca^{2+}. **(B)** Mg^{2+} and Mn^{2+}. **(C)** Ca^{2+} and Mn^{2+}.

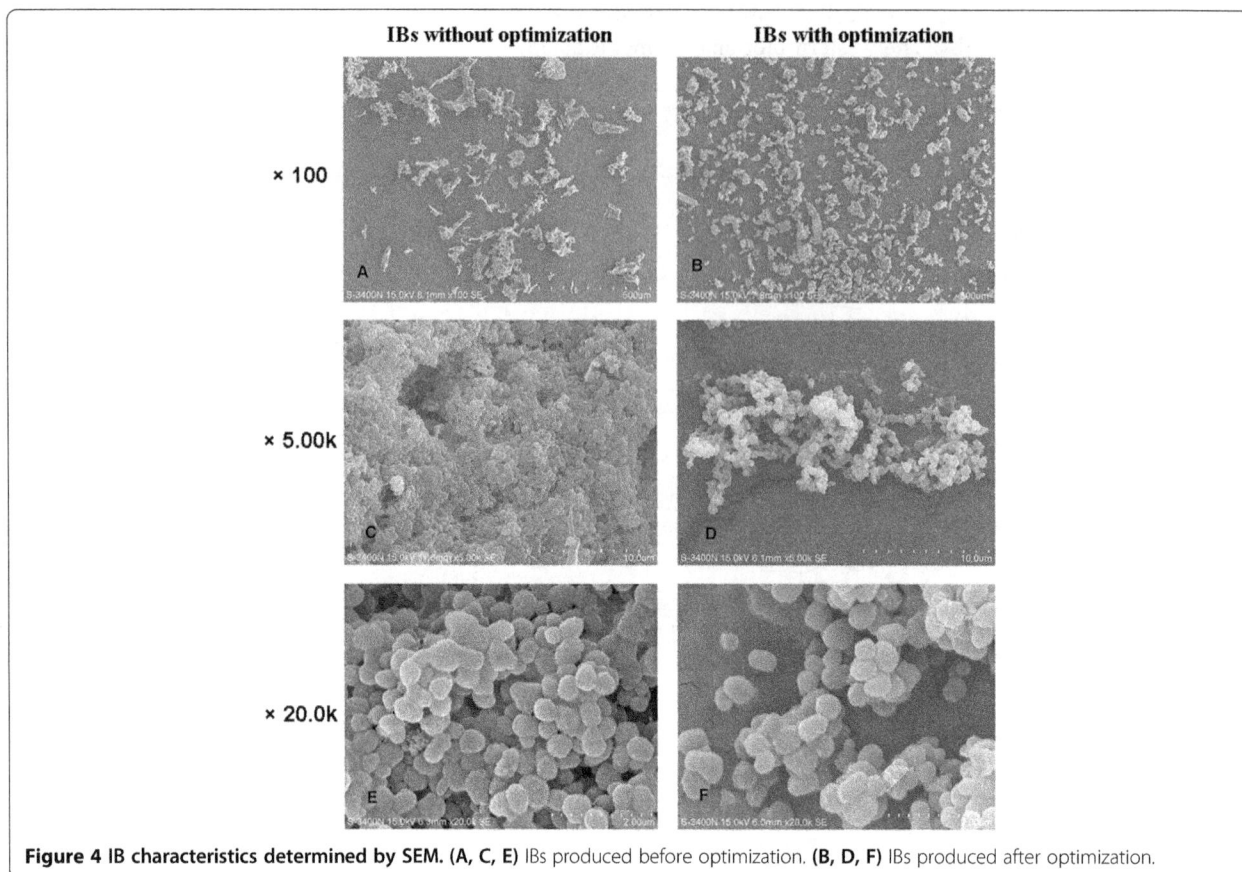

Figure 4 IB characteristics determined by SEM. **(A, C, E)** IBs produced before optimization. **(B, D, F)** IBs produced after optimization.

strain, induction temperature, induction time, and concentration of the inducer [28,29]. We tried to improve the structure of mda-7/IL-24 IBs by optimizing the culture conditions to facilitate the subsequent IB renaturation and purification process. In this study, turbidity was used as a quantitative index of IB solubility. The effects of culture conditions on turbidity are shown in Figure 2. Conditions which could significantly reduce the turbidity of IB suspension without markedly reducing the mda-7/IL-24 protein expression were chosen. As a result, we selected an induction temperature of 30°C, pH 7.0, and 0.1 mM IPTG. Culture additives of Mg^{2+}, Ca^{2+}, and Mn^{2+} had positive effects on reducing the turbidity compared with the control, where the turbidity was decreased by 11%, 16%, and 7.0%, respectively. In contrast, Na^+, NH_4^+, K^+, and Cu^{2+} had no obvious effects, whereas Fe^{3+} increased the turbidity greatly.

To achieve the maximum solubilization of mda-7/IL-24 IBs, we optimized the IB turbidity using the Box-Behnken design, which included metal ion additives, i.e., (A) Mg^{2+}, (B) Ca^{2+}, and (C) Mn^{2+}. The design matrix and the results of the 17 experiments conducted using the Box-Behnken design are described in Table 3, which comprised 12 trials plus five center points. The results obtained were analyzed by ANOVA using the Design-

Expert Software (version 8.0 Stat-Ease Inc.). The regression model used was as follows:

$$Y(OD_{480}) = +16.38667 - 4.28765A - 10.34305B \\ - 3.77760C + 1.01600AB - 0.43800AC \\ + 1.85000BC + 0.70160A^2 \\ + 2.03760B^2 + 1.33040C^2$$

(1)

Table 5 The simplified purification process of the washing steps

Washing step	Total protein (mg)	mda-7/IL-24 (mg)	mda-7/IL-24 yield (%)	mda-7/IL-24 purity (%)
Before washing	105	68	-	65
Buffer I	84	61	89	73
Buffer II	65	56	92	86
Buffer III	53	49	88	92
Deionized water	50	48	98	96

The wet weight of the *E. coli* cells was 3.75 g from 1 L culture medium, and 105 mg IBs was obtained after disruption. mda-7/IL-24 comprised up to 30% of the total protein. The total target protein lost 29% during the washes and 8.4% during the dialysis. The final purity of mda-7/IL-24 was about 95%.

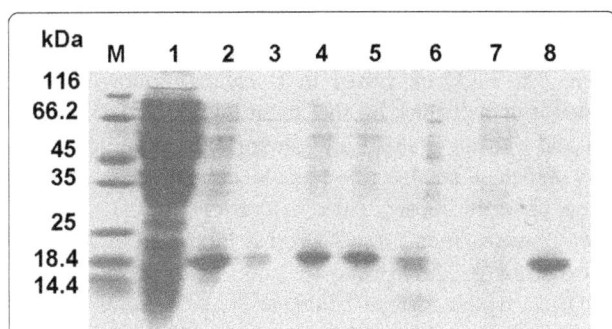

Figure 5 The SDS-PAGE analysis of mda-7/IL-24 after '3 + 1' washing steps. Lane M, protein molecular weight markers; lane 1, supernatant of bacteria lysate; lanes 2, 4, and 5, pellets washed by washing buffers I, II, III, respectively; lanes 3, 6, 7, supernatants after each wash steps (I, II, and III wash steps, respectively); lane 8, the final pellets after '3 + 1' washing steps.

Table 6 Solubilization and refolding before and after optimization

Optimization items	Solubilization		Refolding	
	Absorbance at 480 nm	Increase of solubility (%)	Renatured mda-7/IL-24 (%)	Increase of refolding (%)
Control	3.921	-	13	-
Temperature	3.383	14	16	22
IPTG	3.672	6	14	9.0
pH	Not significant	-	-	-
Metal ion additives	1.94	51	23	84

The results of the ANOVA for the regression model are shown in Table 4. The model F value of 20.89 implied that the model was significant, i.e., there was only a 0.03% chance that a 'model F value' this large could occur due to noise. The coefficient of determination (R^2) was used to check the goodness of fit of the model (Table 4), where R^2 was 0.9641. The value of the adjusted coefficient of determination (adjusted $R^2 = 0.9180$) agreed reasonably well with the predicted R^2 (0.8434). The lack-of-fit value for the regression of Equation 1 was not significant (0.8113), thereby indicating that the model equation was suitable for predicting the turbidity with any combination of values for the variables. 'Adeq Precision' measures the signal-to-noise ratio, where a ratio greater than four is considered to be desirable [30]. The 'Adeq Precision' ratio of 15.336 obtained in the present study indicated an adequate signal. Thus, this model can be used to navigate the design space.

The effects of the addition of metal ions, i.e., (A) Mg^{2+}, (B) Ca^{2+}, and (C) Mn^{2+}, on IB solubilization were visualized using 3D response surface and contour plots in (Figure 3). The 3D graphs were generated based on the response of pairwise combinations of the three factors while keeping the other at its optimum level. The graphs are shown to demonstrate the effects of various factors on the turbidity. The interaction effect of Mg^{2+} and Ca^{2+} is shown in Figure 3A. Ca^{2+} exhibited a quadratic response and it was the major factor, while Mg^{2+} also exhibited a quadratic response. These data suggested that the Mg^{2+} and Ca^{2+} had significant effects on turbidity, and they also had significant interaction effects on each other. The interaction effect of Mg^{2+} and Mn^{2+} is shown in Figure 3B. The effects of Mn^{2+} and its quadratic response were not significant, while Mg^{2+} and Mn^{2+} also had non-significant interaction effects on each other. However, Ca^{2+} and Mn^{2+} had significant interaction effects on each other, where the

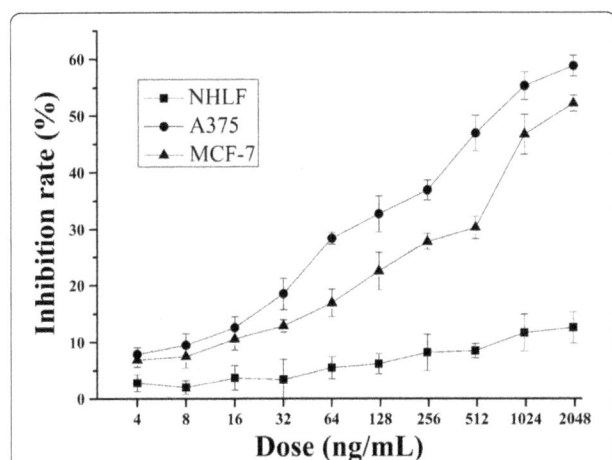

Figure 6 The activity of renatured mda-7/IL-24 (±SD). mda-7/IL-24 inhibits the survival of A375 and MCF-7 cells in a concentration-dependent manner, whereas it has little effect on the growth of NHLFs.

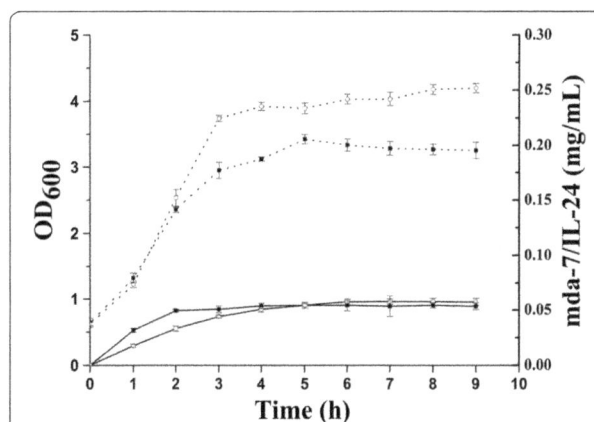

Figure 7 Bacteria growth (OD$_{600}$) and mda-7/IL-24 protein production (mg/mL) before and after optimization. E. coli were grown in a 30 mL culture medium. Dotted lines represent values of mean OD$_{600}$ (±SD), without optimization (...■...) and with optimization (...○...). Solid lines represent values of protein production (±SD), without optimization (—■—) and after optimization (—○—).

effect of Ca^{2+} was enhanced in the presence of Mn^{2+} (Figure 3C).

A validation was performed using the conditions predicted by the model. The optimized levels predicted by the model were as follows: Mg^{2+} (15.7 mM), Ca^{2+} (16.6 mM), and Mn^{2+} (3.0 mM). The predicted turbidity was 1.92 (suspension measured at 480 nm), and the actual obtained was 1.94. The exact correlation between the experimental and predicted values validated this model.

IB characteristics determined by SEM
The structures of the IBs were also examined by SEM (Figure 4). It was clear that IBs without optimization were much larger, and the particles stuck together tightly (Figure 4A,C,E). While IBs with optimization had a more regular shape and appeared to be much looser, even single particles could be observed (Figure 4B,D,F).

Simplified purification process
In our study, IBs comprised up to approximately 100% of the total mda-7/IL-24 even after these optimizations. mda-7/IL-24 IBs were barely dissolved in both 8 M urea and 6 M guanidine hydrochloride. A '3 + 1' IB washing process was designed. High concentration of urea and Triton X-100 could effectively remove the impurities, while the pyknotic mda-7/IL-24 IBs were not completely removed in the washing steps. The purity of mda-7/IL-24 after '3 + 1' steps was 96% (Table 5, Figure 5). This was a simplified purification process, which required no additional procedures, such as affinity chromatography purification.

Bioactivity of mda-7/IL-24
The results showed that mda-7/IL-24 had negligible inhibitory effects on NHLF cells, while it had cytotoxicity effects on MCF-7 cells and A375 cells in a concentration-dependent manner, and the 50% inhibitory concentrations were about 1.0 and 0.7 μg/mL, respectively (Figure 6).

Conclusions
Recombinant protein is prone to form IBs when expressed in *E. coli*. Recombinant human mda-7/IL-24 IBs are difficult to refold due to its low solubility [14]. Some studies tried to obtain soluble expression or easy refolding of mda-7/IL-24 using genetic methods [15,16,31]. However, no previous study has attempted to improve the structure of mda-7/IL-24 IBs by optimizing the culture conditions to facilitate the subsequent IB renaturation and purification. In this study, culture conditions were optimized, which increased the solubility of mda-7/IL-24 IBs by approximately 51% (Table 6). We speculated that the optimized culture conditions improved the IB structure, which was validated by SEM and the increase of solubility.

The renatured protein almost had the same activity as the commercial product of mda-7/IL-24 (Cloud-Clone Corp., Houston, TX, USA) expressed in *E. coli*. Therefore, we confirmed the activities of the renatured mda-7/IL-24 protein and did not evaluate the proportion of the active proteins. Efficient solubilization could contribute to the performance of the subsequent renaturation steps [17,18], and the renatured mda-7/IL-24 protein increased by 84% in our experiments (Table 6).

The '3 + 1' washing steps could gradually remove the miscellaneous proteins and also make the IB particles loose, which benefited the subsequent refolding process. As a result, the purity of mad-7/IL-24 was 96% after '3 + 1' washing steps. Therefore, the '3 + 1' washing approach based on these IB structural characters was a simplified purification process.

With the optimized conditions, the mda-7/IL-24 formation rate was slower during the early stage and faster during the late stage (Figure 7). This may be due to some proteins being synthesized in the early stage that facilitated the correct folding of the mda-7/IL-24 protein, thereby allowing the IBs to possess a greater proportion of native-like secondary structure [32]. Therefore, the structure of IBs was improved and the solubility of the IBs increased. It is an important method to optimize the renaturation and purification process by manipulating the fermentation parameters [33]. This approach may also be applied to the expression of other proteins with similar structural characteristics.

Competing interests
The authors declare that they have no competing interests.

Authors' contributions
XJW and CB carried out the molecular genetic studies, participated in the sequence alignment, and drafted the manuscript. XJW and JZ participated in the design of the study and performed the statistical analysis. AS, XDW, and DW conceived of the study, participated in its design and coordination, and helped draft the manuscript. All authors read and approved the final manuscript.

Acknowledgements
This work was supported by a grant from the Ministry of Science and Technology (National Major Science and Technology Projects of China (No. 2012ZX09304009)).

References
1. Fisher PB (2005) Is mda-7/IL-24 a "magic bullet" for cancer? Cancer Res 65:10128–10138
2. Jiang H, Lin JJ, Su ZZ, Goldstein NI, Fisher PB (1995) Subtraction hybridization identifies a novel melanoma differentiation associated gene, mda-7, modulated during human melanoma differentiation, growth and progression. Oncogene 11:2477–2486
3. Caudell EG, Mumm JB, Poindexter N, Ekmekcioglu S, Mhashilkar AM, Yang XH, Retter MW, Hill P, Chada S, Grimm EA (2002) The protein product of the tumor suppressor gene, melanoma differentiation-associated gene 7, exhibits immunostimulatory activity and is designated IL-24. J Immunol 168:6041–6046
4. Chada S, Sutton RB, Ekmekcioglu S, Ellerhorst J, Mumm JB, Leitner WW, Yang HY, Sahin AA, Hunt KK, Fuson KL, Poindexter N, Roth JA, Ramesh R,

Grimm EA, Mhashilkar AM (2004) mda-7/IL-24 is a unique cytokine-tumor suppressor in the IL-10 family. Int Immunopharmacol 4:649–667

5. Su Z-Z, Madireddi MT, Lin JJ, Young CS, Kitada S, Reed JC, Goldstein NI, Fisher PB (1998) The cancer growth suppressor gene mda-7 selectively induces apoptosis in human breast cancer cells and inhibits tumor growth in nude mice. Proc Natl Acad Sci 95:14400–14405

6. Ekmekcioglu S, Ellerhorst J, Mhashilkar AM, Sahin AA, Read CM, Prieto VG, Chada S, Grimm EA (2001) Down-regulated melanoma differentiation associated gene (mda-7) expression in human melanomas. Int J Cancer 94:54–59

7. Lebedeva IV, Sarkar D, Su Z-Z, Kitada S, Dent P, Stein C, Reed JC, Fisher PB (2003) Bcl-2 and Bcl-x(L) differentially protect human prostate cancer cells from induction of apoptosis by melanoma differentiation associated gene-7, mda-7/IL-24. Oncogene 22:8758–8773

8. Su Z, Emdad L, Sauane M, Lebedeva IV, Sarkar D, Gupta P, James CD, Randolph A, Valerie K, Walter MR (2005) Unique aspects of mda-7/IL-24 antitumor bystander activity: establishing a role for secretion of mda-7/IL-24 protein by normal cells. Oncogene 24:7552–7566

9. Ramesh R, Mhashilkar AM, Tanaka F, Saito Y, Branch CD, Sieger K, Mumm JB, Stewart AL, Boquio A, Dumoutier L (2003) Melanoma differentiation-associated gene 7/interleukin (IL)-24 is a novel ligand that regulates angiogenesis via the IL-22 receptor. Cancer Res 63:5105–5113

10. Dent P, Yacoub A, Grant S, Curie DT, Fisher PB (2005) mda-7/IL-24 regulates proliferation, invasion and tumor cell radiosensitivity: a new cancer therapy? J Cell Biochem 95:712–719

11. Tong AW, Nemunaitis J, Su D, Zhang Y, Cunningham C, Senzer N, Netto G, Rich D, Mhashilkar A, Parker K (2005) Intratumoral injection of INGN 241, a nonreplicating adenovector expressing the melanoma-differentiation associated gene-7 (mda-7/IL24): biologic outcome in advanced cancer patients. Mol Ther 11:160–172

12. Azab B, Dash R, Das SK, Bhutia SK, Shen XN, Quinn BA, Sarkar S, Wang XY, Hedvat M, Dmitriev IP (2012) Enhanced delivery of mda-7/IL-24 using a serotype chimeric adenovirus (Ad. 5/3) in combination with the Apogossypol derivative BI-97C1 (Sabutoclax) improves therapeutic efficacy in low CAR colorectal cancer cells. J cell Physio 227:2145–2153

13. Chen R (2012) Bacterial expression systems for recombinant protein production: E. coli and beyond. Biotechnol Adv 30:1102–1107

14. Ma Q, Jiang H, Liu K, Zhu J, Zheng X (2004) Cloning of human interleukin 24 (mda-7/IL-24) gene and its fusion protein expression in E. coli. Jun Shi Yi Xue Ke Xue Yuan Yuan Kan 28:221–224

15. Yang J, Zhang W, Liu K, Jing S, Guo G, Luo P, Zou Q (2007) Expression, purification, and characterization of recombinant human interleukin 24 in Escherichia coli. Protein Expression Purif 53:339–345

16. Liu J-J, Zhang B-F, Yin X-X, Pei D-S, Yang Z-X, Di J-H, Chen F-F, Li H-Z, Xu W, Wu Y-P (2012) Expression, purification, and characterization of RGD-mda-7, a HIS-TAGGED mda-7/IL-24 mutant protein. J Immunoassay Immunochem 33:352–368

17. Singh SM, Panda AK (2005) Solubilization and refolding of bacterial inclusion body proteins. J Biosci Bioeng 99:303–310

18. Petrides D, Cooney CL, Evans LB, Field RP, Snoswell M (1989) Bioprocess simulation: an integrated approach to process development. Comput Chem Eng 13:553–561

19. Kang H, Sun AY, Shen YL, Wei DZ (2007) Refolding and structural characteristic of TRAIL/Apo2L inclusion bodies from different specific growth rates of recombinant Escherichia coli. Biotechnol Prog 23:286–292

20. Chura-Chambi RM, Cordeiro Y, Malavasi NV, Lemke LS, Rodrigues D, Morganti L (2013) An analysis of the factors that affect the dissociation of inclusion bodies and the refolding of endostatin under high pressure. Process Biochem 48:250–259

21. Hayashi M, Iwamoto S, Sato S, Sudo S, Takagi M, Sakai H, Hayakawa T (2013) Efficient production of recombinant cystatin C using a peptide-tag, 4AaCter, that facilitates formation of insoluble protein inclusion bodies in Escherichia coli. Protein Expression Purif 88:230–234

22. Senthil K, Gautam P (2010) Expression and single-step purification of mercury transporter (merT) from Cupriavidus metallidurans in E. coli. Biotechnol Lett 32:1663–1666

23. Ferreira SC, Bruns R, Ferreira H, Matos G, David J, Brandao G, da Silva EP, Portugal L, Dos Reis P, Souza A (2007) Box-Behnken design: an alternative for the optimization of analytical methods. Anal Chim Acta 597:179–186

24. Mao X, Shen Y, Yang L, Chen S, Yang Y, Yang J, Zhu H, Deng Z, Wei D (2007) Optimizing the medium compositions for accumulation of the novel FR-008/Candicidin derivatives CS101 by a mutant of Streptomyces sp. using statistical experimental methods. Process Biochem 42:878–883

25. Singh SM, Upadhyay AK, Panda AK (2008) Solubilization at high pH results in improved recovery of proteins from inclusion bodies of E. coli. J Chem Technol Biotechnol 83:1126–1134

26. Patra AK, Mukhopadhyay R, Mukhija R, Krishnan A, Garg LC, Panda AK (2000) Optimization of inclusion body solubilization and renaturation of recombinant human growth hormone from Escherichia coli. Protein Expression Purif 18:182–192

27. Welinder C, Ekblad L (2011) Coomassie staining as loading control in Western blot analysis. J Proteome Res 10:1416–1419

28. Correa A, Oppezzo P (2011) Tuning different expression parameters to achieve soluble recombinant proteins in E. coli: advantages of high-throughput screening. Biotechnol J 6:715–730

29. Vincentelli R, Cimino A, Geerlof A, Kubo A, Satou Y, Cambillau C (2011) High-throughput protein expression screening and purification in Escherichia coli. Methods 55:65–72

30. Anderson MJ, Whitcomb PJ (2005) RSM simplified: optimizing processes using response surface methods for design of experiments. Productivity Press, Oxon

31. Xiao B, Li W, Yang J, Guo G, Mao XH, Zou QM (2009) RGD-IL-24, a novel tumor-targeted fusion cytokine: expression, purification and functional evaluation. Mol Biotechnol 41:138–144

32. Shivu B, Seshadri S, Li J, Oberg KA, Uversky VN, Fink AL (2013) Distinct β-sheet structure in protein aggregates determined by ATR–FTIR spectroscopy. Biochemistry 52:5176–5183

33. Shen YL, Xia XX, Zhang Y, Liu JW, Wei DZ, Yang SL (2003) Refolding and purification of Apo2L/TRAIL produced as inclusion bodies in high-cell-density cultures of recombinant Escherichia coli. Biotechnol Lett 25:2097–2101

20

High-throughput system for screening of *Monascus purpureus* high-yield strain in pigment production

Jun Tan, Ju Chu[*], Yonghong Wang, Yingping Zhuang and Siliang Zhang

Abstract

Background: An economical and integrated high-throughput primary screening strategy was developed for high-aerobic microbe *Monascus purpureus* cultivation. A novel and effective mixture culture method was proposed and used to realize the whole mutant library being high-throughput screened after mutagenesis.

Results: The good correlation of fermentation results between differing-scale cultivations confirmed the feasibility of utilizing the 48-deep microtiter plates (MTPs) as a scale-down tool for culturing high-aerobic microbes. In addition, the fluid dynamics of 24-, 48-, and 96-deep MTPs and 500-mL shake flask were studied respectively using the computational fluid dynamic (CFD) tool ANSYS CFX 11.0 to get better understanding of their turbulent regimes.

Conclusions: The by-product citrinin production had no significant change while the pigment production had improved. As a result, the high-yield strain T33-6 was successfully screened out and the pigment was more than 50% higher than that of the parental strain in the shake flask.

Keywords: *Monascus*; High-throughput screening; Mixture cultivation; Computational fluid dynamics

Background

Natural colorants derived from plants and microorganisms have recently gained popularity over synthetic coloring agents because of undesirable toxic effects including mutagenicity and potential carcinogenicity in some cases [1]. *Monascus* pigments produced by *Monascus purpureus* (*M. purpureus*) fermentation are now used in processed seafood, sausages, and sauces in Asia to replace some food additives such as cochineal, potassium nitrate, and nitrites [2-4]. However, *M. purpureus* is also a toxigenic strain that can produce the nephrotoxic and hepatotoxic mycotoxin citrinin which greatly limits the wide application of the *Monascus*-related products [5-7]. *Monascus* pigments are safe, especially characterized with high protein adhesion and heat stability and can be used in a wide pH range [8,9]. Hence, the investigations on the improvement of the pigment production are of commercial importance in the food coloration market.

The high-yield strain is a key factor for the production of fermented foods. However, most traditional fermentation experiments were performed in shake flasks. The extraction of *Monascus* pigments was time consuming and inconvenient for treatment of large numbers of samples. A large amount of material was required, so it was uneconomic and impractical for high-throughput system. Therefore, a simple high-throughput primary screening strategy that can culture and evaluate a large number of isolates simultaneously with a fair degree of accuracy and reproducibility becomes imperative [10]. Microtiter plates (MTPs) are attractive for high-throughput cultivation due to its small working volume and high degree of parallelization [11,12]. Oxygen mass transfer rate (OTR) and mixing studies in MTPs have been extensively reported [13,14]. It proved that the square-shaped vessel can provide higher OTR. Forty-eight-deep-well microtiter plates (48-deep MTPs) were used in the present high-throughput strategy, and the correlation was good with shake flasks. Furthermore, the success of any strain improvement program mainly depends on the number of positive isolates that can be screened after mutagenic treatment [15]. However, only a small part of mutants

* Correspondence: juchu@ecust.edu.cn
State Key Laboratory of Bioreactor Engineering, East China University of Science and Technology, 130 Meilong Road, P.O. Box 329#, Shanghai 200237, People's Republic of China

selected randomly have the chance to be screened in traditional screening method [16,17].

In this paper, the mixture cultivation was effectively proposed to realize the whole mutant library of high-aerobic microbe *M. purpureus* being high-throughput screened after mutagenesis integrating the subsequent high-throughput pigment microassay. The high-yield mixture in one well of the microtiter plate was screened out first. The subsequent isolation of the desirable high-yield colonies was further selected and screened from this specific high-yield mixture. The method greatly increased screening efficiency compared with conventional method.

Methods
Microorganisms

M. purpureus M-403 used in this study was a parent strain preserved in our laboratory at 4°C on peptone-malt extract agar slants (g/L): peptone 30.0, malt extract 12.0, and agar 20.0.

Seed medium adjusted to pH 4.2 by lactic acid was composed of (g/L): starch 30.0, $NaNO_3$ 2.5, KH_2PO_4 2.5, $MgSO_4 \cdot 7H_2O$ 1.3, soybean meal 10.0, and corn steep liquor 15.0. Fermentation medium adjusted to pH 4.5 was composed of (g/L): glucose 40.0, peptone 5.0, $NaNO_3$ 3.0, KH_2PO_4 1.5, and $MgSO_4 \cdot 7H_2O$ 1.0. Media were sterilized by autoclaving for 25 min at 121°C.

Culture of microorganism
Traditional culture in shake flask

M. purpureus M-403 and its mutants were precultured in 500-mL shake flask containing 50 mL seed medium at 33°C on a rotary shaker at 220 rpm for 42 h (50-mm shaking diameter, 30/300, ZHWY-3212, Zhicheng Analytical Instrument Manufacturing Co., Ltd., Shanghai, China). Nine percent (v/v) precultures were inoculated in 500-mL shake flask containing 50 mL fermentation medium at 33°C, 220 rpm. The samples were taken every 6 h. All the experiments were carried out in triplicate at least.

Microculture in 48-deep MTPs

M. purpureus M-403 and its mutants were inoculated with toothpicks in 48-deep MTPs containing 1.0 mL seed medium at 33°C on a rotary shaker at 220 rpm for 42 h. Nine percent (v/v) precultures were inoculated in the corresponding wells of new 48-deep MTPs containing 1.0 mL fermentation medium. The other culture conditions were the same as shake flasks.

High-throughput system for screening
Mutagenesis procedure

M. purpureus (M-403) was subjected to mutagenesis using UV irradiation and LiCl treatment. M-403 was grown on agar slant at 33°C for 7 days. A piece of 2 × 2 × 3 (cm × cm × mm) slant was dug into the homogenizer, gently homogenized with 10 mL sterile water to obtain homogeneous hypha suspension. All the hypha suspension was collected to expose to UV irradiation at 254 nm for 3 min at a distance of 20 cm in petri dish, and the isometric 10 mL 1.2% LiCl solution was added in the petri dish, mixed, and contacted for 20 min, which resulted in 90% kill rate at least.

Mixture cultivation and high-throughput screening

Firstly, screen the high-yield mixture from the mutant library: 20 mL treated hypha suspension was equally allotted into the 48-deep MTPs, roughly 30 plates were needed. Fifteen-microliter mutant suspension called a mixture, was cultured in a well containing 1.0 mL seed medium shaking at 220 rpm for 42 h at 33°C. Nine percent (v/v) seed culture was inoculated in another sterile corresponding well of 48-deep MTPs at 220 rpm for 48 h; the remaining seed cultures were preserved in 4°C refrigerator with parafilm until assay results were obtained. The pigment was determined by high-throughput determination. It can be speculated that the more high-yield mutants dominate in the mixture, the higher pigment production could be observed from the corresponding wells. The subsequent isolation of the desirable colonies from the high-yield mixture should have much higher probability than that only isolated from the parts of treated hypha suspension directly by traditional dilution-plate method.

Secondly, isolate the high-yield strain from the high-yield mixture: The preserved seed of high-yield mixture, approximately 300 μl, was inoculated on the fresh slant for 7 days at 33°C. Single colonies were isolated by a traditional dilution-plate method. The integrated procedure chart is shown in Figure 1.

Extraction of pigments
Traditional determination

Extraction of pigments from 500-mL shake flask: 1 mL of the culture broth was mixed with 9 mL of 70% (v/v) ethanol in a test tube, rest for 15 min; the supernatant was filtered by a filter paper (45 μm, Xinhua Paper Industry Co., Ltd., Hangzhou, China); and the pigment concentrations were measured by a spectrophotometer at 517 nm after dilution. The pigment production (U/mL) = OD_{517} × dilution factor.

High-throughput determination

Extraction of pigments from 48-deep MTPs: 0.1 mL of the culture broth was mixed with 0.9 mL of 70% (v/v) ethanol in new 48-deep MTPs, rest for 15 min and the supernatant was obtained by MTP centrifuge (TDZ5-WS, Jiachuang Biotechnology Co., Ltd., Shanghai, China; 3,000×g, 5 min) and pipetted 300 μl into each well of a new 96-well microtiter plate. The pigment concentrations were measured by a microplate reader (Multiskan Go, Thermo Fisher Scientific Inc., Massachusetts, MA, USA) at 517 nm after dilution. The pigment production (U/mL) = OD_{517} × dilution factor.

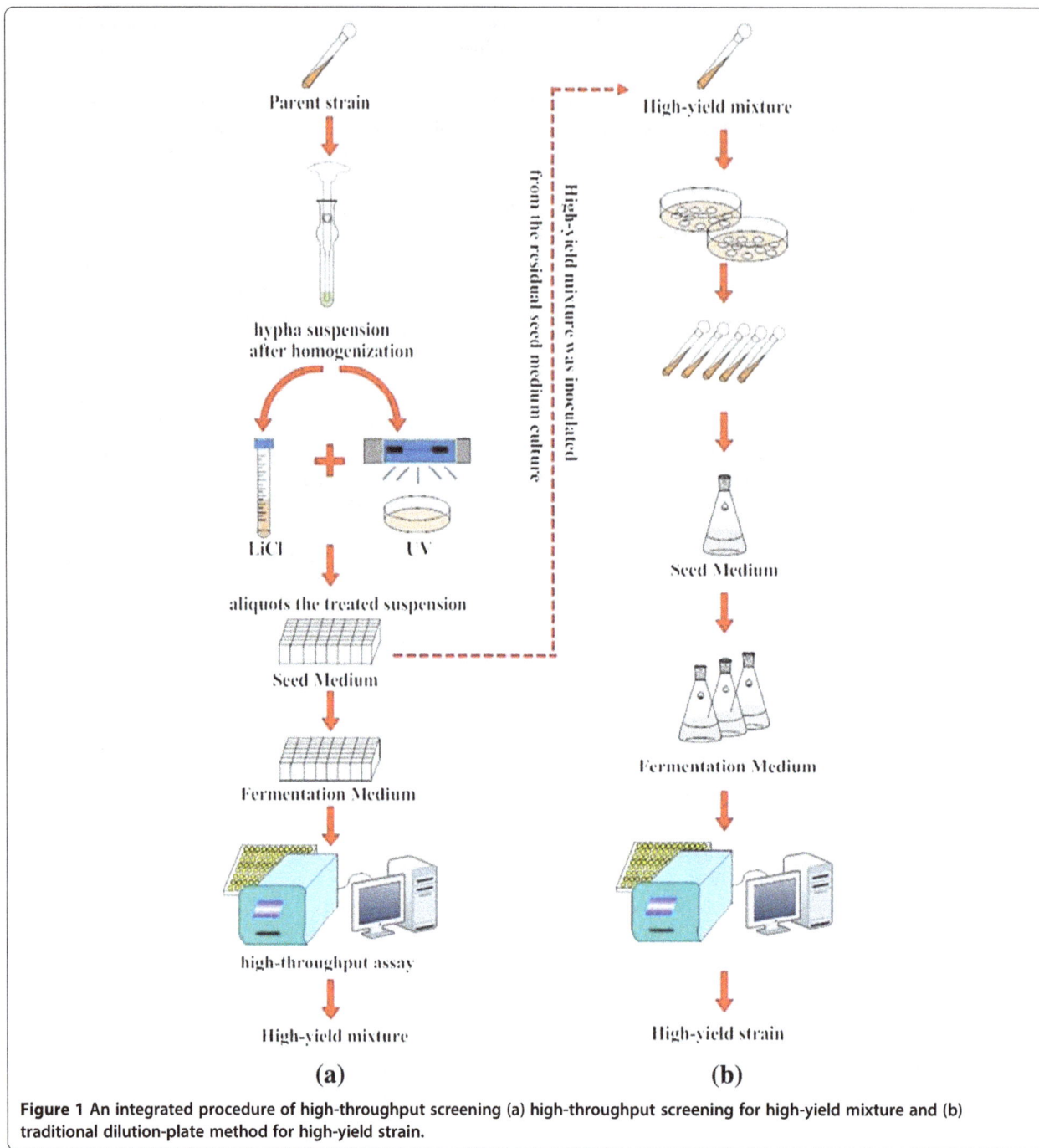

Figure 1 An integrated procedure of high-throughput screening (a) high-throughput screening for high-yield mixture and (b) traditional dilution-plate method for high-yield strain.

Citrinin analysis

Pure citrinin was purchased from Sigma Chemical Company, St. Louis, MO, USA. All solvents used were high-performance liquid chromatography (HPLC) grade. Citrinin was extracted by methanol, 1 mL fermentation broth mixed fiercely with the same volume of methanol for 10 min and centrifuged at $15,000 \times g$ for 20 min. The supernatant was determined by HPLC (Agilent 1100 Series, Shanghai, China). The samples were separated on a TSK-gel ODS-100S C18 column at a flow rate of 1 mL/min, 28°C, automatic injector

50 µl. Fluorescence detection was performed with a FP-920 fluorescence detector (Jasco Corp., Tokyo, Japan) set at 331-nm excitation wavelength and 500-nm emission wavelength. The mobile phase consisted of acetonitrile/water (35/65, v/v), and the pH was 2.5 adjusted by phosphoric acid.

$k_L a$ value measurements

The sulfite oxidation method [18] was employed to determine the $k_L a$ in MTPs and shake flask. The sulfite solution system contained 0.5 M sodium sulfite, 10^{-7} M cobalt sulfate, 0.012 M Na_2HPO_4/NaH_2PO_4 phosphate buffer (pH 8.0), and 2.4×10^{-5} M bromothymol blue. The initial pH was adjusted to 8.0 using 2.0 M sulfuric acid. The pH of the sulfite solution remained at an almost constant value of pH 8.0 before dropping sharply to a value of pH 5.0 at the end of the reaction. The time of the color changes from blue to yellow was measured accurately to calculate the $k_L a$ value of MTPs and shake flask. All the experiments were performed at 220 rpm, five parallelisms at least.

Results and discussion
Performance evaluation of MTPs and shake flask
Computational fluid dynamic (CFD) methods based on the Navier-Stokes equation have become a powerful tool to predict the fluid flow and homogenization in stirred tanks [19,20]. Many papers published the simulation methods of mixing [21], mass transfer [22], and shear environment [23] in different kinds of stirred tanks. The fluid dynamics of 24-, 48-, and 96-deep MTPs and 500-mL shake flask were studied respectively using the CFD tool ANSYS CFX 11.0 to get better understanding of their turbulent regimes. Power consumption, shear strain, and oxygen mass transfer coefficient were calculated by numerical simulation, and the results were verified by experimental data. The parameter values are shown in Table 1. The shear strain presented the shear force in MTPs and shake flask, the result revealed that the shear effect of 24- and 48-deep MTPs was superior or comparable to shake flask, while 96-deep MTPs were unsatisfied. Figure 2 shows the distribution of shear force in MTPs and shake flask. The maximum shear force distributed near the wall of 24- and 48-deep MTPs and 500-mL shake flask, while the maximum shear force mainly distributed at the bottom of the 96-deep MTPs. It indicated that the liquid mixing was insufficient in the 96-deep MTPs, directly influenced the oxygen transfer. The properties of 24- and 48-deep MTPs were comparable with 500-mL shake flask. Furthermore, the power consumption of 24- and 48-deep MTPs was only 5% of the shake flask. These parameters illustrated that the MTPs were very practical equipment for high-throughput screening. Considering the oxygen transfer capability and the screening throughput, 48-deep MTPs were selected for M. purpureus cultivation.

High-throughput screening system
The general method for supernatant collection was time consuming and laborious for the sample treatment, especially for the samples from 48-deep MTPs which was impractical for rapid and high-throughput screening. In order to develop a simple, rapid, and high-throughput screening strategy, it is preferential to establish a simple and accurate analytical method. Therefore, 48-deep MTPs, MTP centrifuge, and microplate reader were employed to solve the problem. Thousands of samples could be determined simultaneously in a short time, meanwhile, could also save more material consumption.

Commercially available standard pigment (98.2%, Sigma) was used to determine the maximum absorbance by microplate reader from 300 to 700 nm; the pigments have maximum absorbance at 517 nm.

Twenty mutants derived from M. purpureus M-403 were fermented in shake flasks, the Monascus pigment was determined by a spectrophotometer and microplate reader. The data based on a microplate reader were higher than the results from a spectrophotometer (Figure 3a), but the high correlation coefficient (0.95) was obtained by statistical analysis. This high-throughput assay could be effectively used for the determination of Monascus pigment instead of traditional methods. The comparisons of differing-scale cultivations are assessed as shown in Figure 4b. The strain M-403 was fermented in shake flasks and in 48-deep MTPs, respectively, both the Monascus pigments were determined by a microplate reader. The comparison results between differing-scale cultivations were generally good (Figure 3b). Both Monascus pigments in 48-deep MTPs and shake flasks were increased to their maximum values at 42 h and were kept constant until 78 h. Although the pigment productions in shake flasks were higher than that in 48-deep MTPs, the results had the same tendency. The correlation coefficient was 0.98 by statistical analysis. The data suggested that the 48-deep MTPs could be used as a scale-down tool for high-aerobic microbe screening applications. The data also illustrated that this scale-down system could be effectively used to determine pigment production. Thus, we had developed an integrated high-throughput screening strategy combined the 48-deep MTPs culture system with the microplate assay. All experiments were performed in triplicate.

Table 1 The performance evaluation of MTPs and shake flask

	Filling volume (mL)	$k_L a$ (1/h)[a]	P/V (W/m³)[b]	SSR (1/s)[c]	$k_L a$ (1/h)[d]
24-deep MTPs	1.2	176	79.1	76.8	148.8
48-deep MTPs	0.7	170	73.8	85.3	144.3
96-deep MTPs	0.5	71.4	7.3	27.8	58.77
500 mL SF	30	130.1	1348.9	61.4	150.7

[a]$k_L a$: oxygen mass transfer coefficient calculated by numerical simulation; [b]P/V: power consumption per unit volume; [c]SSR: shear strain; [d]$k_L a$: oxygen mass transfer coefficient determined by experiment.

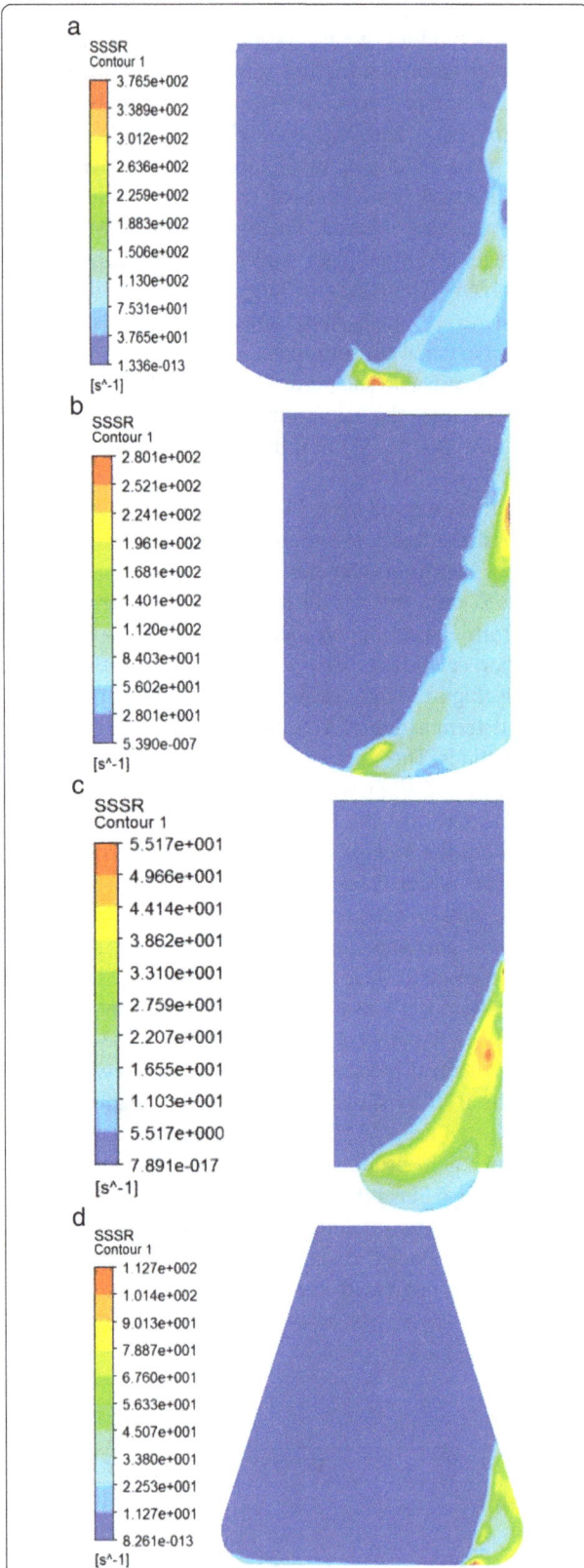

Figure 2 Shear force simulation profile of MTP and shake flask: (a) 24-deep MTPs, (b) 48-deep MTPs, (c) 96-deep MTPs, and (d) shake flask.

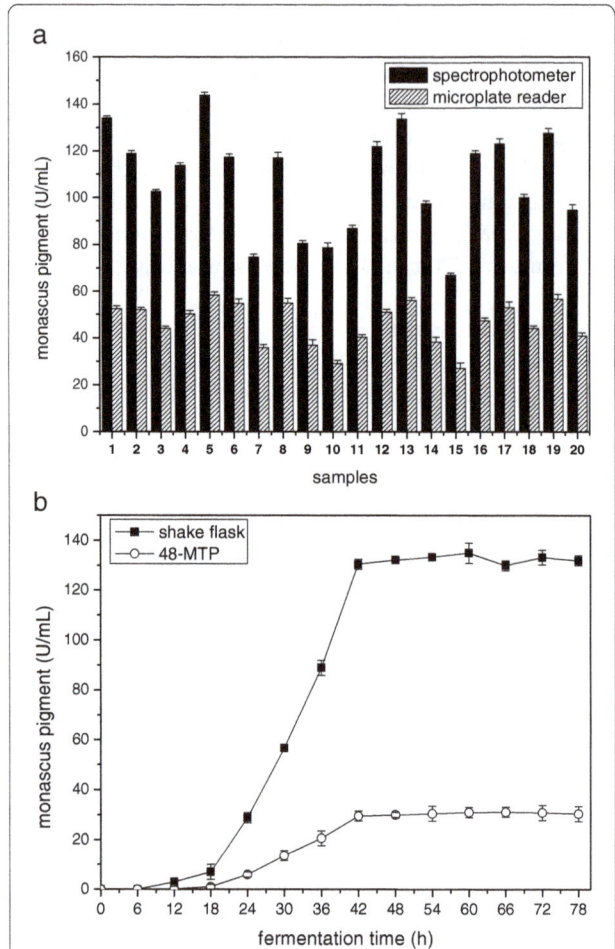

Figure 3 *Monascus* pigment determined by spectrophotometer and microplate reader and comparison of *Monascus* pigment fermented by *M. purpureus* M-403. *Monascus* pigment determined by spectrophotometer and microplate reader **(a)** and comparison of *Monascus* pigment fermented by *M. purpureus* M-403 between shake flasks and 48-deep MTPs **(b)**.

High-throughput screening for *M. purpureus* high-yield strain
Figure 4a shows the results of mutagenic treatment, out of 1,400 mixtures in 48-deep MTPs, 125 mixtures could be determined the pigment production by a microplate reader. The dashed line stands for the pigment production of parent strain M-403 (29.2 U/mL). The pigment productions of 20% mixtures were comparable to that of the parent strain; 15 mixtures were found to possess pigment production at least 20% higher than that of the parent strain. Among these mixtures, the production of the high-yield mixture T33 was 44.5 U/mL, nearly 50% higher than that of the parent strain. The corresponding seed culture well of the T33 preserved in 4°C refrigerator was traced back, and all the remaining seed culture of T33 was inoculated on the fresh slant timely for further single colony isolation.

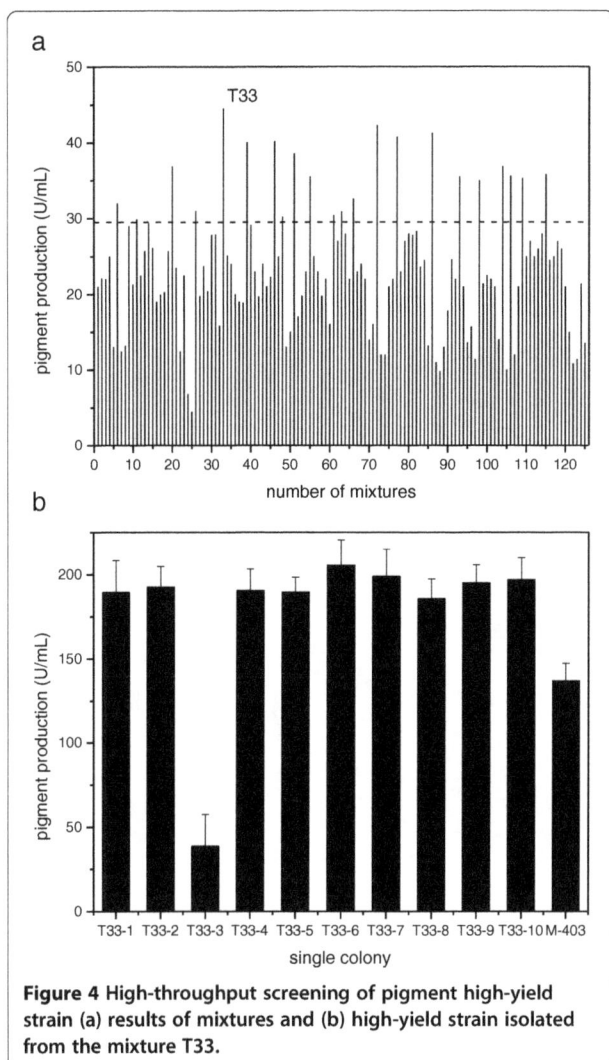

Figure 4 High-throughput screening of pigment high-yield strain (a) results of mixtures and (b) high-yield strain isolated from the mixture T33.

Table 2 Amount of citrinin detected in the samples of *M. purpureus* M-403 and T33-6

	Parent strain M-403	High-yield strain T33-6
Citrinin (mg/L)	12.7	10.5
RSD (%)	4.21	3.72

RSD, relative standard deviation.

Comparison of citrinin production between mutant *M. purpureus* T33-6 and parent strain *M. purpureus* M-403

The parent strain and the high-yield strain were fermented in shake flasks; the amounts of citrinin are summarized in Table 2. The amounts of citrinin produced by *M. purpureus* M-403 and T33-6 were 12.7 mg/L and 10.5 mg/L, respectively; there was no significant change while the pigment production was improved, so this high-yield strain was satisfactory.

Conclusions

In summary, a novel high-throughput screening method was established for screening the whole mutant library of *M. purpureus* after mutagenesis. The high-yield strain T33-6 was screened out by mixture cultivation successfully. The production was 205.5 U/mL fermented in shake flasks, nearly 50% higher than that of parent strain. The amount of citrinin had no significant change produced by *M. purpureus* T33-6 (11.9 mg/L) and the parent strain (12.7 mg/L) while the pigment production had improved, so this high-throughput strategy was feasible. High-throughput technology played an important role in strain improvement screening, significantly increased throughput and reduced assay volume, efficaciously solve the low-throughput problem encountered in conventional strain screening. High-throughput screening field will continue to be promising and dynamic in the future.

Competing interests
The authors declare that they have no competing interests.

Authors' contributions
JT, JC, and YHW were in charge of the experiments and paper writing. YPZ and SLZ directed the study as the tutors. All authors read and approved the final manuscript.

Acknowledgements
This work was financially supported by a grant from the National High Technology Research and Development Program of China (863 program), No. 2012AA021201; the Major State Basic Research Development Program of China (973 program), No. 2012CB721006; National Major Scientific and Technological Special Project, No. 2012YQ15008709; and Research Fund for the Doctoral Program of Higher Education of China, No. 20110074110015.

Among the colonies isolated from the high-yield mixture T33, ten available colonies were selected by our practical experience and their submerged cultivations in shake flasks were carried out with the parent strain (M-403) as the control (Figure 4b). Except colony T33-3, the productions of other nine colonies were almost 190.0 U/mL averagely, higher than that of the parent strain (137.0 U/mL). There was only a small part of low-yield mutants mixed in the high-yield mixture, so there was a high possibility to isolate the desired strain from the high-yield mixture compared with traditional shake flask screening. Repeated studies showed that the production of single colony T33-6 remained at the same level after several consecutive generations by submerged fermentation, three parallelisms for shake flasks at least. To the best of our knowledge, this is the first report to buildup an integrated HTS strategy for *M. purpureus* applications.

References
1. Mapari SAS, Nielsen KF, Larsen TO, Frisvad JC, Meyer AS, Thrane U (2005) Exploring fugal biodiversity for production of water-soluble pigments as potential natural food colorants. Curr Opin Biotechnol 16:231–238

2. Fabre CE, Santerre AL, Loret MO, Baberian R, Paresllerin A, Goma G, Blanc PJ (1993) Production and food applications of the red pigment of *Monascus rubber*. J Food Sci 58:1099–1110

3. Lee YK, Chen DC, Chauvatcharin S, Seki T, Yoshida T (1995) Production of *Monascus* pigments by a solid-liquid state culture method. J Ferment Bioeng 2:21–26

4. Kim HL, Kim HJ, Oh HJ, Shin CS (2002) Morphology control of *Monascus* cell and scale up of pigment fermentation. Process Biochem 38:649–655

5. Blanc PJ, Loret MO, Goma G (1995) Production of citrinin by various species of *Monascus*. Biotechnol Lett 17:291–294

6. Pisareva E, Savov V, Kujumdzieva A (2005) Pigments and citrinin biosynthesis by fungi belonging to genus *Monascus*. Z Naturforsch C 60:116–120

7. Blanc PJ, Laussac JP, Lebars J, Lebars P, Loret MO, Pareilleux A, Prome D, Prome JC, Santerre AL, Goma G (1995) Characterization of monascidin A from *Monascus* as citrinin. Int J Food Microbiol 27:201–213

8. Li ZQ, Guo F (2004) A further studies on the species of *Monascus*. Mycosystema 23:1–6

9. Kim JY, Kim HJ, Oh JH, Lee IH (2010) Characteristics of *Monascus* sp. isolated from *Monascus* fermentation products. Food Sci Biotechnol 19:1151–1157

10. Huang L, Wei PL, Zang R, Xu ZN, Cen PL (2010) High-throughput screening of high-yield colonies of *Rhizopus oryzae* for enhanced production of fumaric acid. Ann Microbiol 60:287–292

11. Betts JI, Doig SD, Baganz F (2006) Characterization and application of a miniature 10 mL stirred-tank bioreactor, showing scale-down equivalence with a conventional 7 L reactor. Biotechnol Prog 22(3):681–688

12. Gao H, Liu M, Zhou XL, Liu JT, Zhuo Y, Gou ZX, Xu B, Zhang WQ, Liu XY, Luo AQ, Zheng CS, Chen XP, Zhang LX (2010) Identification of avermectin-high-producing strains by HTS methods. Appl Microbiol Biotechnol 85:1219–1225

13. Harms P, Kostov Y, French JA, Soliman M, Anjanappa M, Ram A, Rao G (2006) Design and performance of a 24-station high throughput microbioreactor. Biotechnol Bioeng 93:6–13

14. Chen A, Chitta R, Chang D, Amanullah A (2008) Twenty-four well plate miniature bioreactor system as a scale-down model for cell culture process development. Biotechnol Bioeng 102(1):148–160

15. Buchs J (2001) Introduction to advantages and problems of shaken cultures. Biochem Eng J 7:91–98

16. Du Toit EA, Rautenbach M (2000) A sensitive standardised micro-gel well diffusion assay for the determination of antimicrobial activity. J Microbiol Methods 42:159–165

17. Kumar MS, Kumar PM, Sarnaik HM, Sadhukhan AK (2000) A rapid technique for screening of lovastatin-producing strains of *Aspergillus terreus* by agar plug and *Neurospora crassa* bioassay. J Microbiol Methods 40:99–104

18. Hermann R, Lehmann M, Büchs J (2003) Characterization of gas-liquid mass transfer phenomena in microtiter plates. Biotechnol Bioeng 81(2):178–186

19. Bujalski JM, Jaworski Z, Bujalski W, Nienow AW (2002) The influence of the addition position of a tracer on CFD simulated mixing times in a vessel agitated by a Rushton turbine. Chem Eng Res Des 80:824–831

20. Lane GL, Schwarz MP, Evans GM (2005) Numerical modelling of gas-liquid flow in stirred tanks. Chem Eng Sci 60:2203–2214

21. Min J, Gao ZM, Shi LT (2005) CFD simulation of mixing in a stirred tank with multiple hydrofoil impellers. Chin J Chem Eng 13:583–588

22. Hristov HV, Mann R, Lossev V, Vlaev SD (2004) A simplified CFD for three dimensional analysis of fluid mixing, mass transfer and bioreaction in a fermenter equipped with triple novel geometry impellers. Food Bioprod Process 82:21–34

23. Raimondi MT, Moretti M, Cioffi M, Giordano C, Boschetti F, Lagana K, Pietrabissa R (2006) The effect of hydrodynamic shear on 3D engineered chondrocyte systems subject to direct perfusion. Biorheology 43:215–222

21

Separation of ursodeoxycholic acid by silylation crystallization

Xiaolei Ma and Xuejun Cao[*]

Abstract

Background: Ursodeoxycholic acid is an important clinical drug in the treatment of liver disease. In our previous work, ursodeoxycholic acid was prepared by electroreduction of 7-ketolithocholic acid. The separation of ursodeoxycholic acid from the electroreduction product (47% (*w/w*) ursodeoxycholic acid) by silylation crystallization is described herein.

Results: *N,N*-dimethylformamide was used as the solvent, whereas hexamethyldisilazane was the reaction agent. The optimal material ratio of electroreduction product/*N,N*-dimethylformamide/hexamethyldisilazane was found to be 1:10:2 (*w/v/v*). The reaction proceeded for 2 h at 60°C, and the corresponding silylation derivative was separated by crystallization and pure ursodeoxycholic acid was recovered by 5% acid hydrolysis at 50°C for 0.5 h. The maximum recovery and purity of ursodeoxycholic acid were 99.8% and 99.5%, respectively.

Conclusion: Ursodeoxycholic acid with high purity and high recovery can be prepared directly. The developed method offers a potential application for large-scale production of ursodeoxycholic acid.

Keywords: Ursodeoxycholic acid; Silylation; Separation

Background

Ursodeoxycholic acid (3α, 7β-2-hydroxy-5β-cholanic acid, UDCA) is an important clinical drug used in the treatment of liver disease, such as gallstones [1], alcoholic fatty liver [2], nonalcoholic fatty liver [3], viral hepatitis [4], primary biliary cirrhosis [5], primary sclerosing cholangitis [6], and cholestatic [7].

UDCA was originally separated from the black bear [8]. In our previous work, UDCA was prepared by electroreduction of 7-ketolithocholic acid (3α-hydroxy-7-oxo-5β-cholanic acid, 7K-LCA) [9]. UDCA and its epimer chenodeoxycholic acid (3α, 7α-2-hydroxy-5β-cholanic acid, CDCA) were both reduction products of 7K-LCA; thus, the product of this electrochemistry conversion was a mixture of 7K-LCA, UDCA, and CDCA, which are difficult to separate. This problem limits its application in the production of UDCA. The structures of UDCA, 7K-LCA, and CDCA are shown in Figure 1.

Researchers have focused on the separation and purification of UDCA. Guillemette et al. [10] described the purification of UDCA by reacting an aqueous alkali metal salt solution of UDCA in the presence of chloroform with an acid to recover crystalline UDCA. Bonaldi et al. [11] prepared high-purity UDCA, starting from cholic acid by forming the tris-trimethylsilyl derivative acid thereof, reducing the acid by the Wolff-Kishner method into UDCA, and the total impurities were less than 1.3%. Xu et al. [12] separated UDCA from its isomeric mixture using a core-shell molecular imprinting polymer, and the separation factor of the molecular imprinting polymer with acrylamide for UDCA was 2.20. Tian et al. [9] produced UDCA by catalytic transfer hydrogenation of 7K-LCA with Raney nickel, then UDCA was purified by column chromatography and recrystallized. Ninety-seven percent UDCA was obtained via this last method.

---

* Correspondence: caoxj@ecust.edu.cn
State Key Laboratory of Bioreactor Engineering, Department of Bioengineering, East China University of Science & Technology, 130 Meilong Rd., Shanghai 200237, China

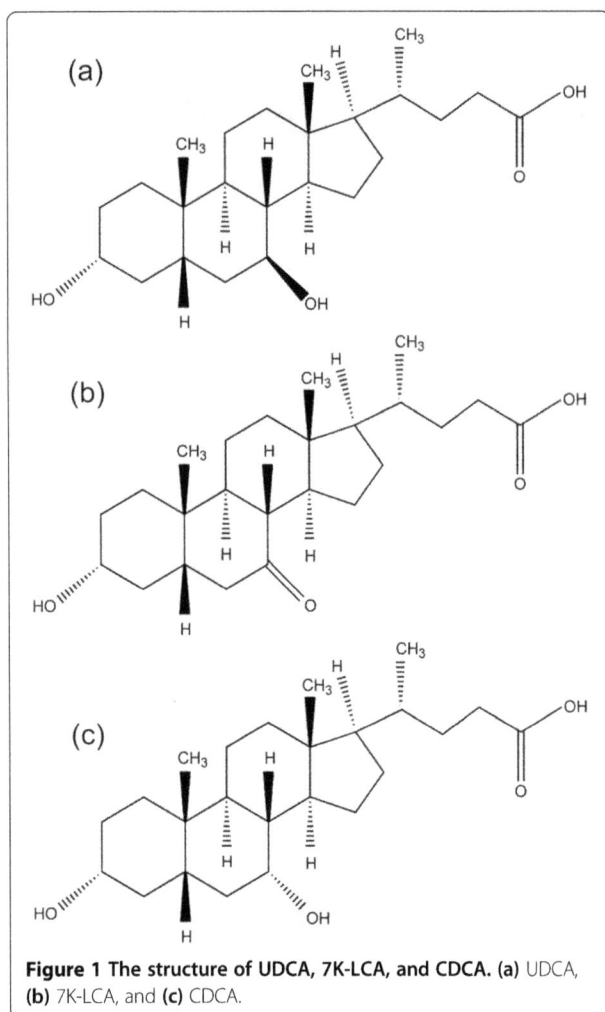

Figure 1 The structure of UDCA, 7K-LCA, and CDCA. (a) UDCA, **(b)** 7K-LCA, and **(c)** CDCA.

Methods

General methods

The standard samples of UDCA, CDCA, and silylating reagents were purchased from Aladdin Chemistry Co. Ltd (Shanghai, China). 7K-LCA was prepared according to our previous work [9]. Acetonitrile and methanol were of high-performance liquid chromatography (HPLC) grade and purchased from Shanghai Xingke Biochemistry Co. Ltd (Shanghai, China). All other reagents were of analytical grade. HPLC (LC-20A, Shimadzu Corporation, Kyoto, Japan) with a UV detector (SPD-20A) using a C18 column (Welchrom-C18, 4.6 × 150 mm, 5 µm, Welch Materials Inc. Shanghai, China) was used for quantification of the reaction products. Electron impact ionization time-of-flight mass spectrometry (EI-TOF-MS) (Micromass GCTTM, Micromass UK Ltd, Lancas, UK) and Fourier transform infrared spectroscopy (FTIR) (Magna-IR 550, Thermo Nicolet Ltd, Wisconsin, USA) were used for the characterization of the product and provided by the Analysis and Test Center, East China University of Science and Technology.

Analytical methods

(1) HPLC was used to analyze the product. The mobile phase was a mixture of phosphate acid buffer (pH 3.0) and acetonitrile (50:50, v/v) at a flow rate of 1.0 ml/min at 25°C. Detection was performed using a UV detector at 208 nm. UDCA was quantified by an external standard.
(2) EI-TOF-MS was used to identify the molecular weight of the derivatives.
(3) FTIR spectra with KBr pellets were used to analyze the structure of the product and the standard UDCA.

Preparation of UDCA

The synthesis procedure followed Yuan's method [23]. UDCA was synthesized in a divided electrolytic cell by direct electroreduction of 7K-LCA. A titanium ruthenium mesh electrode was used as the anode, and a high-purity lead plate was used as the cathode. Under the optimized process conditions, the content of UDCA was 47%.

Silylation crystallization

Silylation crystallization experiments were carried out as below. One gram of the electroreduction product (47% (w/w) of UDCA) was added into a 100-ml conical flask containing 10 ml N,N-dimethylformamide (DMF) at 30°C, and following the dissolution of the material under magnetic stirring at 150 rpm, 1 ml of HMDS was added into the solution. The conical flask was sealed with thread seal tape. The reaction was carried out for

Silylation is a powerful tool to improve the production process and the quality of the product in modern pharmaceutical and organic synthesis, and has been applied in increasing the volatility [13], changing the solubility in organic solvents [14], and the protection of sensitive functional groups such as hydroxyl and carboxyl moieties [15]. Silylation of alcohols and phenols with hexamethyldisilazane (HMDS) has been achieved using various types of catalysts [16-21]. The reaction conditions were nearly neutral, and the corresponding silyl ethers yields were high. Mormann et al. [22] reported the silylation of cellulose by HMDS in liquid ammonia and the reaction gave high degrees of silylation.

In this work, UDCA was isolated from the electroreduction product with analogues by silylation crystallization and UDCA. The process afforded UDCA with high purity and high recovery via a direct method.

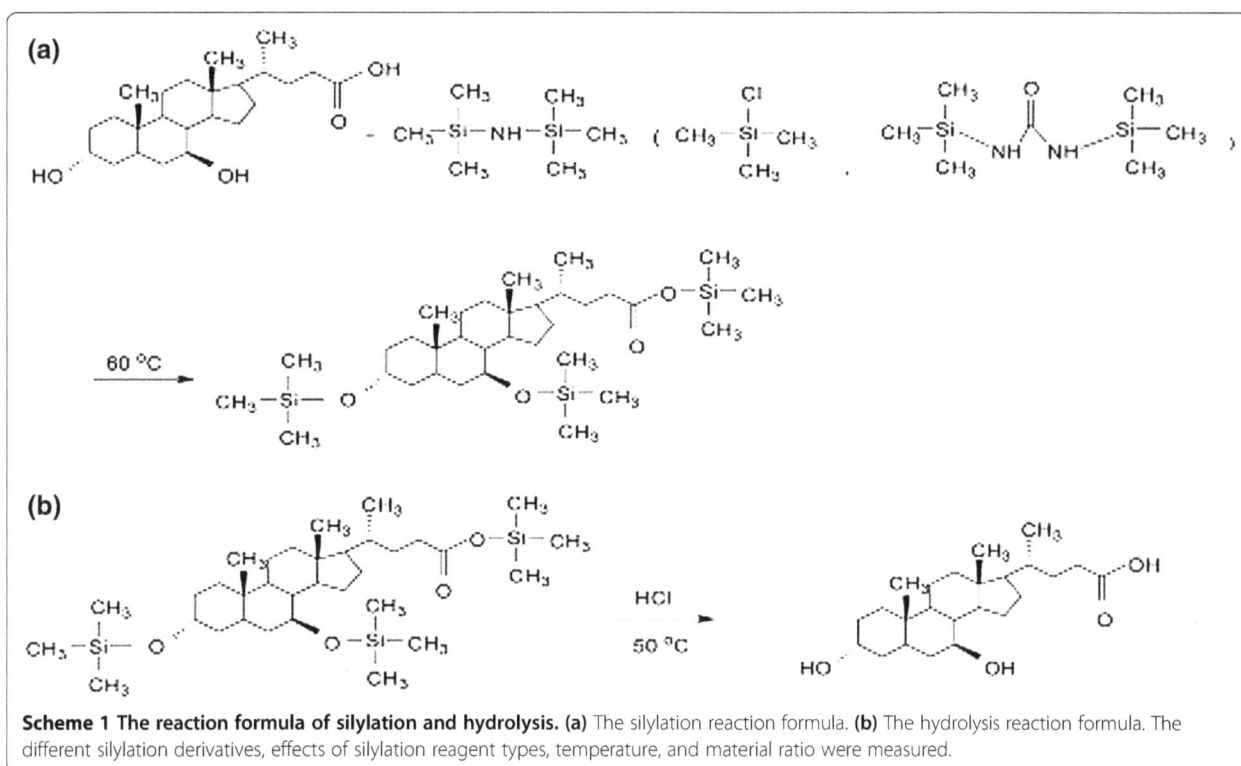

Scheme 1 The reaction formula of silylation and hydrolysis. (a) The silylation reaction formula. **(b)** The hydrolysis reaction formula. The different silylation derivatives, effects of silylation reagent types, temperature, and material ratio were measured.

2 h at 30°C at a speed of 150 rpm. After the reaction was complete, the flask was moved to glacial water and maintained at 0°C for 24 h. The crystalline material was collected by filtration and washed with the same silylating reagent and dried in a vacuum oven.

The silylation derivative of UDCA was then added to a 5% solution of hydrochloric acid, and the suspension was warmed to 50°C for 0.5 h. After being cooled, the precipitate was collected by filtration, then washed with deionized water and dried in a vacuum oven. The recovery and purity of the product were measured. The reaction formula is shown in Scheme 1.

Results and discussion

Selection of silylating reagents

There are many kinds of silylation reagents. In this study, hexamethyldisilazane (HMDS), trimethylchlorosilane (TMCS), and 1,3-bis (trimethylsilyl) urea (BSU) were chosen as the silylating reagents.

The effect of different silylating reagents on UDCA recovery and purity is presented in Figure 2. These results indicate that the silylating reagent has an important effect on the recovery and purity of UDCA. In this study, when HMDS was used as the silylating reagent, maximal recovery and purity of UDCA were achieved with values of 87.7% and 90.4%, respectively.

Since the boiling point of TMCS is 57.7°C, the reaction took place intensively after TMCS was added to the system under the experimental temperature (30°C), and TMCS evaporated. Consequently, the contact between reactants was reduced, and this made the final product yield

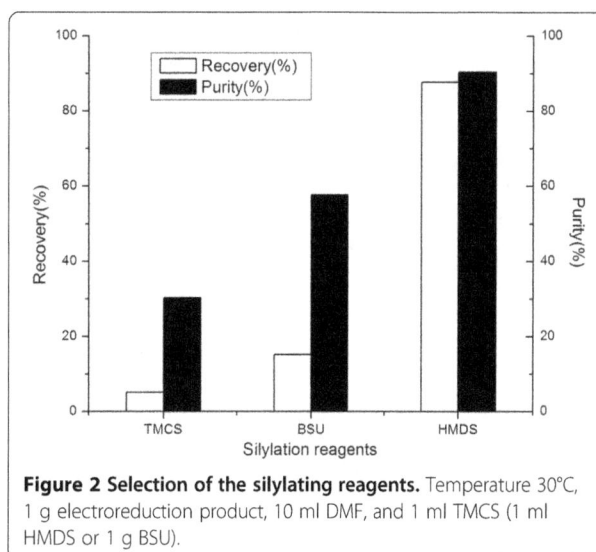

Figure 2 Selection of the silylating reagents. Temperature 30°C, 1 g electroreduction product, 10 ml DMF, and 1 ml TMCS (1 ml HMDS or 1 g BSU).

lower. BSU is a silylation reagent that can increase the solubility of the derivative of UDCA in DMF, so it is difficult to form a precipitant. HMDS, the commercially available reagent for trimethylsilylation of reactive hydrogen, is stable and yields ammonia as the only by-product, which is simple to remove from the reaction medium. Therefore, HMDS was chosen for the following study.

Difference between the silylation derivatives

Since a silylation reagent can replace the hydroxyl groups in UDCA, CDCA, and 7K-LCA, the difference between their corresponding silylation derivatives was the key to the subsequent success of the separation step. As such, silylation was conducted using standard samples (UDCA, CDCA, 7K-LCA, and their mixtures at different proportions) as the silylation reaction substrates and HMDS as the silylation reagent.

The difference between the silylation derivatives of UDCA, CDCA, and 7K-LCA is presented in Table 1. The results showed that crystallization was only observed when UDCA was present in the sample. When CDCA and 7K-LCA individually or together were used, no crystallization was observed. The tris-trimethylsilyl derivative of UDCA has very low solubility in organic solvents, whereas the derivatives of 7K-LCA and CDCA are highly soluble in the same solvents. This suggested that HMDS can be used to isolate UDCA from CDCA and 7K-LCA.

Influence of temperature

The silylation reaction is a strongly exothermic reaction. Thus, temperature could affect the reaction to some extent. Thus, silylation was conducted over the temperature range of 20°C to 70°C.

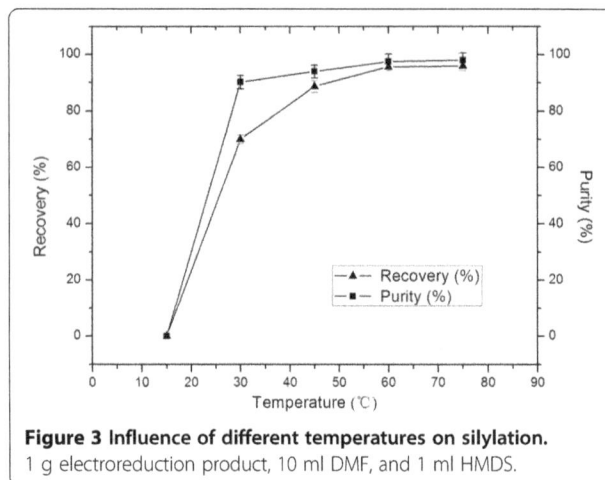

Figure 3 Influence of different temperatures on silylation.
1 g electroreduction product, 10 ml DMF, and 1 ml HMDS.

The effect of different temperatures on UDCA yield is shown in Figure 3. In the range of 20°C to 70°C, the yield of UDCA increased as the reaction temperature increased. This indicated that the reaction was an endothermic reaction. When the temperature was 60°C, the recovery and purity of UDCA were 95.6% and 97.5%, respectively. The use of higher temperatures did not lead to further increases in yield. Consequently, 60°C was chosen for the following study. Under this temperature, the reaction was mild and only a small amount of ammonia was produced as a by-product.

Influence of the material ratio

The ratio of reactants needs to be controlled, so silylation was conducted at different material ratios of electroreduction product, DMF, and HMDS. Table 2 presents the effect of different material ratios on the yield and purity of UDCA.

Table 1 Difference between the silylation derivatives

Groups	UDCA (g)	CDCA (g)	7K-LCA (g)	Crystallization
1	1.0	0	0	Yes
2	0	1.0	0	No
3	0	0	1.0	No
4	0.8	0.2	0	Yes
5	0.6	0.4	0	Yes
6	0.4	0.6	0	Yes
7	0.2	0.8	0	Yes
8	0.8	0	0.2	Yes
9	0.6	0	0.4	Yes
10	0.4	0	0.6	Yes
11	0.2	0	0.8	Yes
12	0	0.5	0.5	No

Temperature 30°C, 1 g electroreduction product, 10 ml DMF, and 1 g HMDS.

Table 2 Influence of different material ratios

Groups	DMF (ml)	Electroreduction product (g)	HMDS (ml)	Recovery (%)	Purity (%)
1	10.0	1.0	1.0	60.4	92.5
2	10.0	1.0	2.0	99.8	99.5
3	10.0	1.0	3.0	87.6	94.6
4	10.0	2.0	1.0	40.3	90.1
5	10.0	2.0	2.0	45.6	91.2
6	10.0	2.0	3.0	50.8	92.8
7	15.0	1.0	1.0	30.2	93.3
8	20.0	1.0	1.0	26.4	90.6

Temperature 60°C; silylation reagent, HMDS.

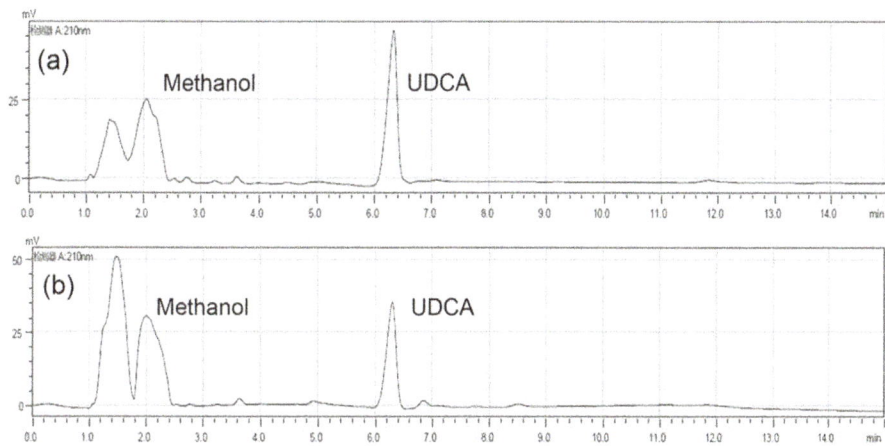

Figure 4 The HPLC spectrum of the product and standard spectrum of UDCA. Mobile phase was a mixture of acetonitrile and phosphate acid buffer (pH 3.0) with a volume ratio of 50:50 at a flow rate of 1.0 ml/min at 25°C. **(a)** Standard UDCA and **(b)** product.

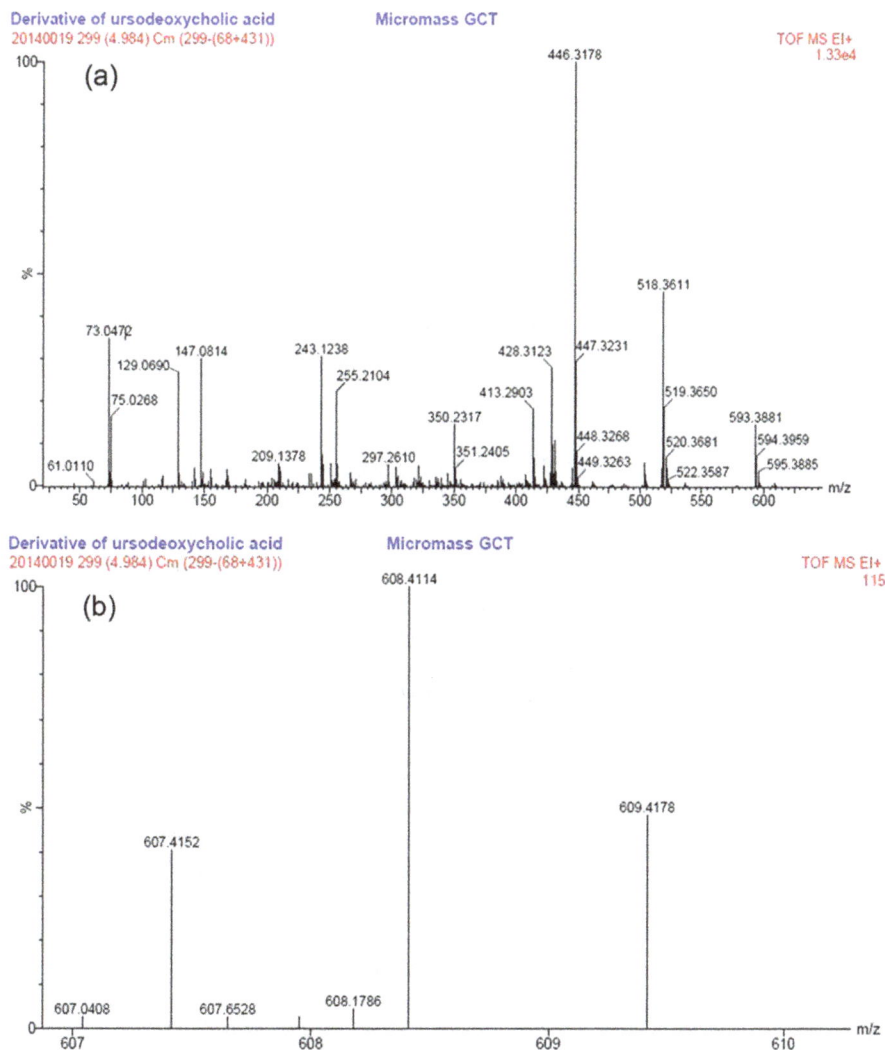

Figure 5 The EI-TOF-MS spectra of the derivative of UDCA. (a) The m/z ranged from 0 to 600. **(b)** The m/z ranged from 605 to 610.

Table 3 Formula evaluated by multiple mass analysis

| Minimum | 4.00 | | | | −1.5 | |
| Maximum | 100.00 | | 5.0 | 5.0 | 50.0 | |
Mass	Relative abundance	Calculated mass	mDa	PPM	Double bond equivalents	Formula
73.0472	34.67	73.0474	−0.2	−2.1	0.5	C_3H_9Si
147.0814	30.08	147.0841	−2.7	−18.6	0.5	$C_6H_{15}O_2Si$
243.1238	4.43	243.1205	3.3	13.5	7.5	$C_{15}H_{19}OSi$
350.2317	14.64	350.2277	4.0	11.4	5.0	$C_{20}H_{34}O_3Si$
413.2903	18.26	413.2876	2.7	6.6	7.5	$C_{26}H_{41}O_2Si$
428.3123	28.07	428.3111	1.2	2.9	7.0	$C_{27}H_{44}O_2Si$
446.3178	100.00	446.3216	−3.8	−8.6	6.0	$C_{27}H_{46}O_3Si$
518.3611	45.94	518.3612	−0.1	−0.1	6.0	$C_{30}H_{53}O_3Si_2$
608.4114	100.00	608.4112	0.2	0.3	5.0	$C_{33}H_{64}O_4Si_3$

Multiple mass analysis, 37 mass(es) processed. Tolerance = 5.0 mDa/double bond equivalents: min = −1.5, max = 50.0. Isotope cluster parameters: separation = 1.0, abundance = 1.0%.

When the material ratio of electroreduction product/DMF/HMDS was 1:10:2 (*w/v/v*), the recovery and purity of UDCA were 99.8% and 99.5%, respectively. With 1 ml HMDS present, the reaction did not complete; however, when 3 ml of HMDS was used, the silylation derivative of UDCA did not readily form a crystal because of the increase of HMDS in the solution. A comparison of group 2 with groups 4, 5, and 6 showed that excessive crude UDCA made the reaction system too viscous to react.

Characterization of the product

After being hydrolyzed and dried, the product was determined by HPLC, EI-TOF-MS, and FTIR.

Characterization by HPLC

As presented in Figure 4a, the retention time of the product was 6.3 min and consistent with the standard HPLC spectrum of UDCA (Figure 4b). In the range of 0 to 20 mg/ml, the peak area of the standard UDCA sample and its concentration gave a linear correlation, with the standard curve of UDCA being $A = 226,034.6C + 18,928.2$ (*A* is the for peak area, *C* stands for concentration, $R^2 = 0.9997$). The sample was run under identical conditions as the standard UDCA material, and the recovery and purity were calculated according to an external standard method.

Characterization by EI-TOF-MS

The EI-TOF-MS spectrum of the tris-trimethylsilyl derivative of UDCA is shown in Figure 5 and Table 3.

The spectrum showed its characteristic fragmentation ions at *m/z* 608 (100), 593 (14), 518 (45), 446 (100), 428 (28), 413 (18), 350 (14), 243 (30), 147 (30), and 73 (34). The evaluation formulas corresponding to these fragments are listed in Table 3. As observed, the ion at *m/z* 608 corresponds to the tris-trimethylsilyl derivative of UDCA ($C_{33}H_{64}O_4Si_3$). This result proved that the reaction proceeded and indicated that the active hydrogen atoms of the three hydroxyl groups on UDCA were substituted by silylation.

Characterization by FTIR

The FTIR spectrum is shown in Figure 6. It indicated that -Si-O- (839.36 cm^{-1}, 1,080.8 cm^{-1}), -Si-C- (1,250 cm^{-1}), -C-H (2,867.6 cm^{-1}, 2,943.7 cm^{-1}), and -CH$_3$ (1,380.0 cm^{-1}) existed in the structure of the derivative of UDCA. Thus, the silylation reaction was confirmed, and the FTIR spectrum of the product is consistent with the standard FTIR spectrum of UDCA.

Conclusions

Ursodeoxycholic acid was purified by silylation crystallization in this study. By optimizing process conditions, recovery and purity of ursodeoxycholic acid was up to 99.8% and 99.5%, respectively. HPLC, EI-TOF-MS, and FTIR analysis showed that silylation is a highly efficient purification method of ursodeoxycholic acid. Compared with previous methods, the UDCA preparation methods presented herein (i) gave higher purity and recovery, (ii) avoided cumbersome procedures, (iii) was more cost efficient, and

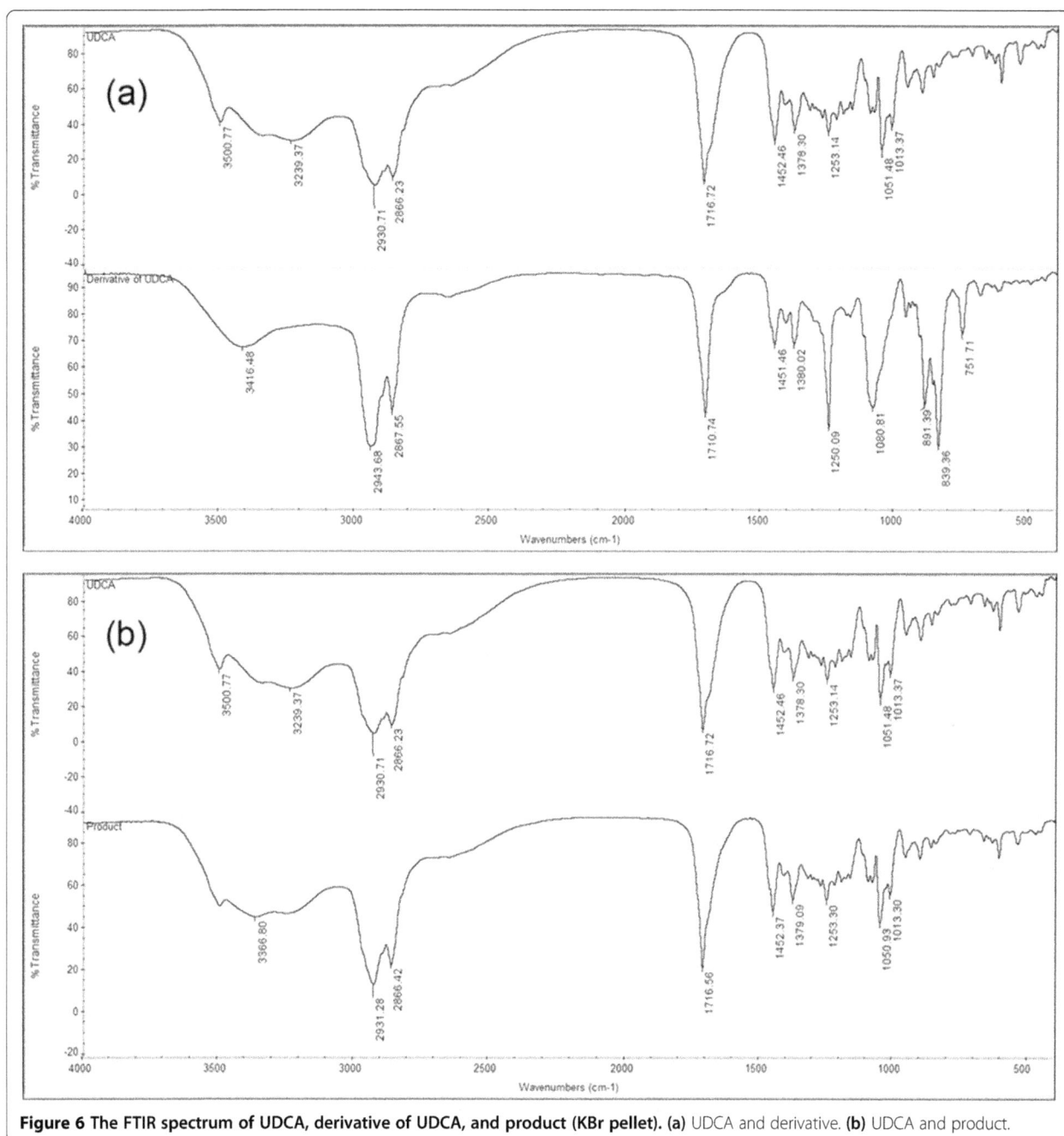

Figure 6 The FTIR spectrum of UDCA, derivative of UDCA, and product (KBr pellet). (a) UDCA and derivative. **(b)** UDCA and product.

(iv) did not require harsh reaction conditions. The silylation crystallization approach was easy to operate, economic, and timesaving, and the by-product (i.e., ammonia) was simple to remove from the reaction medium.

Abbreviations

BSU: 1,3-bis(trimethylsilyl)urea; CDCA: chenodeoxycholic acid; DMF: *N,N*-dimethylformamide; EI-TOF-MS: electron impact ionization time-of-flight mass spectrometry; FTIR: Fourier transform infrared spectroscopy; HMDS: hexamethyldisilazane; HPLC: high-performance liquid chromatography; TMCS: trimethylchlorosilane; UDCA: ursodeoxycholic acid; 7K-LCA: 7-ketolithocholic acid.

Competing interests

The authors declare that they have no competing interests.

Authors' contributions

XLM designed the study; collected, processed, and analyzed the data; and wrote the article. XJC contributed to study design and article corrections. Both authors have read and approved the final manuscript.

References

1. Maton P, Murphy G, Dowling R (1977) Ursodeoxycholic acid treatment of gallstones: dose–response study and possible mechanism of action. Lancet 310(8052):1297–1301

2. Lukivskaya OY, Maskevich AA, Buko VU (2001) Effect of ursodeoxycholic acid on prostaglandin metabolism and microsomal membranes in alcoholic fatty liver. Alcohol 25(2):99–105

3. Matteoni CA, Younossi ZM, Gramlich T, Boparai N, Liu YC, McCullough AJ (1999) Nonalcoholic fatty liver disease: a spectrum of clinical and pathological severity. Gastroenterology 116(6):1413–1419

4. De Lalla F (1999) Effect of ursodeoxycholic acid administration in patients with acute viral hepatitis: a pilot study. Aliment Pharmacol Therapeut 13(9):1187–1193

5. Poupon R, Poupon R, Calmus Y, Chrétien Y, Ballet F, Darnis F (1987) Is ursodeoxycholic acid an effective treatment for primary biliary cirrhosis? Lancet 329(8537):834–836

6. Pardi DS, Loftus EV Jr, Kremers WK, Keach J, Lindor KD (2003) Ursodeoxycholic acid as a chemopreventive agent in patients with ulcerative colitis and primary sclerosing cholangitis. Gastroenterology 124(4):889–893

7. Paumgartner G, Beuers U (2002) Ursodeoxycholic acid in cholestatic liver disease: mechanisms of action and therapeutic use revisited. Hepatology 36(3):525–531

8. Shoda M (1927) Über die Ursodesoxycholsäure aus Bärengallen und ihre physiologische Wirkung. J Biochem 7(3):505–517

9. Tian H, Zhao HB, Cao XJ (2012) Catalytic transfer hydrogenation of 7-ketolithocholic acid to ursodeoxycholic acid with Raney nickel. J Ind Eng Chem 19:606–613

10. Guillemette A, Francois A (1981) Novel purification process. United States Patent and Trademark Office, US Patent 4,282,161. 4 Aug 1981

11. Bonaldi A, Molinari E (1982) Process for preparing high purity ursodeoxycholic acid. United States Patent and Trademark Office, US Patent 4,316,848. 23 Feb 1982

12. Xu ZY, Wan JF, Liang S, Cao XJ (2008) Separation of ursodeoxycholic acid from its isomeric mixture using core–shell molecular imprinting polymer. Biochem Eng J 41(3):280–287

13. Halket JM, Zaikin VG (2003) Derivatization in mass spectrometry–1: silylation. Eur J Mass Spectrom 9(1):1–21

14. Goussé C, Chanzy H, Excoffier G, Soubeyrand L, Fleury E (2002) Stable suspensions of partially silylated cellulose whiskers dispersed in organic solvents. Polymer 43(9):2645–2651

15. Yamamoto K, Takemae M (1989) The utility of t-butyldimethylsilane as an effective silylation reagent for the protection of functional groups. B Chem Soc Jpn 62(6):2111–2113

16. Karimi B, Golshani B (2000) Mild and highly efficient method for the silylation of alcohols using hexamethyldisilazane catalyzed by iodine under nearly neutral reaction conditions. J Org Chem 65(21):7228–7230

17. Khazaei A, Zolfigol MA, Rostami A, Choghamarani AG (2007) Trichloroisocyanuric acid (TCCA) as a mild and efficient catalyst for the trimethylsilylation of alcohols and phenols with hexamethyldisilazane (HMDS) under heterogonous conditions. Catal Commun 8(3):543–547

18. Kadam ST, Kim SS (2009) Mild and efficient silylation of alcohols and phenols with HMDS using Bi(OTf)$_3$ under solvent-free condition. J Organomet Chem 694(16):2562–2566

19. Shirini F, Mamaghani M, Atghia SV (2012) A mild and efficient method for the chemoselective trimethylsilylation of alcohols and phenols and deprotection of silyl ethers using sulfonic acid-functionalized ordered nanoporous Na$^+$-montmorillonite. Appl Clay Sci 58:67–72

20. Zareyee D, Ghandali MS, Khalilzadeh MA (2011) Sulfonated ordered nanoporous carbon (CMK-5-SO3H) as an efficient and highly recyclable catalyst for the silylation of alcohols and phenols with hexamethyldisilazane (HMDS). Catal lett 141(10):1521–1525

21. Lee SH, Kadam ST (2011) Cross-linked poly (4-vinylpyridine/styrene) copolymer-supported bismuth(III) triflate: an efficient heterogeneous catalyst for silylation of alcohols and phenols with HMDS. Appl Organomet Chem 25(8):608–615

22. Mormann W, Wagner T (2000) Silylation of cellulose with hexamethyldisilazane in liquid ammonia. Carbohyd Polym 43(3):257–262

23. Yuan XX, Ma XL, Cao XJ (2014) Preparation of ursodeoxycholic acid by direct electro-reduction of 7-ketolithocholic acid. Korean J Chem Eng 1–5. doi:10.1007/s11814-013-0245-y

22

Optimization of algal methyl esters using RSM and evaluation of biodiesel storage characteristics

Annam Renita A[1*], Nurshaun Sreedhar[1] and Magesh Peter D[2]

Abstract

Background: This paper deals with the production of biodiesel from the brown seaweed *Sargassum myriocystum*, a third-generation biodiesel from the Gulf of Mannar, Rameshwaram, India. The optimization of reaction parameters was done using Design-Expert software version 8.0.7.1. Algal oil was transesterified using methanol and sodium hydroxide. The effect of oil:alcohol ratio, catalyst amount, temperature, and time on biodiesel yield was investigated by response surface methodology using central composite design.

Results: It was found that the maximum biodiesel yield was obtained at 60°C for 1:6 (*v/v*) oil:alcohol ratio, 0.4 (*w/w*) catalyst amount, and 120 min. The R^2, adjusted R^2, and predicted R^2 values are 0.9977, 0.9956, and 0.9923, respectively, which implies that experimental values are in good agreement with predicted values. The fatty acid profile of *S. myriocystum* biodiesel was determined using gas chromatography. Algal biodiesel was stored in dark and light conditions. Fuel properties like kinematic viscosity and acid value were determined. It was found that the samples exposed to light led to an increase in kinematic viscosity and acid value with some sediment formation.

Conclusions: The acid value and kinematic viscosity of the samples stored in the dark environment had only marginal increase in fuel properties which were within the range specified by the American Society of Testing Materials (ASTM D6751).

Keywords: Brown seaweed; Transesterification; Biodiesel; Optimization; RSM; Fatty acid profile; Storage properties

Background

India consumes almost five times more diesel fuel than gasoline and burns about 450 million barrels a year of diesel [1]. So in India, search for alternatives is an important criterion to meet current and future energy requirements. India's tropical climate is ideal for the growth of algal species which serves as an advantage over other countries. Approximately 841 species of marine algae are found in both intertidal and deep water regions of the Indian coast [2]. Less than 100 species have been explored for biodiesel extraction. There are two global biorenewable liquid transportation fuels that might replace gasoline and diesel fuel in the future. These are bioethanol and biodiesel [3]. Biodiesel is a processed fuel derived from the esterification and transesterification of free fatty acids and triglycerides, respectively, that occur naturally in renewable biological sources such as plant oils and animal fats

[4]. Different countries use the oil which is abundant in that particular country. India, being a peninsula, has a rich source of seaweeds. Algae are classified as microalgae and macroalgae. The potential value of microalgal fuel production is widely recognized [5-11]. Macroalgae, which are also known as seaweeds, have comparatively low lipid content. Nevertheless, they are a potential raw material for biofuel production. Biodiesel has good biodegradability, and it is expected to degrade over 98% biologically within 3 weeks while commercial diesel will only degrade 50% biologically within the same period [12]. In Indian waters, species of *Sargassum* are the most abundant algae [13]. The seaweeds detach themselves from the rocky substrates and are frequently washed ashore. Though details of the phytoprofile [2], antioxidant activity, free scavenger property [14], and fatty acid profile of *Sargassum* have been reported, optimization of biodiesel yield and storage properties have not been reported regarding its suitability as a raw material for biodiesel. Hence, this paper deals with the production of biodiesel from the brown seaweed *Sargassum myriocystum* as a potential resource for biodiesel.

* Correspondence: reniriana@gmail.com
[1]Department of Chemical Engineering, Sathyabama University, Jeppiaar Nagar, Rajiv Gandhi Salai, Chennai TN 119, India
Full list of author information is available at the end of the article

Algal oil extracted from the brown seaweed *S. myriocystum* was transesterified to biodiesel. Analysis of the biodiesel sample was done using gas chromatography-mass spectrometry (GC-MS). The GC-MS of the biodiesel sample indicated the presence of low amounts of unsaturated fatty acids [15]. Since the fatty acid profile predicted low concentration of unsaturated fatty acids, the biodiesel sample was tested for storage properties. Biodiesel production from algae is hindered by its production cost. This can be overcome by optimizing the reaction parameters to achieve maximum production of biodiesel with minimum cost of raw materials. For this research work, optimization of reaction conditions was done by Design-Expert software version 8.0.7.1 (Stat-Ease Inc., Minneapolis, MN, USA). Central composite design (CCD) was applied to optimize the catalyzed transesterification reaction variables like oil:alcohol ratio, catalyst amount, time, and temperature. Regression analysis and analysis of variance (ANOVA) tested the significance of the model. *S. myriocystum* is also rich in proteins and sugars. It is being used in food and pharmaceutical industries [16]. By applying the biorefinery concept, all valuable end products can be extracted from *S. myriocystum*, thus making the cost of algal biodiesel economically viable. This study will enable the seaweed to be considered as a potential option as a biorefinery crude.

Methods
Raw materials
The brown seaweed *S. myriocystum* belonging to the class Phaeophyceae was obtained from the Gulf of Mannar, Rameshwaram, India. The seaweed species was chosen because of its availability and lipid profile. It is available all through the year and shows high productivity during the months of July and September. The samples were first washed with seawater and the debris was removed manually. Then they were washed with distilled water, shade dried under the sun, and finally dried in an oven at 60°C. Ten kilograms of wet algae was collected, and after drying, 8.8 kg of dry algae was obtained. Hexane (≥85%), chloroform (≥98%), methanol (≥99%), and sodium hydroxide (≥98%) used for the experiments were of analytical grade.

Experimental method
The dried algae, which were chopped into fine pieces, were ground into fine powder which passes through 200 mesh size. Two solvent extraction systems were tested to find out the best suitable solvent for extraction of lipids. In one system, the solvent system was hexane and the second solvent system was chloroform:methanol (2:1 *v/v*). The solvent extraction was done by a Soxhlet apparatus, and the extracted algal oil was vacuum evaporated in a rotary evaporator to obtain pure algal oil. The chloroform:methanol solvent system extracted lipids better than hexane. Using the chloroform:methanol solvent

system, 5.2 kg of algal oil was extracted. The viscosity of the algal oil has to be reduced in order to be an acceptable fuel for running an engine. The most possible method of reducing the viscosity and the commercially accepted method is the chemical transesterification method. In this process, the pure algal oil was reacted with methanol in the presence of sodium hydroxide catalyst to produce fatty acid methyl esters (FAME), which more closely resemble petroleum-based diesel fuel, and glycerol is obtained as a co-product. Sodium hydroxide was dissolved in methanol and stirred for 20 min prior to transesterification to ensure the complete dissolution of sodium hydroxide pellets. Algal oil and premixed mixture of catalyst and alcohol are kept in a round-bottom flask fitted with a reflux condenser to maintain isothermal operation. The round-bottom flask was mounted on top of a magnetic stirrer, and the reaction was carried out with continuous stirring. Experiments were done with different feed ratios of oil:alcohol, catalyst amounts, temperatures, and times to obtain the highest yield of biodiesel. Experiments on volume basis for oil:alcohol ratio were done on trial and error analysis. The volume ratio was chosen for ease of application in commercial processes. All the experiments were done in triplicate to ensure the reproducibility of the data. The error of analysis for the experimental procedure is approximately ±5%. The product mixture is biodiesel and glycerol. Both phases were separated using a gravity separator. The biodiesel which was obtained as the upper layer was washed with hot water and cold water alternately to remove impurities like sodium hydroxide, methanol, and foam. The sample was fan dried and stored for analysis. The yield of biodiesel is estimated from Equation 1:

$$\text{Yield of biodiesel} = \text{Weight of algal oil/Weight of} \\ \text{raw material} \times 100$$

(1)

Statistical analysis
Design-Expert software version 8.0.7.1 was used for the optimization of reaction parameters. CCD of response surface methodology (RSM) was used to optimize the parameters affecting the transesterification so as to achieve maximum yield of biodiesel. RSM can be used to find the relationship among process variables and response in an efficient manner using minimum number of experiments [17]. Several parameters affect transesterification: catalyst concentration, methanol concentration, temperature, reaction time, pressure, and type of oil [18]. The process of transesterification is affected by various factors: free fatty acids, moisture, catalyst type, catalyst concentration, molar ratio of alcohol to oil, type of alcohol, reaction time, reaction temperature, mixing

Table 1 Coded levels for variables used in the experimental design

Factors	Symbol	Coded levels		
		−1	0	+1
Oil:alcohol ratio (v/v)	A	4	6	8
Catalyst amount (w/w)	B	0.1	0.3	0.5
Time (min)	C	30	90	150
Temperature (°C)	D	50	70	90

intensity, and organic solvents [19]. The CCD of the RSM was employed to evaluate the effect of reaction time, reaction temperature, catalyst amount, and oil:alcohol ratio on biodiesel yield. The coded values for the variables are shown in Table 1.

The empirical formula to find the optimal biodiesel yield is given by Equation 2:

$$Y = \beta_0 + \beta_1 A + \beta_2 B + \beta_3 C + \beta_4 D + \beta_5 AB + \beta_6 AC$$
$$+ \beta_7 AD + \beta_8 BC + \beta_9 BD + \beta_{10} CD + \beta_{11} A^2 + \beta_{12} B^2$$
$$+ \beta_{13} C^2 + \beta_{14} D^2$$

$$(2)$$

where Y is the measured response in percentage yield (biodiesel yield). A, B, C, and D are the coded independent inputs. β_0 is the intercept term and β_1, β_2, β_3, and β_4 are the coefficients showing the linear effects. β_5, β_6, β_7, β_8, β_9, and β_{10} are the cross-product coefficients showing the interaction effects. β_{11}, β_{12}, β_{13}, and β_{14} are the quadratic coefficients showing the squared effects.

Table 2 Central composite design matrix for *Sargassum myriocystum* biodiesel

Run number	Oil:alcohol ratio (A)	Catalyst amount (B)	Time (C)	Temperature (D)	Experimental yield (%)	Predicted yield (%)
1	4	0.50	150	50	78.4	79.54
2	4	0.10	30	90	59.35	54.09
3	4	0.30	90	90	68.57	69.18
4	6	0.30	90	60	82.11	83.54
5	4	0.10	150	50	72.11	73.05
6	8	0.50	30	90	73.23	71.79
7	6	0.50	90	70	81.75	82.88
8	4	0.50	30	90	61.62	60.48
9	8	0.10	150	50	62.54	64.8
10	6	0.30	90	60	82.11	83.54
11	6	0.30	90	70	81.71	82.88
12	8	0.50	150	50	65.70	69.54
13	8	0.50	30	50	63.53	64.8
14	4	0.10	150	90	71.47	68.88
15	8	0.10	30	50	55.28	54.09
16	8	0.50	150	90	69.85	71.79
17	6	0.30	90	50	78.85	69.54
18	4	0.10	30	50	55.32	57.7
19	6	0.30	90	90	81.45	82.88
20	6	0.30	90	70	82.11	83.54
21	8	0.30	90	70	81.65	80.22
22	6	0.30	30	70	69.82	75.52
23	6	0.30	90	60	83.54	80.88
24	8	0.10	30	90	63.95	69.73
25	6	0.30	90	70	82.82	83.1
26	4	0.50	150	90	67.57	65.25
27	8	0.10	150	90	66.59	69.87
28	6	0.30	150	60	78.23	82.14
29	4	0.50	30	50	58.25	52.57
30	6	0.10	90	70	49.27	52.88

Evaluation of storage properties

A good fuel should have low amounts of unsaturated fatty acids. The fatty acid profile was determined by GC-MS. Since *S. myriocystum* biodiesel has low amounts of unsaturated fatty acids, the fuel was investigated for its storage properties. Fuel properties like kinematic viscosity and acid value were determined at different time intervals to determine the shelf life of *S. myriocystum* biodiesel. The samples were stored for a period of 5 months from March to July, which is the average summer period in India. The samples were stored in two different conditions. Two liters of sample was kept in a stoppered amber glass bottle and exposed to sunlight for a period of 5 months, and properties like kinematic viscosity and acid value were determined in an interval of 1 month. Another 2 l of sample was kept in a stoppered amber glass bottle in a dark room for a period of 5 months, and properties like kinematic viscosity and acid value were determined at an interval of 1 month.

Results and discussion
Response surface methodology

The Design-Expert software developed a 2^4 factorial CCD resulting in 30 runs with eight axial points and six replicates. The second-order polynomial equation predicted by the model for maximum algal biodiesel is given by Equation 3:

$$\text{Percentage yield} = 92.88 + 0.52A + 1.33B + 3.43C$$
$$+1.96D + 1.61AB - 2.22AC$$
$$+1.30AD - 1.45BC + 0.025BD$$
$$-1.21CD - 3.19A^2 - 2.94B^2 - 8.94C^2$$
$$-3.24D^2$$
$$(3)$$

where A, B, C, D are the coded forms of oil:alcohol ratio, catalyst amount, time, and temperature, respectively. AB, AC, AD, BC, BD, and CD are the interaction terms, and A^2, B^2, C^2, and D^2 are the squared terms of the independent variables. The CCD matrix is represented by Table 2. Table 3 shows ANOVA.

The results of ANOVA for the predicted values fit well with the experimental values. The goodness of fit of the model is predicted by the determination coefficient value of 0.9977 and the adjusted R^2 values of 0.9956. The predicted value is also in good agreement with adjusted R^2 value implying the significance of the model. Only 0.23% of the sample variance could not be satisfactorily explained by the model. Hence, the predictability of the model is at 99% confidence interval. Lack of fit is also not significant, indicating the goodness of fit of the model. The p value serves as a tool to check the significance of

Table 3 ANOVA for *Sargassum myriocystum* biodiesel and regression coefficient estimate

Source	Sum of squares	df	Mean square	F value	p value probability > F	
Model	2,480.56	14	177.18	465.69	<0.0001	Significant
A	4.80	1	4.80	12.63	0.0029	
B	32.00	1	32.00	84.10	<0.0001	
C	212.18	1	212.18	557.67	<0.0001	
D	68.84	1	68.84	180.92	<0.0001	
AB	41.60	1	41.60	109.34	<0.0001	
AC	79.21	1	79.21	208.19	<0.0001	
AD	27.04	1	27.04	71.07	<0.0001	
BC	33.64	1	33.64	88.42	<0.0001	
BD		1		0.026	0.8734	
CD	23.52	1	23.52	61.82	<0.0001	
A^2	36.28	1	36.28	69.08	<0.0001	
B^2	23.32	1	23.32	58.66	<0.0001	
C^2	206.85	1	206.85	543.65	<0.0001	
D^2	27.12	1	27.12	71.27	<0.0001	
Residual	5.71	15	0.38			
Lack of fit	3.37	10	0.34	0.72	0.6909	Not significant
Pure error	2.33	5	0.47			
Correlated total	2,486.27	29				

R^2 0.9977; adjusted R^2 0.9956; predicted R^2 0.9923.

each coefficient. The smaller the p value and the larger the F value, the more significant is the parameter reflecting the relative importance of the term attached to that parameter [20]. ANOVA for the response surface quadratic model gave an F value of 465.69, with a p value of <0.0001, implying its significance.

Two-dimensional (contour) and three-dimensional (surface) plots indicate the effect of interaction of the different test variables on the predicted response. The three-dimensional graphs and contour graphs are the common graphical representation of the regression equation which shows the optimal values of each dependent variable [21]. Four surface and contour plots are shown to indicate the effect of interaction of oil:alcohol ratio, catalyst amount, time, and temperature on biodiesel yield. From the shape of contour plots, one could estimate the significance of the mutual interactions between the independent variables in that an elliptical profile of the contour plots indicates a remarkable interaction between the independent variables [22]. Figures 1a and 2a show the interaction between temperature and time on biodiesel yield. They show the interaction between time and temperature at a catalyst amount of 0.4 (*w/w*) and oil:alcohol ratio of 1:6 (*v/v*). The midpoint of the plots gives the maximum value of 81.79% which is obtained at 1:6 (*v/v*) oil:alcohol ratio

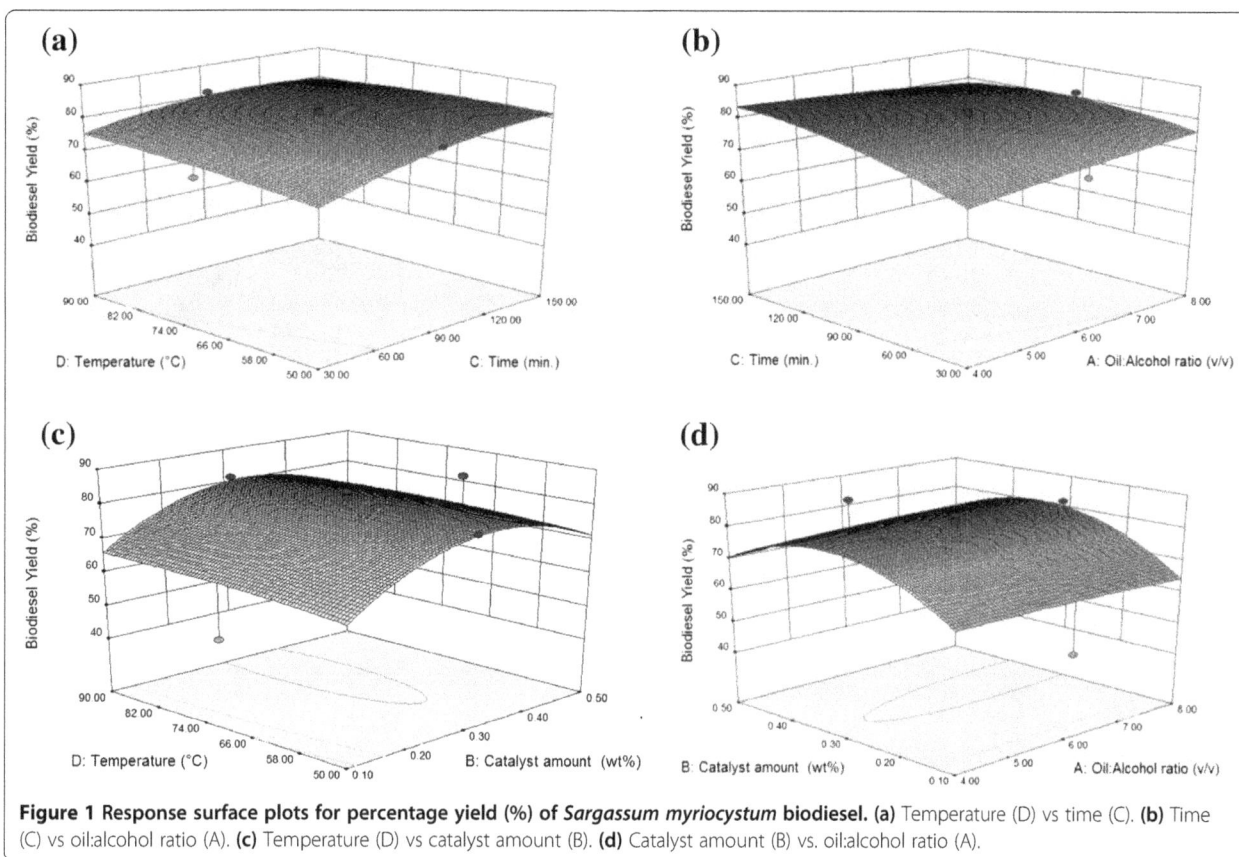

Figure 1 Response surface plots for percentage yield (%) of *Sargassum myriocystum* biodiesel. **(a)** Temperature (D) vs time (C). **(b)** Time (C) vs oil:alcohol ratio (A). **(c)** Temperature (D) vs catalyst amount (B). **(d)** Catalyst amount (B) vs. oil:alcohol ratio (A).

and 120 min. For low oil:alcohol ratio, the yield value is marginally low at 49.18%. High oil:alcohol ratios are not economically and environmentally viable. Moreover, excessive methanol amounts may reduce the concentration of the catalyst in the reactant mixture and retard the transesterification reaction [23]. Figures 1b and 2b show the interaction between time and oil:alcohol ratio on biodiesel yield. The biodiesel yield progressively increases as temperature increases and shows a decline after 60°C. This could be due to the fact that the solvent tends to evaporate after 60°C since it exceeds its boiling point. Figures 1c and 2c show the interaction between temperature and catalyst amount on biodiesel yield. At low catalyst concentration, the biodiesel yield shows a decline till 0.3 wt% and starts increasing from 0.4 to 0.5 wt%. After 0.4 wt%, the washing of the biodiesel layer developed more foam because of the residual catalyst. Figures 1d and 2d show the interaction between catalyst amount and oil: alcohol ratio on biodiesel yield. The biodiesel yield progressively increases as oil:alcohol ratio increases and starts decreasing after 1:7. At low oil:alcohol ratio, phase separation between biodiesel and glycerol layer was not observed, and as the ratio increased, a good phase separation was achieved.

The optimum values of the variables as predicted by the model are 1:6 (*v*/*v*) oil:alcohol ratio, 0.4 (*w*/*w*) catalyst amount, 120 min, and 60°C. Experiments were carried out to validate the optimum conditions indicated by the model. The experimental result obtained was 79.5% biodiesel yield with a 3.6% error to the value of 83.54% predicted by the model. The experiment was done in triplicate. Since the error was not significant, it can be stated that the optimum conditions are 1:6 (*v*/*v*) oil:alcohol ratio, 0.4 (*w*/*w*) catalyst amount, 120 min, and 60°C.

Fatty acid profile

The composition of the algal biodiesel was analyzed using gas chromatography. Of the sample mixed with 10 ml of methyl heptadecanoate solution, 250 mg was injected into a flame ionization detector at 250°C. The capillary column of length 30 m and internal diameter 0.32 m was coated with polyethylene glycol. The carrier gas was helium with a flow of 1 to 2 ml/min at a pressure of 30 to 100 kPa.

Figure 3 shows the composition of *S. myriocystum* biodiesel. It was found that monounsaturated fatty acids (MUFA) is 5.68 g/100 g, polyunsaturated fatty acids (PUFA) is 7.31 g/100 g, and transfat is 3.31 g/100 g. The sample is composed of saturated and unsaturated fatty acids with 10 to 36 carbon atoms. *S. myriocystum* biodiesel has 16.3 g/100 g of unsaturated fatty acids and 82.1 g/100 g

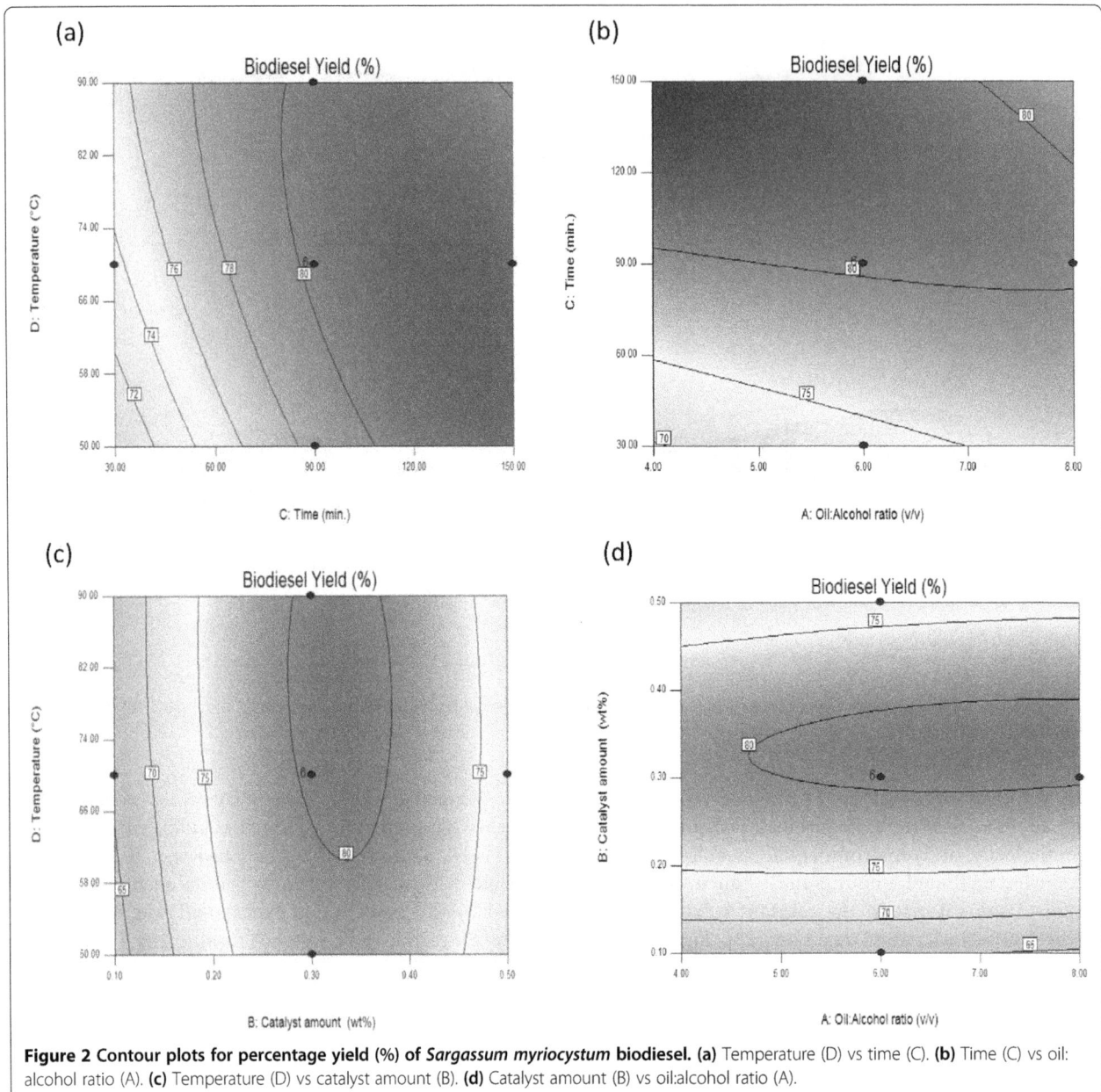

Figure 2 Contour plots for percentage yield (%) of *Sargassum myriocystum* biodiesel. **(a)** Temperature (D) vs time (C). **(b)** Time (C) vs oil: alcohol ratio (A). **(c)** Temperature (D) vs catalyst amount (B). **(d)** Catalyst amount (B) vs oil:alcohol ratio (A).

of saturated fatty acids, indicating it to be a good prospective source for biodiesel [12].

Storage properties

A major disadvantage of widespread biodiesel commercialization is its low oxidation stability [24]. Conventional biodiesel is a hydrocarbon molecule having no oxygen integrated in its molecular structure, whereas algal biodiesel has fuel-bound oxygen atoms; hence, there is a lot of possibility of oxidation problems when it is exposed to air. The degree of unsaturation of biodiesel makes it susceptible to thermal and oxidative degradation leading to problems of injection in fuel systems [25]. Bouaid et al. studied the storage properties of high

oleic sunflower oil, erucic *Brassica carinata* oil, and used frying oil. It was found that acid value, peroxide value, viscosity, and insoluble impurities decreased whereas iodine value increased with increase in storage time [26]. Das et al. studied the storage characteristics of karanja oil methyl ester over a period of 180 days and found that the addition of antioxidants like tert-butylated hydroxytoluene, tert-butylated hydroxyanisol, and propyl gallate increased the oxidative stability [27]. *S. myriocystum* biodiesel has a major advantage of having low amounts of unsaturated fatty acids. The unsaturation influences not only the engine performance but also the emission of pollutants into the environment. Emissions of pollutant gases like nitrogen oxides and carbon monoxides will be

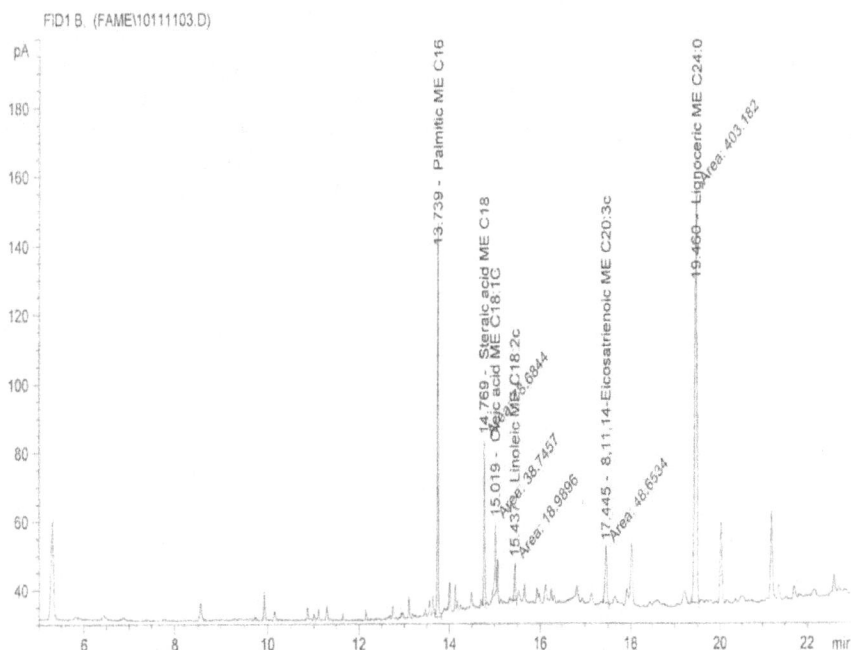

Peak #	Ret time [min]	Type	Width (min)	Area [pA*s]	Area %	Name
1.	4.861		0.0000	0.00000	0.00000	Butyric ME C4
2.	7.226		0.0000	0.00000	0.00000	Caproic ME C6
3.	9.031		0.0000	0.00000	0.00000	Caprylic ME C8
4.	10.482		0.0000	0.00000	0.00000	Capric ME C10
5.	11.124		0.0000	0.00000	0.00000	Undecanoic ME C11
6.	11.734	BB	0.0296	3.71682	0.25220	Lauric ME C12
7.	12.287		0.0000	0.00000	0.00000	Tridecanoic ME C13
8.	12.762	BB	0.0222	38.29747	2.59859	Myristic ME C14
9.	13.124		0.0000	0.00000	0.00000	Myristoleic ME C14:1
10.	13.325		0.0000	0.00000	0.00000	Pentadecanoic ME C15
11.	13.630		0.0000	0.00000	0.00000	Pentadecanoic ME C15:1c
12.	13.782	BB	0.0254	1025.46973	69.58085	Palmitic ME C16
13.	14.129		0.0000	0.00000	0.00000	Palmitoleic ME C16:1
14.	14.327	VP	0.0243	22.76040	1.54435	Heptadecanoic ME C17
15.	14.610		0.0000	0.00000	0.00000	Heptadecanoic ME C17:1c
16.	14.782	MM	0.0487	73. 35067	4.97704	Stearic acid ME C18
17.	15.049	MF	0.0311	48.72641	3.30622	Eladic acid ME C18:1t
18.	15.096	FM	0.0284	35.02279	2.37639	Oleic acid ME C18:1c
19.	15.306		0.0000	0.00000	0.00000	Linoelidic acid ME C18:2t
20.	15.467	BP	0.0297	107.75860	7.31171	Linoleic ME C18:2c
21.	15.854		0.0000	0.00000	0.00000	Linolenic ME C18:3
22.	16.007	BB	0.0330	118.67853	8.05266	Arachidic ME C20:0
23.	16.078		0.0000	0.00000	0.00000	g- Linolenic ME C18:3c
24.	16.381		0.0000	0.00000	0.00000	11-Eicosenoic ME C20:1
25.	16.772		0.0000	0.00000	0.00000	Heneicosanoic ME C21:0
26.	16.956		0.0000	0.00000	0.00000	8,11,14-Eicosatrienoic ME C20:3c
27.	17.388		0.0000	0.00000	0.00000	11,14-Eicosatrienoate ME C20:2
28.	17.623		0.0000	0.00000	0.00000	Arachidonic ME C20:4
29.	17.723		0.0000	0.00000	0.00000	Behenic acid ME C22:0
30.	18.097		0.0000	0.00000	0.00000	Euricic ME C22:1
31.	18.606		0.0000	0.00000	0.00000	5,8,11,14,17 Eicosanpantenoic acid ME C20:2
32.	18.658		0.0000	0.00000	0.00000	Tricosanoic ME C23:0
33.	18.896		0.0000	0.00000	0.00000	11,14,17- Eicosatrieonate ME
34.	19.621		0.0000	0.00000	0.00000	Lignoceric ME C24:0
35.	20.086		0.0000	0.00000	0.00000	Nervonic acid ME C24:1
36.	20.956		0.0000	0.00000	0.00000	4,7,10,13,15,19-decosahexanoicacid ME C22:6

Figure 3 GC-MS of *Sargassum myriocystum* biodiesel.

Table 4 Storage test results in dark and light conditions

Sargassum myriocystum biodiesel	Time (days)					
	0	30	60	90	120	150
In dark condition						
Kinematic viscosity (CSt)	2.82	3.4	3.52	3.96	3.98	3.98
Acid value (mg KOH/g)	0.32	0.33	0.31	0.32	0.34	0.34
In light condition						
Kinematic viscosity (CSt)	2.82	4.02	4.95	5.31	6.94	7.89
Acid value (mg KOH/g)	0.32	0.38	0.49	0.43	0.47	0.52

high if the degree of unsaturation is high. Algal biodiesel has fuel-bound oxygen for combustion to be complete which ensures complete oxidation of pollutant gases.

Table 4 shows the storage tests done on biodiesel stored in dark and light conditions for a period of 150 days. The samples were periodically taken out at an interval of 30 days and were checked for both kinematic viscosity and acid value. Kinematic viscosity is an important property of all fuels. High viscosity results in negative impact on fuel injection system performance. The specified range of viscosity of the American Society of Testing Materials (ASTM D6751) is 2.5 to 6. Kinematic viscosity was measured by an Ostwald viscometer. High viscosity leads to decreased atomization, fuel vaporization, and combustion. Samples stored in dark conditions showed only a marginal increase in kinematic viscosity for all samples which did not exceed the specified limits of ASTM D6751 as indicated by Table 4. For samples stored in light conditions, there was a sharp increase in viscosity values which exceeded the standard specifications. The viscosity rises during storage owing to the formation of polymers and to the hydrolytic cleavage of methyl esters into fatty acids [25]. Hence, the properties of the biodiesel samples stored in dark conditions are comparable to conventional diesel properties. Acid number indicates the corrosive nature of the fuel. The acid number for biodiesel should be lower than 0.50 mg KOH/g according to ASTM D6751. Acid value was measured in the laboratory by mixing the sample with ethanol and heating in water bath and titrating it with potassium hydroxide using phenolphthalein as indicator. For samples stored in dark conditions, the acid values did not exceed the prescribed value after a period of 150 days and showed only a small difference from the initial value. The acid value is low for the samples stored in light conditions for a month but the value exceeds the prescribed limit after 150 days, indicating that the fuel had become very corrosive. This is in agreement with earlier studies [25,27]. There was sediment formation in the glass bottles after a period of 90 days, indicating that it cannot be stored and used as a fuel after 4 months. This could be due to the breakdown of molecular bonds and sample degradation. If the quality of biodiesel is not good and

contains high degree of unsaturated fatty acids, it would tend to polymerize with the lubricating oil, forming sludge, and increasing engine wear [28]. Storage problems will be more for algal biodiesel than commercial diesel and shelf life is much reduced. The problems encountered in storing biodiesel for a long duration are mainly due to auto-oxidation, photo-oxidation, hydrolytic degradation, and oxidative degradation. This could be due to the fact that light can oxidize the oil. Degradation of the fuel has a profound influence on engine performance. The fuel was tested for calorific value before and after the test for both conditions of exposure to light and in the dark. The calorific value tested according to ASTM D6751 before the test was 41,499 kJ/kg. The calorific value of the sample kept in the dark after a period of 150 days is 38,700 kJ/kg, and that of the sample kept in light after a period of 150 days is 24,520 kJ/kg. The calorific value has drastically reduced because of molecular breakdown owing to degradation. A lower calorific value leads to more fuel consumption for the same amount of work done by a reference fuel. The engine will have high specific fuel consumption for the maintenance of a constant power output in maximum load conditions. Hence, it can be suggested that suitable additives can be added to the fuel to increase its shelf life and storage characteristics. Some additives for biodiesel storage found in literature are pyrimidinols [29], butylated hydroxytoluene, butylated hydroxyanisol, tert-butyl hydroquinone, propyl gallate, ethylenediaminetetraacetic acid, citric acid, phosphoric acid, and amino acids [30].

Conclusions

The brown seaweed S. myriocystum is a potential resource for biodiesel production. To make the process of making biodiesel cost effective, optimization of parameters was done using Design-Expert software version 8.0.7.1. RSM gave the optimum values of 120 min, 60°C, 1:6 (v/v) oil:alcohol ratio, and 0.4 (w/w) catalyst amount. Maximum yield of biodiesel was achieved at the optimum conditions. The fuel was further investigated for its storage properties. It was found that the fuel deteriorated when exposed to air after a period of 4 months and started forming deposits with high increase in acid value. The fuel which was not exposed to air did not show a drastic change in kinematic viscosity and acid value but maintained a slow and progressive increase in the properties which did not exceed the limits prescribed by ASTM D6751. The calorific value of the fuel had drastically decreased to a lower value on exposure to light after a period of 150 days. It can be suggested that biodiesel prepared from S. myriocystum seaweed is best stored without any contact with air or suitable additives should be added to enhance the storage properties. In India, S. myriocystum seaweed is used in food and pharmaceutical industries. Since this paper substantiates its biofuel property which also meets ASTM D6751, it is an ideal crude for biorefineries.

Competing interests

The authors declare that they have no competing interests.

Authors' contributions

AR carried out the extraction, reaction, and optimization studies of the paper. NS performed the storage tests. MP identified the species and helped in the extraction of algal oil. All authors read and approved the final manuscript.

Acknowledgements

The authors wish to thank the Director of the National Institute of Ocean Technology, India, for the assistance provided. The authors also thank the editor and reviewers for their suggestions, thus improving the quality of the manuscript.

Author details

[1]Department of Chemical Engineering, Sathyabama University, Jeppiaar Nagar, Rajiv Gandhi Salai, Chennai TN 119, India. [2]Department of Marine Biotechnology, National Institute of Ocean Technology, Velacherry-Tambaram Main Road, Chennai 100, India.

References

1. Shakeel A, Khan R, Mir Z, Hussain PS, Banerjee UL (2009) Prospects of biodiesel production from microalgae in India. Renew Sustain Energy Rev 13:2361–2372
2. Jeyabalan JPP, Johnson Marimuthu A (2012) Preliminary phytochemical analysis of *Sargassum myriocystum* J.Ag. and *Turbinaria ornata* (Turner) J.Ag. from the southern coast of Tamil Nadu, India. Asian Pac J Trop Biomed 2:1–4
3. Fatih Demirbas M (2009) Biorefineries for biofuel upgrading: a critical review. Appl Energy 86:S151–S161
4. Mcneff CV, McNeff LC, Bingwen Y, Nowlan DT, Rasmussen M, Gyberg AE, Krohn BJ, Fedie RL, Hoye TR (2008) A continuous catalytic system for biodiesel production. Appl Catal Gen 343:39–48
5. Sharma R, Chisti Y, Banerjee UC (2001) Production, purification, characterization and applications of lipases. Biotechnol Adv 19:627–62
6. Roessler PG, Brown LM, Dunahay TG, Heacox DA, Jravis EE, Schneider JC (1994) Genetic engineering approaches for enhanced production of biodiesel fuel from microalgae. ACS Symp Ser 566:255–270
7. Sawayama S, Inoue S, Dote Y, Yokoyama SY (1995) CO_2 fixation and oil production through microalga. Energy Convers Manag 36:729–731
8. Hu Q, Sommerfeld M, Jravis E, Ghirardi M, Posewitz M, Seibert M (2008) Microalgal triacylglycerols as feedstocks for biofuel production: perspectives and advances. Plant J 54:621–639
9. Schenk PM, Thomas Hall SR, Stephens E, Marx UC, Mussgnug JH, Posten C (2008) Second generation biofuels: high-efficiency microalgae for biodiesel production. Bioenerg Res 54:621–639
10. Maeda K, Owada M, Kimura N, Omata K, Kraube I (1995) CO_2 fixation from the flue gas on coal-fired thermal power plant by microalgae. Energy Convers Manag 36:717–720
11. Patil V, Tran KQ, Giselrod HR (2008) Towards sustainable production of biofuels from microalgae. Int J Mol Sci 9:1188–1195
12. Williamson AM, Badr O (1998) Assessing the viability of using rape methyl ester (RME) as an alternative to mineral diesel fuel for powering road vehicles in the UK. Appl Energy 59:187–214
13. Sobha V, Surendran M, Vasudevan Nair T (1992) Heavy metal and biochemical studies of different groups of algae form Cape Comorin and Kovalam. Seaweed Res Util 15(1&2):77–85
14. Badrinathan S, Suneeva SC, Shiju TM, Girish Kumar CP, Pragasam V (2011) Exploration of a novel hydroxyl radical scavenger from *Sargassum myriocystum*. J Pharm Res 5(10):1997–2005
15. Annam Renita A, Joshua Amarnath D (2011) Multifaceted applications of marine macro algae *Sargassum myriocystum*. J Pharm Res 4(11):3871–3872
16. Tressler DK (1951) Marine products of commerce. Reinhold, New York
17. Nadyaini WN, Omar W, Aishah N, Amin S (2011) Optimization of heterogeneous biodiesel production from waste cooking palm oil via response surface methodology. Biomass Bioenergy 35:1329–1338
18. Ferella F, Di Celso M, De Michelis I, Stanisci V, Veglio F (2010) Optimization of the transesterification reaction in biodiesel production. Fuel 89(1):36–42
19. Meher LC, Sagar V, Naik SN (2006) Technical aspects of biodiesel production by transesterifcation - a review. Renew Sustain Energy Rev 10:248–268
20. Khuri A, Cornell JA (1987) Response surfaces: design and analysis. Marcel Dekker, New York
21. Jang MG, Kim DK, Park SC, Lee JS, Kim SW (2012) Biodiesel production from crude canola oil by two-step enzymatic process. Renew Energy 42:99–104
22. Chen X, Wei D, Liu D (2008) Response surface optimization of biocatalytic biodiesel production with acid oil. Biochem Eng J 40:423–429
23. Zhang S, Zu YG, Fu YJ, Luo M, Zhang DY, Effreth T (2010) Rapid microwave-assisted transesterification of yellow horn oil to biodiesel using a heteropolyacid solid catalyst. Bioresour Technol 101:931–936
24. Karavalakis G, Hilari D, Givalou L, Karonis D, Stournas S (2011) Storage stability and ageing effect of biodiesel blends treated with different antioxidants. Energy 36:369–374
25. Mittelbach M, Gangl S (2001) Long storage stability of biodiesel made from rapeseed and used frying oil. JAOCS 78(6):573–577
26. Abderrahim B, Martinez M, Aracil J (2007) Long storage stability of biodiesel from vegetable and used frying oils. Fuel 86:2596–2602
27. Das LM, Bora DK, Pradhan S, Malaya K, Naik NSN (2009) Long-term storage stability of biodiesel produced from Karanja oil. Fuel 88:2315–2318
28. Leung DYC, Koo BCP, Guo Y (2006) Degradation of biodiesel under different storage conditions. Bioresour Technol 97:250–256
29. Wijtmans M, Pratt DA, Brinkhorst J, Serwa R, Valgimigli L, Pedulli GF, Porter NA (2004) Synthesis and reactivity of some 6-substituted-2,4-dimethyl-3-pyridinols, a novel class of chain-breaking antioxidants. J Org Chem 69:9215–9223
30. Knothe G (2007) Some aspects of biodiesel oxidative stability. Fuel Process Technol 88:669–677

23

A new process for the rapid and direct vermicomposting of the aquatic weed salvinia (*Salvinia molesta*)

T Ganeshkumar, M Premalatha, S Gajalakshmi[*] and SA Abbasi

Abstract

Background: The concept of high-rate vermicomposting was successfully used to achieve direct vermicomposting of the aquatic weed salvinia - without any precomposting or cow dung supplementation as previously reported processes for the vermicomposting of phytomass had necessitated.

Results: Both the epigeic species of earthworms that were explored, *Eudrilus eugeniae* and *Eisenia fetida*, provided efficient vermicast production with no mortality, persistent gain in body mass, and good fecundity over the 270-day-long course of the reactor operation. In this period, all reactors were pulse-fed at the solid retention time of 15 days and were operated in the pseudo-discretized continuous operation protocol developed earlier by the authors. With this, it was possible to almost completely dampen the influence of natural biodegradation of the feed or grazing by the earthworm born in the vermireactors. This has made it possible to link vermicast production directly to the ability of the earthworm to feed upon, and digest, salvinia. In turn, this enables accurate process monitoring and provides clear pointers on how to improve process efficiency.

Conclusion: The paper establishes the capability of high-rate vermicomposting technology developed earlier by the authors in direct and efficient vermicomposting of salvinia without any precomposting or manure supplementation. The findings have very significant implications in improving process economics and consequently process utility. No previous report exists in primary literature on the vermicomposting of salvinia.

Background

Salvinia (*Salvinia molesta* D. S. Mitchell) is a free-floating aquatic weed, native to South Africa [1]. It is capable of sexual as well as vegetative reproduction and is known for its explosive growth rate. There are instances on record where a few salvinia plants have multiplied so rapidly that even lakes as large as Lake Kariba, with a water-spread of 100 km^2, have been covered by the plant's dense mats in a matter of a few weeks [2,3]. During the last five decades, salvinia has invaded many tropical and subtropical countries of Asia, Africa, and the Pacific regions. It has become a scourge of wetlands, eroding their water-storage space, jeopardizing their water quality, and harming their biodiversity [4,5]. The debris of dead leaves and plants of salvinia undergoes aerobic or anaerobic biodegradation - depending on whether the decomposition occurs in aerobic or anoxic zones of the wetland - generating global warming gases CO_2 and CH_4 [6]. In this manner, colonization of wetlands by salvinia is a major cause of eco-degradation and global warming.

Attempts have been made from time to time to find a use for salvinia as a bioagent for wastewater treatment; as a mulch/fertilizer/animal feed/adsorbent; as raw material for handicraft, paper, chemicals, or biogas; and for nanoparticles synthesis [7-10]. However, none of these potential uses except the one in wastewater treatment have proved economically viable. Moreover, use of salvinia in this manner still leaves the problem of its disposal unsolved.

In recent years, Abbasi and coworkers have introduced the concept of high-rate vermicomposting and have developed associated technology [11-17], which enables direct and efficient vermicomposting of phytomass. As has been

* Correspondence: dr.s.gajalakshmi@gmail.com
Centre for Pollution Control and Environmental Engineering, Pondicherry University, Kalapet, Puducherry 605014, India

reviewed recently [18,19], the use of phytomass as a substrate for generating vermicompost had been very limited prior to the introduction of this know-how and no methods had existed on vermicomposting of phytomass which had the potential for large-scale utilization. This was due to the following reasons:

a) All the past attempts have depended on supplementing the phytomass with large proportions of animal manure, especially cow dung [13,15,18,19]. This has two major disadvantages. First is that to process the very large quantities of phytomass that are available, equally large quantities of animal manure shall be needed. But it is not easy to find so much manure because of numerous competitive uses already in existence [20,21]. The second major disadvantage is that, unlike waste phytomass, animal manure is not available free of cost [21]. Hence dependence on animal manure makes the process economics highly unfavorable.

b) The reported processes have all been very slow, taking 2 months or more to achieve substantial conversion of phytomass to vermicast (unless precomposting had been done). As the rate of any process is directly related to its efficiency, hence economics, this aspect further diminishes the utilizability of the reported processes [18].

c) When precomposting had to be done, it further adds time and cost to the overall process, eroding its economic viability still further [19].

The present paper reports the successful application of the high-rate vermicomposting technology in achieving direct and rapid vermicomposting of salvinia. Two epigeic earthworm species - *Eudrilus eugeniae* and *Eisenia fetida* - which are well-known for their ability to vermicompost animal manure [22-24], were explored and were found to be highly successful in utilizing salvinia as well. No previous report exists in primary literature on the vermicomposting of salvinia.

Methods
Fresh whole plants of salvinia, which are about 1-ft long if measured from tip to toe, were collected from the ponds situated near the Pondicherry University campus. The plants were washed thoroughly to remove adhering soil particles and lightly wiped before subjecting them to vermicomposting.

Circular 7-L plastic containers (diameter 16 cm, effective height 12 cm) were used as vermireactors. The reactors were lined with plastic material to prevent worms from escaping and predation. Jute cloth sheets of 3-mm thickness saturated with water were placed at the bottom of each reactor, and the feed, in the form of 3 kg of

salvinia (dry weight 250 g), was laid over it. The reactors were operated in the *pseudo-discretized continuous reactor operation* (*PDCOP*) mode as described earlier by the authors ([22,25]; Sankar [14]); PDCOP enables an operation which is not really continuous but creates an ambience of a continuous reactor operation. In it, the reactors are started with a certain fixed quantity of substrate and population of adult earthworm. After 15 or 20 days, the contents are removed and the extent of conversion of the substrate to vermicast, change of zoomass of the adult earthworms, and fecundity (in terms of number of juveniles and cocoons generated) are quantified. Within minutes, the reactors are restarted with fresh substrate and the same adult earthworms that were employed initially. In this way, it is possible to record the rate of vermicast production per adult earthworm as a function of time. By removing unconsumed substrate - which would otherwise biodegrade even without the action of the earthworms - the impact of happenings other than ingestion by the earthworms is minimized. Also, the earthworms are always grazing upon totally fresh, or nearly fresh, substrate as they would have in a truly continuous vermireactor. Additionally, since the juveniles that are produced are removed before they grow significantly big to consume significant quantities of substrate, it is possible to dampen their influence on the reactor performance as well.

In the present work, the reactors were started by introducing in them 250 healthy, adult individuals of either *E. fetida* or *E. eugeniae*, picked randomly from cow-dung-fed cultures maintained by the authors. Once every 15 days, the reactor contents were removed and placed in a separate container for the quantification of vermicast, zoomass change in adults, and production of juveniles and cocoons. Within a few minutes, fresh reactors were started with everything else the same as at the start except that from the earthworms removed from the previous run, only the adults were reintroduced into the reactor. All treatments were done in triplicate.

During the course of the experiments, the reactors were kept under the same ambient conditions of $30°C \pm 4°C$ temperature and $60\% \pm 10\%$ relative humidity. Their water content was maintained in the range 60% to 70%. All quantities were adjusted so that the feed and the casting mass reported in this paper represent dry weights (taken after oven-drying at 105°C to constant weight). The earthworm biomass is reported as live weight, taken after rinsing adhering material of the worms and blotting them dry. The castings were sieved through a 3-mm mesh to separate other particles. In this manner, it was possible to assess the vermicast output of 'parent' worms as a function of 1 kg of feed, without competition from offspring. It also ensured that the unutilized feed did not accumulate, and possibly biodegrade, in the reactors.

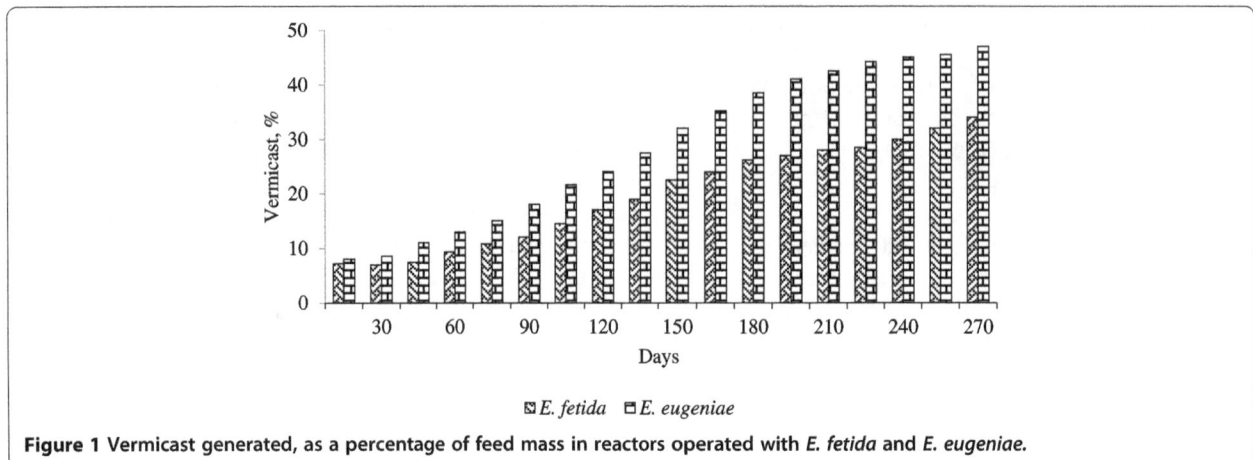

Figure 1 Vermicast generated, as a percentage of feed mass in reactors operated with *E. fetida* and *E. eugeniae*.

Results and discussion

The reactors were operated for 270 days (9 months) and appeared set to go on indefinitely when the experiment was terminated. The extent of vermicast production, in terms of fraction of the feed converted to vermicast, is summarized in Figure 1. Median value of the reactors operated in triplicate have been given, the agreement between the replicates was within a relative error of ±10% which indicates good reproducibility given the heterogeneous nature of the reactor content and natural variability in the feed characteristics.

As is seen in Figure 1, less than 10% of the feed was vermicomposted during the first 30 days. Then the rate began to increase steadily and crossed 40% in the reactors with *E. eugeniae* and 27% in the reactor with *E. fetida* by the 13th run (6½ months of reactor operation). The difference in per capita vermicast output between *E. fetida* and *E. eugeniae* was not statistically significant till the 75th day. Thereafter, it became significant at ≥95% confidence level as revealed by Student's t-test. The rate of vermicast generation continued to increase even after the 13th run but much slowly. The pattern of change in the rate of vermicast production with time is seen to approximately follow a sigmoid pattern - an initial lag phase (30 days), followed by rapid increase during the next 165 days, and a mildly increasing trend thereafter. As the earthworms had been reared on cow dung as the sole feed, upon transfer to vermireactors which had a feed of totally different texture and taste, there might have been an initial sluggishness in feeding. As the animals got acclimated to the new feed as well as the new type of confinements in the vermireactors, their feeding activity picked up. As may be seen from Table 1, the animals steadily gained weight from the 30th day onwards after a zero or negligible weight gain in the first 30 days. The fecundity also increased with time as is reflected in the steady rise in the average number of juveniles produced by

the two species in their respective vermireactors (Figure 2). The trend in the vermiconversion-time curve of Figure 1 indicates that the rate of conversion of salvinia to vermicast was approaching 50% and 35% per fortnight in reactors operated with *E. fetida*. This amounts to 100% conversion in 30 and 45 days, respectively, for the two species. Had the juveniles produced during the first 120 days not been removed as we have done, the rate of

Table 1 Increase in earthworm zoomass with time

	E. fetida		*E. eugeniae*	
	Zoomass, g, of 250 adults	Increase with respect to initial mass, %	Zoomass, g, of 250 adults	Increase with respect to initial mass, %
0	112.1	0	134.3	0
15	112.3	0.2	134.3	0
30	112.8	0.6	135.3	0.7
45	113	0.8	136.6	1.7
60	113.9	1.6	137.9	2.7
75	114.8	2.4	139.2	3.6
90	116	3.5	140.8	4.8
105	117.1	4.5	142.3	6.0
120	118.2	5.4	143.8	7.0
135	119.3	6.4	145.3	8.2
150	120.1	7.1	146.6	9.2
165	121.3	8.2	148.5	10.6
180	122.5	9.3	150.7	12.2
195	124	10.6	152.6	13.6
210	125.2	11.7	154.9	15.3
225	126.7	13.0	156.7	16.7
240	127.9	14.1	158.9	18.3
255	129.1	15.2	160.8	19.7
270	130.8	16.7	163.3	21.6

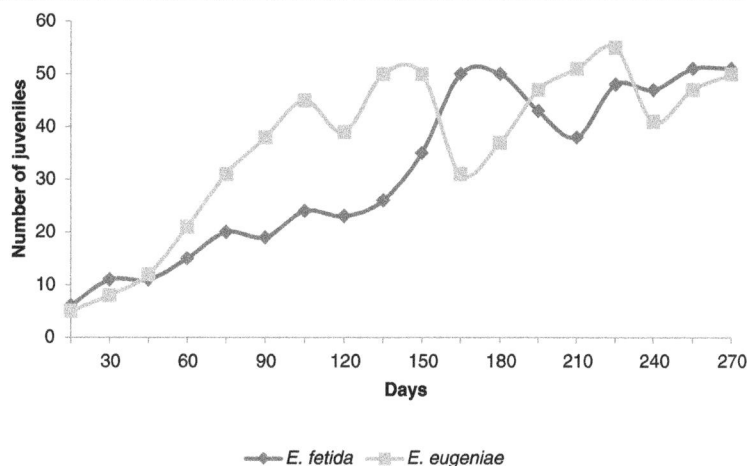

Figure 2 Juveniles produced in vermireactors operated with *E. fetida* and *E. eugeniae*. AJ5 *Salvinia molesta* 31.5.14_3.

vermiconversion would have been much higher by the 270th day as, by then, the offspring produced during the first few months would have reached adulthood and caused a much higher rate-of-feed utilization. Indeed, given the high rate of juvenile production, which averaged 2.1 and 2.4 individuals per day in *E. fetida* and *E. eugeniae*, respectively (Figure 2), the rate of feed utilization is likely to have been near quantitative at the 15-day SRT had the offspring been retained in the reactor.

Even though, on an average, each adult of *E. eugeniae* was generating about 15% more vermicast than each individual of *E. fetida,* on a unit zoomass basis, the latter was nearly as efficient a vermicomposter as the former - both animals generating about 0.7 g of vermicast per gram of body weight towards the closing stages of the experiment.

The results also indicate that if the reactors are operated in a normal pulse-fed fashion, wherein some feed biodegradation will occur naturally as well as earthworms born in the reactors will hasten feed utilization, it may be possible to achieve near 100% process efficiency at solid retention times of 15 to 20 days. Given that conventional vermicomposting requires 3 to 4 months for complete feed conversion, this may amount to a threefold to fourfold gain in process efficiency and consequently in process economics. Indeed, the rate of vermicast production can be increased further by employing a larger number of earthworms from the outset - up to a limit that has been defined by us earlier as the 'highest sustainable earthworm density' [11,13]. Hence the process efficiency can be made substantially higher than achieved in the experiments.

Ongoing, yet-to-be-published, studies on the effect of salvinia vermicast on the germination and growth of botanical plants, in comparison to chemical fertilizers of identical macronutrient and micronutrient content, have revealed that salvinia vermicast is a superior fertilizer. This is possibly due to the presence of plant-friendly enzymes and hormones that vermicast, in general, are known to possess [24].

Conclusions

Whole plants of the aquatic weed salvinia (*S. molesta*) were directly vermicomposted - without precomposting or any other form of elaborate pretreatment and without any fortification with animal manure - by the application of the concept of high-rate vermicomposting developed earlier by the authors. With it, vermireactors were operated, separately, with two common earthworm species *E. fetida* or *E. eugeniae* for 270 days. The PDCOP earlier developed by the authors was used to assess the efficiency of the conversion of the feed to vermicast in a manner that almost completely dampened the possible effect of the natural biodegradation of the feed. PDCOP similarly dampened the contribution of the earthworms that were born in the reactors on the process. With this, it became possible to establish the fact that the application of the high-rate vermicomposting concept and the associated know-how enables the vermicomposting of salvinia in a highly efficient and sustainable manner. The results also indicate that if the reactors are operated in a normal pulse-fed fashion, wherein some feed biodegradation will occur naturally as well as earthworms born in the reactors will hasten feed utilization, it may be possible to achieve a near 100% process efficiency at solid retention times of 15 to 20 days. Given that vermicomposting of phytomass like salvinia in conventional vermicomposting systems requires 3 to 4 months for complete feed conversion - that, too, after pretreatment and liberal supplementation of animal manure - the present process is several times faster, besides being significantly more frugal.

Competing interests
The authors declare that they have no competing interests.

Authors' contributions
All authors have contributed equally. All authors read and approved the final manuscript.

Acknowledgements
SG, PL and SAA, thank the Department of Biotechnology, Government of India for support. TGK is grateful to the University Grants Commission for Rajiv Gandhi National Fellowship.

References
1. Abbasi SA, Nipaney PC (1986) Infestation by aquatic weeds of the fern genus *Salvinia*: its status and control. Environ Conserv 13:235–241
2. Room PM, Thomas PA (1986) Population growth of the floating weed *Salvinia molesta*: field observation and a global model based on temperature and nitrogen. J Appl Ecol 23(3):1013–1028
3. Abbasi SA, Nipaney PC (1993) World's worst weed (salvinia): its impact and utilization. International book Distributors, Dehradun, p 226
4. Abbasi SA, Nipaney PC (1994) Potential of aquatic weed *Salvinia molesta* (Mitchell) for water-treatment and energy recovery. Ind J Chem Technol 1(4):204–213
5. Abbasi SA, Nipaney PC (1995) Productivity of aquatic weed salvinia (*Salvinia molesta*, Mitchell) in natural waters. Ecol Environ Conserv 1(1–4):11–12
6. Abbasi T, Tauseef SM, Abbasi SA (2012) Biogas energy, Springer, New York & London, xiv+169 Pages; ISBN 978-1-4614-1039-3
7. Abbasi SA, Nipaney PC (1991) Effect of temperature on biogas production from aquatic fern *Salvinia*. Ind J Technol 29:306–309
8. Abbasi SA, Nipaney PC, Soni R (1988) Aquatic weeds - distribution, impact, and control. J Sci Ind Res 47(11):650–661
9. Abbasi SA, Nipaney PC, Ramasamy EV (1992) Use of aquatic weed salvinia (*Salvinia molesta*, Mitchell) as full partial feed in commercial biogas digesters. Ind J Technol 30(9):451–457
10. Anuradha J, Abbasi T, Abbasi SA (2014) An eco-friendly method of synthesizing gold nanoparticles using on otherwise worthless weed pistia (*Pistia stratiotes* L.). *J Adv Res*, http://dx.doi.org/10.1016/j.jare.2014.03.006.
11. Gajalakshmi S, Ramasamy EV, Abbasi SA (2002) Vermicomposting of different forms of water hyacinth by the earthworm *Eudrilus engeniae*, Kinberg. Biorescour Technol (Elsevier) 82:165–169
12. Gajalakshmi S, Ganesh PS, Abbasi SA (2005) A highly cost-effective simplification in the design of fast-paced vermireactor. Biochem Eng J (Elsevier) 22:111–116
13. Abbasi T, Gajalakshmi S, Abbasi SA (2009) Towards modeling and design of vermicomposting systems: mechanisms of composting/vermicomposting and their implications. Ind J Biotechnol 8:177–182
14. Ganesh PS, Gajalakshmi S, Abbasi SA (2009) Vermicomposting of the leaf litter of acacia (*Acacia auriculiformis*): possible roles of reactor geometry, polyphenols, and lignin. Bioresour Technol (Elsevier) 100:1819–1827
15. Abbasi T, Tauseef SM, Abbasi SA (2011) The inclined parallel stack continuously operable vermireactor. Official J Patent Off 22:9571
16. Tauseef SM, Abbasi T, Banupriya D, Vaishnavi V, Abbasi SA (2013a) HEVSPAR: a novel vermireactor system for treating paper waste. Official J Patent Off 24:12726
17. Tauseef SM, Abbasi T, Banupriya G, Banupriya D, Abbasi SA (2013b) A new machine for clean and rapid separation of vermicast, earthworms, and undigested substrate in vermicomposting systems, *Journal of Environmental Science and Engineering*, communicated.
18. Abbasi SA, Nayeem-Shah M, Abbasi T (2014) Vermicomposting of phytomass: limitations of the past approaches and the promise of the clean and efficient high-rate vermicomposting technology. J Cleaner Product, in press.
19. Nayeem-Shah M (2014) Exploration of methods for gainful utilization of phytomass-based biowaste. PhD Thesis, Pondicherry University. pp. 92.
20. Abbasi T, Tauseef SM, Abbasi SA (2013) Energy recovery from wastewaters with high-rate anaerobic digesters. Renewab Sustain Energ Rev 19:704–741
21. Tauseef SM, Premalatha M, Abbasi T, Abbasi SA (2013) Methane capture from livestock manure. J Environ Manage 117:187–207
22. Gajalakshmi S, Abbasi SA (2004) Earthworms and vermicomposting. Ind J Biotechnol 3:486–494
23. Gajalakshmi S, Abbasi SA (2008) Solid waste management by composting: state of the art. Critic Rev Environ Sci Technol (CRC Press) 38:311–400
24. Edwards CA, Arancon NQ, Sherman R (2011) Vermiculture technology. CRC Press, London, New York. pp. xix+601
25. Gajalakshmi S, Abbasi SA (2003) High-rate vermicomposting systems for recycling paper waste. Ind J Biotechnol 2:613–615

Permissions

List of Contributors

Zhihui Zhong
State Key Laboratory of Bioreactor Engineering, East China University of Science and Technology, 130 Meilong Road, Shanghai 200237, People's Republic of China

Longfei Liu
State Key Laboratory of Bioreactor Engineering, East China University of Science and Technology, 130 Meilong Road, Shanghai 200237, People's Republic of China

Jiajia Zhou
State Key Laboratory of Bioreactor Engineering, East China University of Science and Technology, 130 Meilong Road, Shanghai 200237, People's Republic of China

Lirong Gao
State Key Laboratory of Bioreactor Engineering, East China University of Science and Technology, 130 Meilong Road, Shanghai 200237, People's Republic of China

Jiajie Xu
State Key Laboratory of Bioreactor Engineering, East China University of Science and Technology, 130 Meilong Road, Shanghai 200237, People's Republic of China

Shuilin Fu
State Key Laboratory of Bioreactor Engineering, East China University of Science and Technology, 130 Meilong Road, Shanghai 200237, People's Republic of China

Heng Gong
State Key Laboratory of Bioreactor Engineering, East China University of Science and Technology, 130 Meilong Road, Shanghai 200237, People's Republic of China

Saptarshi Ghosh
Department of Pharmaceutical Technology (Biotechnology), National Institute of Pharmaceutical Education and Research, Sector- 67, S.A.S. Nagar, Punjab 160062, India

Harish Pawar
Department of Pharmaceutical Technology (Biotechnology), National Institute of Pharmaceutical Education and Research, Sector- 67, S.A.S. Nagar, Punjab 160062, India

Omkar Pai
Department of Pharmaceutical Technology (Biotechnology), National Institute of Pharmaceutical Education and Research, Sector- 67, S.A.S. Nagar, Punjab 160062, India

Uttam Chand Banerjee
Department of Pharmaceutical Technology (Biotechnology), National Institute of Pharmaceutical Education and Research, Sector- 67, S.A.S. Nagar, Punjab 160062, India

Anish Kumari Bhuwal
Department of Microbiology, Kurukshetra University, Kurukshetra, Haryana 136119, India.

Gulab Singh
Department of Microbiology, Kurukshetra University, Kurukshetra, Haryana 136119, India.

Neeraj Kumar Aggarwal
Department of Microbiology, Kurukshetra University, Kurukshetra, Haryana 136119, India.

Varsha Goyal
Department of Microbiology, Kurukshetra University, Kurukshetra, Haryana 136119, India.

Anita Yadav
Department of Biotechnology, Kurukshetra University, Kurukshetra 136119, India.

Anvarsadat Kianmehr
Genetic and Metabolism Research Group, Department of Biochemistry, Pasteur Institute of Iran, 13164 Tehran, Iran.
Department of Medical Biotechnology, School of Advanced Medical Sciences, Tabriz University of Medical Sciences, 13164 Tabriz, Iran.

Maryam Pooraskari
Department of Cell and Molecular Biology, Islamic Azad University, East Tehran Branch, 13164 Tehran, Iran.

Batoul Mousavikoodehi
Department of Microbiology, Islamic Azad University, North Tehran Branch, 13164 Tehran, Iran.

Seyede Samaneh Mostafavi
Department of Cell and Molecular Biology, Faculty of Science, Islamic Azad University, Sanandaj Branch, 13164 Sanandaj, Iran.

Veeranjaneya Reddy Lebaka
Department of Microbiology, Yogi Vemana University, Kadapa, Andhra Pradesh 516003, India.

Hwa-Won Ryu
School of Biological Sciences and Technology, Chonnam National University, Gwangju 500-757, Korea.

Young-Jung Wee
Department of Food Science and Technology, College of Natural Resources, Yeungnam University, Gyeongbuk 712-749, Korea.

Chalisa Jaturapaktrarak
Department of Microbiology, Faculty of Science, Chulalongkorn University, Phayathai Road, Patumwan, Bangkok 10330, Thailand

Suchada Chanprateep Napathorn
Department of Microbiology, Faculty of Science, Chulalongkorn University, Phayathai Road, Patumwan, Bangkok 10330, Thailand

Maria Cheng
Department of Biotechnology, Graduate School of Engineering, Osaka University, 2-1 Yamadaoka, Suita, Osaka 565-0871, Japan

Kenji Okano
Department of Biotechnology, Graduate School of Engineering, Osaka University, 2-1 Yamadaoka, Suita, Osaka 565-0871, Japan

Hisao Ohtake
Department of Biotechnology, Graduate School of Engineering, Osaka University, 2-1 Yamadaoka, Suita, Osaka 565-0871, Japan

Kohsuke Honda
Department of Biotechnology, Graduate School of Engineering, Osaka University, 2-1 Yamadaoka, Suita, Osaka 565-0871, Japan

Yalan Liu
State Key Laboratory of Bioreactor Engineering, Department of Food Science and Engineering, East China University of Science and Technology, 130# Meilong Rd., P.O. Box 283, Shanghai 200237, People's Republic of China.

Lujia Zhang
State Key Laboratory of Bioreactor Engineering, Department of Food Science and Engineering, East China University of Science and Technology, 130# Meilong Rd., P.O. Box 283, Shanghai 200237, People's Republic of China.

Mingrong Guo
State Key Laboratory of Bioreactor Engineering, Department of Food Science and Engineering, East China University of Science and Technology, 130# Meilong Rd., P.O. Box 283, Shanghai 200237, People's Republic of China.

Hongxi Wu
Zhejiang Key Lab of Exploitation and Preservation of Coastal Bio-resource, Wenzhou 325005, People's Republic of China.

Jingli Xie
State Key Laboratory of Bioreactor Engineering, Department of Food Science and Engineering, East China University of Science and Technology, 130# Meilong Rd., P.O. Box 283, Shanghai 200237, People's Republic of China.

Dongzhi Wei
State Key Laboratory of Bioreactor Engineering, Department of Food Science and Engineering, East China University of Science and Technology, 130# Meilong Rd., P.O. Box 283, Shanghai 200237, People's Republic of China.

Carlos Luna
Department of Organic Chemistry, University of Cordoba, Campus de Rabanales, Bldg. Marie Curie, 14014, Cordoba, Spain.

Cristóbal Verdugo
Crystallographic Studies Laboratory, Andalusian Institute of Earth Sciences, CSIC, Avda. Las Palmeras, n°4, 18100, Armilla, Granada, Spain.

Enrique D Sancho
Department of Microbiology, University of Cordoba, Campus de Rabanales, Ed. Severo Ochoa, 14014, Cordoba, Spain.

Diego Luna
Department of Organic Chemistry, University of Cordoba, Campus de Rabanales, Bldg. Marie Curie, 14014, Cordoba, Spain.
Seneca Green Catalyst S.A., Bldg Centauro, Technological Science Park of Cordoba, Rabanales XXI, 14014, Córdoba, Spain.

Juan Calero
Department of Organic Chemistry, University of Cordoba, Campus de Rabanales, Bldg. Marie Curie, 14014, Cordoba, Spain.

Alejandro Posadillo
Seneca Green Catalyst S.A., Bldg Centauro, Technological Science Park of Cordoba, Rabanales XXI, 14014, Córdoba, Spain.

Felipa M Bautista
Department of Organic Chemistry, University of Cordoba, Campus de Rabanales, Bldg. Marie Curie, 14014, Cordoba, Spain.

Antonio A Romero
Department of Organic Chemistry, University of Cordoba, Campus de Rabanales, Bldg. Marie Curie, 14014, Cordoba, Spain.

Jersson Plácido
Department of Biological and Agricultural Engineering, Texas A&M University, Room 201 Scoates Hall, TAMU 2117, College Station, Texas 77841, USA

Sergio Capareda
Department of Biological and Agricultural Engineering, Texas A&M University, Room 201 Scoates Hall, TAMU 2117, College Station, Texas 77841, USA

Huizhong Dong
State Key Laboratory of Bioreactor Engineering, R&D Center of Separation and Extraction Technology in Fermentation Industry, East China University of Science and Technology, Shanghai 200237, China
Shanghai Collaborative Innovation Center for Biomanufacturing Technology, Shanghai 200237, China

Yaosong Wang
State Key Laboratory of Bioreactor Engineering, R&D Center of Separation and Extraction Technology in Fermentation Industry, East China University of Science and Technology, Shanghai 200237, China
Shanghai Collaborative Innovation Center for Biomanufacturing Technology, Shanghai 200237, China

Liming Zhao
State Key Laboratory of Bioreactor Engineering, R&D Center of Separation and Extraction Technology in Fermentation Industry, East China University of Science and Technology, Shanghai 200237, China
Shanghai Collaborative Innovation Center for Biomanufacturing Technology, Shanghai 200237, China

Jiachun Zhou
State Key Laboratory of Bioreactor Engineering, R&D Center of Separation and Extraction Technology in Fermentation Industry, East China University of Science and Technology, Shanghai 200237, China
Shanghai Collaborative Innovation Center for Biomanufacturing Technology, Shanghai 200237, China

Quanming Xia
State Key Laboratory of Bioreactor Engineering, R&D Center of Separation and Extraction Technology in Fermentation Industry, East China University of Science and Technology, Shanghai 200237, China
Shanghai Collaborative Innovation Center for Biomanufacturing Technology, Shanghai 200237, China

Lihua Jiang
State Key Laboratory of Bioreactor Engineering, R&D Center of Separation and Extraction Technology in Fermentation Industry, East China University of Science and Technology, Shanghai 200237, China
Shanghai Collaborative Innovation Center for Biomanufacturing Technology, Shanghai 200237, China

Liqiang Fan
State Key Laboratory of Bioreactor Engineering, R&D Center of Separation and Extraction Technology in Fermentation Industry, East China University of Science and Technology, Shanghai 200237, China
Shanghai Collaborative Innovation Center for Biomanufacturing Technology, Shanghai 200237, China

Folasade M Olajuyigbe
Department of Biochemistry, Federal University of Technology, Akure 340001, Nigeria

Ayodele M Falade
Department of Biochemistry, Federal University of Technology, Akure 340001, Nigeria

Bin Lei
State Key Laboratory of Bioreactor Engineering, East China University of Science and Technology, Shanghai 200237, People's Republic of China

Xu Zhang
State Key Laboratory of Bioreactor Engineering, East China University of Science and Technology, Shanghai 200237, People's Republic of China

Minglong Zhu
State Key Laboratory of Bioreactor Engineering, East China University of Science and Technology, Shanghai 200237, People's Republic of China

Wensong Tan
State Key Laboratory of Bioreactor Engineering, East China University of Science and Technology, Shanghai 200237, People's Republic of China

Atika Hadiati
Genetics of Prokaryotes, Center for Biotechnology, Faculty of Biology, Bielefeld University, Universitaetsstraße 25, Bielefeld 33615, Germany

Irene Krahn
Genetics of Prokaryotes, Center for Biotechnology, Faculty of Biology, Bielefeld University, Universitaetsstraße 25, Bielefeld 33615, Germany

Steffen N Lindner
Genetics of Prokaryotes, Center for Biotechnology, Faculty of Biology, Bielefeld University, Universitaetsstraße 25, Bielefeld 33615, Germany

Volker F Wendisch
Genetics of Prokaryotes, Center for Biotechnology, Faculty of Biology, Bielefeld University, Universitaetsstraße 25, Bielefeld 33615, Germany

Wentao Kong
Department of Bioengineering, University of Illinois at Urbana-Champaign,1304 W Springfield Avenue, Urbana, IL 61801, USA.
Institute for Genomic Biology, University of Illinois at Urbana-Champaign, 1206 W Gregory Drive, Urbana, IL 61801, USA.

Venhar Celik
Department of Bioengineering, University of Illinois at Urbana-Champaign,1304 W Springfield Avenue, Urbana, IL 61801, USA.
Institute for Genomic Biology, University of Illinois at Urbana-Champaign, 1206 W Gregory Drive, Urbana, IL 61801, USA.

Chen Liao
Department of Bioengineering, University of Illinois at Urbana-Champaign,1304 W Springfield Avenue, Urbana, IL 61801, USA.
Institute for Genomic Biology, University of Illinois at Urbana-Champaign, 1206 W Gregory Drive, Urbana, IL 61801, USA.

Qiang Hua
State Key Laboratory of Bioreactor Engineering, East China University of Science and Technology, 130 Meilong Road, Shanghai 200237, People's Republic of China.

Ting Lu
Department of Bioengineering, University of Illinois at Urbana-Champaign,1304 W Springfield Avenue, Urbana, IL 61801, USA.
Institute for Genomic Biology, University of Illinois at Urbana-Champaign, 1206 W Gregory Drive, Urbana, IL 61801, USA.
Department of Physics, University of Illinois at Urbana-Champaign, 1110 W Green Street, Urbana, IL 61801, USA.

Chinnathambi Pothiraj
PG and Research Department of Botany, Alagappa Government Arts College (Alagappa University), Karaikudi, Tamilnadu 630003, India.

Ramasubramanian Arumugam
PG and Research Department of Botany, Alagappa Government Arts College (Alagappa University), Karaikudi, Tamilnadu 630003, India.

Muthukrishnan Gobinath
Department of Microbiology, VHNSN College, Virudhunagar, Tamilnadu 626 001, India.

Ming-Yang Li
State Key Laboratory of Bioreactor Engineering, East China University of Science and Technology, 130 Meilong Road, Shanghai 200237, People's Republic of China

Xue-Dong Wang
State Key Laboratory of Bioreactor Engineering, East China University of Science and Technology, 130 Meilong Road, Shanghai 200237, People's Republic of China

Priya Banerjee
Department of Environmental Science, University of Calcutta, 51/2 Hazra Road, Kolkata 700 026, India.

Mantosh Satapathy
Biotechnology Department, National Institute of Technology, Durgapur 713209, India.

Aniruddha Mukhopahayay
Department of Environmental Science, University of Calcutta, 51/2 Hazra Road, Kolkata 700 026, India.

Papita Das
Department of Chemical Engineering, Jadavpur University, Kolkata 700 032, India.

Fu-Liang Du
Laboratory of Biocatalysis and Synthetic Biotechnology, State Key Laboratory of Bioreactor Engineering, East China University of Science and Technology, Shanghai 200237, P. R. China

Hui-Lei Yu
Laboratory of Biocatalysis and Synthetic Biotechnology, State Key Laboratory of Bioreactor Engineering, East China University of Science and Technology, Shanghai 200237, P. R. China

Jian-He Xu
Laboratory of Biocatalysis and Synthetic Biotechnology, State Key Laboratory of Bioreactor Engineering, East China University of Science and Technology, Shanghai 200237, P. R. China

Chun-Xiu Li
Laboratory of Biocatalysis and Synthetic Biotechnology, State Key Laboratory of Bioreactor Engineering, East China University of Science and Technology, Shanghai 200237, P. R. China

Xiaojuan Wang
State Key Laboratory of Bioreactor Engineering, East China University of Science and Technology, 130 Meilong Road, Shanghai 200237, People's Republic of China

Chaogang Bai
State Key Laboratory of Bioreactor Engineering, East China University of Science and Technology, 130 Meilong Road, Shanghai 200237, People's Republic of China

Jian Zhang
State Key Laboratory of Bioreactor Engineering, East China University of Science and Technology, 130 Meilong Road, Shanghai 200237, People's Republic of China

Aiyou Sun
State Key Laboratory of Bioreactor Engineering, East China University of Science and Technology, 130 Meilong Road, Shanghai 200237, People's Republic of China

Xuedong Wang
State Key Laboratory of Bioreactor Engineering, East China University of Science and Technology, 130 Meilong Road, Shanghai 200237, People's Republic of China

Dongzhi Wei
State Key Laboratory of Bioreactor Engineering, East China University of Science and Technology, 130 Meilong Road, Shanghai 200237, People's Republic of China

Jun Tan, Ju Chu
State Key Laboratory of Bioreactor Engineering, East China University of Science and Technology, 130 Meilong Road, P.O. Box 329#, Shanghai 200237, People's Republic of China

Yonghong Wang
State Key Laboratory of Bioreactor Engineering, East China University of Science and Technology, 130 Meilong Road, P.O. Box 329#, Shanghai 200237, People's Republic of China

Yingping Zhuang
State Key Laboratory of Bioreactor Engineering, East China University of Science and Technology, 130 Meilong Road, P.O. Box 329#, Shanghai 200237, People's Republic of China

Siliang Zhang
State Key Laboratory of Bioreactor Engineering, East China University of Science and Technology, 130 Meilong Road, P.O. Box 329#, Shanghai 200237, People's Republic of China

Xiaolei Ma
State Key Laboratory of Bioreactor Engineering, Department of Bioengineering, East China University of Science & Technology, 130 Meilong Rd., Shanghai 200237, China

Xuejun Cao
State Key Laboratory of Bioreactor Engineering, Department of Bioengineering, East China University of Science & Technology, 130 Meilong Rd., Shanghai 200237, China

Annam Renita
Department of Chemical Engineering, Sathyabama University, Jeppiaar Nagar, Rajiv Gandhi Salai, Chennai TN 119, India.

Nurshaun Sreedhar
Department of Chemical Engineering, Sathyabama University, Jeppiaar Nagar, Rajiv Gandhi Salai, Chennai TN 119, India.

Magesh Peter
Department of Marine Biotechnology, National Institute of Ocean Technology, Velacherry-Tambaram Main Road, Chennai 100, India.

T Ganeshkumar
Centre for Pollution Control and Environmental Engineering, Pondicherry University, Kalapet, Puducherry 605014, India

M Premalatha
Centre for Pollution Control and Environmental Engineering, Pondicherry University, Kalapet, Puducherry 605014, India

S Gajalakshmi
Centre for Pollution Control and Environmental Engineering, Pondicherry University, Kalapet, Puducherry 605014, India

SA Abbasi
Centre for Pollution Control and Environmental Engineering, Pondicherry University, Kalapet, Puducherry 605014, India

www.ingramcontent.com/pod-product-compliance
Lightning Source LLC
Chambersburg PA
CBHW080648200326
41458CB00013B/4780